# Resilient Cities

## Re-thinking Urban Transformation

**Series Editors**

Nicola Tollin
SDU Civil and Architectural Engineering
University of Southern Denmark
Odense, Denmark

Jordi Morató
UNESCO Chair on Sustainability
Technical University of Catalunya
Terrassa, Spain

Ernesto DR Santibanez Gonzalez
Centre Environmental Sciences
Federal University of Bahia
Bahia, Brazil

The RESILIENT CITIES book series aims to analyse the challenges faced by cities and provide an up-to-date body of knowledge, including a systematic collection of global cutting-edge best practices, fundamental to managing the urban transition toward resilience. The best practices will be collected and analysed following a common format, enabling the reader to understand the solutions adopted and clearly highlighting the parameters and possibilities for replication and up-scaling. The best practices are taken from a global city base including, Barcelona, Medellin, Adelaide, Copenhagen, Seoul, and Accra. The distinctiveness of the RESILIENT CITIES book series is its international dimension, coupled with a multidisciplinary and a cross sectorial approach. The RESILIENT CITIES book series will be a unique and fundamental resource for practitioners, policy makers and scientists involved in planning and governing the transition of cities. It presents the latest and up-to-date systematized information on research, practices and policies development, defining clear means and pathways for replication and up-scaling.

More information about this series at http://www.springer.com/series/14171

Grazia Brunetta • Ombretta Caldarice
Nicola Tollin • Marti Rosas-Casals
Jordi Morató
Editors

# Urban Resilience for Risk and Adaptation Governance

## Theory and Practice

Springer

*Editors*
Grazia Brunetta
Interuniversity Department of Regional and Urban Studies and Planning
Politecnico di Torino
Torino, Italy

Ombretta Caldarice
Interuniversity Department of Regional and Urban Studies and Planning
Politecnico di Torino
Torino, Italy

Nicola Tollin
SDU Civil and Architectural Engineering
University of Southern Denmark
Odense, Denmark

Marti Rosas-Casals
Sustainability Measurement and Modelling Lab (SUMMLab)
Universitat Politècnica de Catalunya (UPC) – Barcelona Tech
Terrassa, Spain

Jordi Morató
UNESCO Chair on Sustainability
Technical University of Catalunya
Terrassa, Spain

ISSN 2524-5988 ISSN 2524-5996 (electronic)
Resilient Cities
ISBN 978-3-030-08330-4 ISBN 978-3-319-76944-8 (eBook)
https://doi.org/10.1007/978-3-319-76944-8



Printed on acid-free paper

This Springer imprint is published by the registered company Springer International Publishing AG part of Springer Nature.
The registered company address is: Gewerbestrasse 11, 6330 Cham, Switzerland

# Foreword

"Urban Resilience for Risk and Adaptation Governance. Theory and Practice" is the first volume of the book series "Resilient City: Re-thinking Urban Transformation" published by Springer. The Series has been envisioned to analyse the contemporary challenges faced by cities and provide an up-to-date body of knowledge, including a systematic collection of global cutting-edge best practices, fundamental to managing the urban transition toward resilience.

This Book is the first result of the ongoing research collaboration between the "Recycling City Network – RECNET" and the "Responsible Risk Resilience Centre – R3C" (Politecnico di Torino), also in the frame of RESURBE International Program on Urban Resilience, in order to merge different approaches and competencies on urban resilience in an international scenario. In other word, while RECNET aims at supporting the key challenges of adaptation and mitigation of climate change related to intangible heritage and circular economy, R3C is the first Italian research centre that internationally-leading investigates risk vulnerabilities and provide resilient solutions for policies and practices using an interdisciplinary perspective in order to strengthen and safeguard the European's cultural and natural heritage.

In this perspective, "Urban Resilience for Risk and Adaptation Governance. Theory and Practice" aims at collecting a series of theory and practice based articles on risk management in urban contexts within a resilient perspective. It proposes a transversal approach regarding the role of spatial planning in promoting and fostering risk management as well as institutions' challenges and perspective for governing risk, particularly in relation to new forms of multi-level governance that include stakeholders and citizen engagement. The different contributions are based on five topics:

- *Concepts and definitions of risk adaptation and urban resilience*. The topic aims to analyse and benchmark different definitions, approaches and concepts of risk, risk management and adaptation in cities. It also includes the understanding of the key characteristics of urban resilience in relation to risk;

- *Disaster risk prevention and reduction*. The topic aims at defining approaches related to urban risk prevention and reduction, with reference to both slow and rapid on-setting disasters and crises. It includes issues related with the prevision capacity and approaches to foresee and forecast future risk, in relation to prevention and preparedness;
- *Community based initiatives for urban risk management and adaptation*. The topic aims to collect experiences on community initiatives and organisational process, based on social inclusion and capacity building for urban risk management. It includes analyses on how local communities and citizens can enhance urban resilience through risk management and adaptation;
- *Institutional frameworks and multi-level governance for urban resilience*. The topic analyses institutional and governance frameworks, with specific focus on multi-level governance, local communities' participation, and bottom-up approaches for urban risk management and adaptation. It includes strategies and frameworks for stakeholders participation and support to participatory and inclusive decision- making;
- *The integration of urban resilience in spatial planning*. This topic aims at discussing in which way risk management and adaptation can be integrated and deployed in spatial planning theories and practices. It also includes the potential of planning for controlling uncertainty for the prevention against and preparedness to occurring events.

We are extremely grateful to the authors of the chapters and all those who have participated in this long-lasting dialogue since the beginning of this research project. All chapters were double-refereed. Our special thanks go to the referees of the chapters Maria Dolores Alvarez, Valentina Antoniucci, Stefano Aragona, Ana María Buritica, Luigi Buzzacchi, Alessandra Casu, Stefano Cozzolino, Arabella Fraser, Boris Lazzarini, Josep Mercadè, Ezio Micelli, Stefano Moroni, and Rocio Perez Campana. We also acknowledge the support given by the Technical University of Antioquia for providing technical assistance to the book series.

Torino, Italy — Grazia Brunetta, Ombretta Caldarice

Odense, Denmark — Nicola Tollin

Terrassa, Spain — Marti Rosas-Casals, Jordi Moratò

15 March 2018

recnet.

R3C Responsible Risk Resilience Centre

# Contents

# Chapter 1
# The Challenge of Urban Resilience: Operationalization

**Ombretta Caldarice, Grazia Brunetta, and Nicola Tollin**

**Abstract** Urban resilience is a structural property of urban systems and it is related to their capacity to continually self-organising and adapt in the face of ongoing and unpredicted changes and risks. Although academic debate on urban resilience is deeply understood, several methodological challenges remain both related to the theoretical and the practical domain of the concept. Operationalising resilience is arguably one of the most impactful global issues for the future research as it implies to link the concept about what urban resilience is and what urban resilience ought to be. To this end, this chapter proposes to deal with the principal issues related to the operationalization of urban resilience, investigating current problems and future perspectives about meanings and values of urban resilience and discussing the main contents of the book.

**Keywords** Urban resilience · Resilience meaning · Adaptive capacity · Adaptation in spatial planning · Resilience in practice

## 1.1 Understanding Resilience

*"The world breaks everyone and afterward many are strong at the broken places"*
*Ernest Hemingway, A farewell to Arms (1929)*

In 2013, one of the most read articles in the *Guardian* newspaper was searching for the real meaning of 'resilience' criticising the common use of the term by recipient governments, implementing agencies, and donors. The article questioned if resilience stills a mere catchword that development organisations are using to

O. Caldarice (✉) · G. Brunetta
Interuniversity Department of Regional and Urban Studies and Planning, Politecnico di Torino, Torino, Italy
e-mail: ombretta.caldarice@polito.it; grazia.brunetta@polito.it

N. Tollin
SDU Civil and Architectural Engineering, University of Southern Denmark, Odense, Denmark
e-mail: nto@iti.sdu.dk

 
G. Brunetta et al. (eds.), *Urban Resilience for Risk and Adaptation Governance*, Resilient Cities, https://doi.org/10.1007/978-3-319-76944-8_1

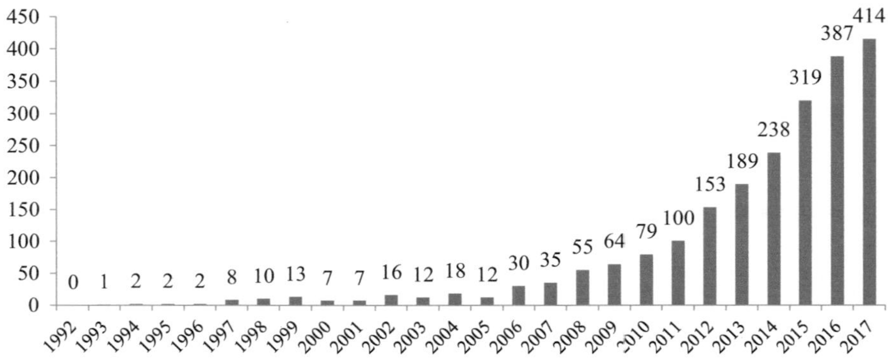


**Fig. 1.1** The rapid rise of urban resilience research (1992–2017) (*Source:* Authors' elaboration based on the Scopus database. The research query represents the number of publications for each year from 1992 (UN Conference on Environment & Development, Rio de Janeiro) to 2017 with the terms "urban resilience" in the title, abstract, or keywords)

rebrand their actions or instead if it has a positive force to drive innovation, creativity, adaptation, and technological evolution. In the same year, *Time* magazine declared 'resilience' the buzzword of 2013 because of its conceptual vagueness and malleability (Brown 2016). Almost five years on, urban resilience is here to stay as an organic concept both to respond to major global challenges as urbanization, development, climate change and sustainability, and to implement global policies, including the 2030 Agenda-Sustainable Development Goals, the Sendai Framework on Disaster Risk Reduction, the Paris Agreement, and the New Urban Agenda. Similarly, scholars have increasingly questioned and debated the concept of urban resilience (Boin et al. 2010), as shown from the rapidly rising and proliferating number of scientific articles published particularly during the last decade (Fig. 1.1).

The concept of resilience has been developed and used with very different understandings and meanings for over 40 years within different disciplines, such as ecology, biology, and phycology, and in most recent years also in engineering, urban studies, and economy. For example, in the contemporary environmental-change literature, resilience is commonly closed to the management of new and emergent risks in urban areas - i.e., globalisation, climate change, spread urbanisation, demographic pressures, resource shortages, and ageing infrastructure. In the risk domain, urban resilience is interpreted as the property of cities to mediate the outcomes of the systemic global phenomenon of climate change oriented to influence the city's capacity to adapt in the face of extreme events (Davoudi et al. 2009). In other words, new urban challenges put increasing pressure on cities that seek to react to them so that the shared concept of urban resilience has emerged as an important lens and a policy response in an era of public concern about disasters and risks (Hayward 2013). In this scenario, cities urgently need to explore further the meaning of urban resilience, which is beyond the simple sum of other disciplinary approaches and requiring a new understanding able to integrate systemically its various dimensions: individual, societal, economic, environmental, political, and infrastructural. At the same time, cities need to understand and define the full potential for the operationalization of resilience, beyond the simple disaster risk reduction and adaptation, but through a systemic framework generating co-benefit by its integration with mitigation, and ultimately with innovation and development.

In this scenario, "*while the academic debate on describing resilience continues, governments around the world have developed plans and programmes that aim to guide cities, communities and authorities towards achieving it*" (Weichselgartner and Kelman 2015: 254). In Europe and worldwide[1], cities are gradually envisaging climate change adaptation strategies so that more than 650 European cities[2] are currently involved in different initiatives related to address the increasing climatic vulnerability and ultimately to reduce risks. At the same time, the great majority of these cities, significantly over 90%, have insufficient capacity to take long-term and potentially climate-impacted decisions in order to achieve significant economic, social and ecological objectives[3]. As outlined in the Ricardo-AEA report (2013), only a few international cities have developed a real capacity for adaptation and consequently can fully implement measures against climate change. In other words, despite theoretical debate on urban resilience is deeply understood, several methodological challenges remain both related to the theoretical and the practical domain of the concept (Crowe et al. 2016). In this perspective, 'operationalising resilience' is arguably one of the most impactful global issues for the future research on this topic (Brunetta and Caldarice 2017). This question implies that urban resilience will be framed as a powerful agent and driver of change for future development, ultimately to re-think the way in which cities are designed, planned, managed and lived facing current major global challenges.

## 1.2 Operationalising Urban Resilience

Resilience "*is about having the capacity to continue to learn, self-organise, and developed in dynamic environments faced with true uncertainty and the unexpected, like steering a vessel in turbulent waters*" (Folke 2016: 7). This book falls under this heading and, starting from these remarks, deals with the principal issue linked to the 'operationalization of urban resilience' in urban contexts within a multidimensional perspective. It brings together a series of theory and practice essays that aim to unpick the complexity of urban resilience. After this brief introduction (Caldarice, Brunetta, and Tollin), which presents an overall implication about meanings and values of urban resilience, the contributions are divided into two main parts.

---

[1] See the report "Sustainable Urbanization in the Paris Agreement' – a comparative review of Nationally Determined Contributions for Urban Content" by UN-Habitat

[2] For a detailed picture of the American situation, see Meerow and Mitchell (2017), Woodruff and Stults (2016), and Hanna et al. (2014).

[3] See the Report "Urban adaptation to climate change in Europe 2016 - Transforming cities in a changing climate" by the European Environmental Agency.

*Part I – From Risk Management to Urban Resilience. Towards a Theoretical Framework* reports some ground-breaking considerations about urban resilience's common understanding. The concept is approached concerning to different geographical contexts (Normandin *et al.*; Giaimo *et al.*; Torres-Lima *et al.*) and diverse disciplinary approaches (Brunetta and Caldarice; Barrico and Castro; Pede). Normandin, Therrien, Pelling and Paterson investigate how London and Montreal have interpreted resilience in their urban policies assuming different references, i.e. paradigm change, governance model to better manage complex issues, and destination to reach. "Putting Resilience into Practice. The Spatial Planning Response to Urban Risks" by Brunetta and Caldarice focuses our attention on what resilience means in the 'doing'. The contribution reports a qualitative comparison between the local climate adaptation strategies of two European metropolitan areas, namely Bologna in Italy and London in the UK discussing the gap between theory and practice of spatial planning in achieving a resilient approach to urban risks through adaptation. Torres-Lima, Lee Pinel and Conway-Gómez's paper is based on a recent literature review on theories and paradigms of adaptive management and collaboration for resilience in the context of socially complex peri-urban landscapes. The paper reports future themes of research into adaptive governance for resilience in peri-urban areas in developing countries stressing on both structure and decision-making and economic processes in the context of demographic, economic, and climate change. In Chap. 5, Giamo, Barbieri and Salata recall the experimentation of the European project LIFE SAM4CP based on an innovative methodological approach that provides the development of appropriate tools for sustainable territorial governance aimed at mitigating disasters risk and increasing the safety of urban systems through the integration of ecological, economic and socio-political aspects. Barrico and Castro's paper focused on the unsustainable urbanisation of the XXI century as the primary challenge that contemporary cities need to face to be climate-resilient. The paper relates climate change to urban sprawl underling how urban expansion intensifies heat island effect, watershed run-off, and stream flow. Finally, "Improving Resilience through Cross-scale Knowledge Sharing" by Pede focuses on the uncertainty of contemporary society that becomes more reliant to complex threats and risks. The paper interprets the cross-scale knowledge sharing as a key theme for urban resilience fostering the systemic perspective in resilience theories and practices through the exploration of the risk society implications.

In the *Part II - From Risk Management to Urban Resilience. Experiences and Best Practices* a number of case studies are presented related to fire risk (Ingaramo *et al.*), flood risk (Ferrari *et al.*; Moreno-Mata *et al.*; Chundeli and Visakha), landslide risk (Ferreira and Abiko), seismic risk (Aguilar-Meléndez *et al.*; Resendiz-Vazquez; Platt) and more general urban transformation dynamics (Hosseinioon). The paper by Hosseinioon reports the effect of formalisation in the informalised development in three different neighbourhoods in the city of Golestan (Iran) underlining how informality and urban resilience have common features because they are considered both as solutions and as survival strategies. "Ecosystem Service valuation for forest landscape resilience: managing fire risk" by Ingaramo, Salizzoni and Voghera identify a new evaluation and mapping methodology to enhance resilience in the crucial challenge against fire risk. Using the concept of ecosystem services, the paper frames the Forest Ecosystem Services valuation method as a useful tool to foster adaptive management to fire risk. The Chap. 10 by Ferrari, Oliveira, Pautasso and Zêzere frames the European project MOVE (Methods for the Improvement of Vulnerability assessment in Europe) in two case studies, i.e. Turin in Italy and Porto-Vila Nova de Gaia in Portugal. The project describes the territorial vulnerability along three dimensions – i.e. exposure, susceptibility and lack of resilience – and suggests some practical actions against flood risk. Moreno-Mata, Villasís-Keever, and Jordi Morató's paper investigates the relationships between climate change, rainwater management and flood risk in the Metropolitan Area of San Luis Potosí pursuing an integrated methodological approach to identify problems, clarify dynamics and present tools for a better understanding of the systemic urban environment. "Urban Resilience and landslide risk management: the case of Santos (Brazil)" by Ferreira and Abiko aims to identify the Brazilian landslide risk management tools and to analyse the role of institutional tools in promoting urban resilience in the case of Santos. Chapter 13 "Probabilistic Assessment of Seismic Risk of Dwelling Buildings of Barcelona. Implications for the City Resilience" proposes a new assessment of the seismic risk of dwelling buildings for the city of Barcelona fostering the possible uses of its results in the Barcelona City scenarios on seismic risk. The Reséndiz-Vázquez's paper relates resilience to post-disaster reconstruction phase in which the societal conditions must be restored. This methodological paper is based on the comparison between the later French reconstruction after World War II and the Mexican reconstruction after the earthquakes of September 1985. "Operational Mitigation – Transient Housing: the case of Jammu, India" by Chundeli and Visakha describes the project of rehabilitation led to intense flooding of the Jhelum River in Jammu focusing on the temporary housing set for the emergency phase. Finally, Chap. 16 "Planning recovery and reconstruction after the 2010 Maule earthquake and tsunami in Chile" by Platt analyses the urban planning process and the disaster recovery strategies adopted, both at the national and local level, after the Maule Chilean earthquake of 27 February 2010. In particular, the paper focuses on how Chile balanced the need for speed with building back better and how effective was the transition from temporary relief to long-term resilience.

This collection of both theoretical and practical essays attempts to contribute to the literature on urban resilience nevertheless it does not exhaust the topics. The book opens new lines of enquiry focusing in particular on three research themes on main challenges and debates related to:

- The *structural challenges* for urban resilience thinking and practice. This includes the different and conflicting ways in which urban resilience is interpreted and the implications for the practice of the broad dichotomy between more functional and more dynamic interpretations of it.
- The *technical challenges* around the spatial aspects, the resilience measurement, and how to assess the costs and benefits of resilience-building.
- The *political challenges* focused on critiques of urban resilience practices. It includes the attention to the potential use of resilience by cities for managing complex issues through a multi-level governance approach.

## References

Boin, A., Comfort, L. K., & Demchak, C. (2010). The rise of resilience. In L. K. Comfort, A. Boin, & C. Demchak (Eds.), *Designing resilience: Preparing for extreme events* (pp. 1–12). Pittsburgh: University of Pittsburgh Press.

Brown, K. (2016). *Resilience, development and global change*. London/New York: Routledge.

Brunetta, G., & Caldarice, O. (2017). Governing for climate change adaptation: Approaches and challenges for designing the resilient city. *Urbanistica, 160* (forthcoming).

Crowe, P. R., Foley, K., & Collier, M. J. (2016). Operationalizing urban resilience through a framework for adaptive co-management and design: Five experiments in urban planning practice and policy. *Environmental Science & Policy, 62*, 112–119.

Davoudi, S., Crawford, J., & Mehmood, A. (Eds.). (2009). *Planning for climate change: Strategies for mitigation and adaptation for spatial planners*. London: Earthscan.

Folke, C. (2016). *Resilience, in Oxford research encyclopedia of environmental science*. https://doi.org/10.1093/acrefore/9780199389414.013.8

Hanna, K., Seasons, M., Dale, A., et al. (2014). Planning for climate change: The state of current practice in Canada. *Plan Canada, 54*(2), 28–36.

Hayward, B. (2013). Rethinking resilience: Reflections on the earthquakes in Christchurch, New Zealand, 2010 and 2011. *Ecology and Society, 18*(4), 37.

Meerow, S., & Mitchell, C. L. (2017). Weathering the storm: The politics of urban climate change adaptation planning. *Environment and Planning A, 49*(11), 2619–2627.

Weichselgartner, J., & Kelman, I. (2015). Geographies of resilience: Challenges and opportunities of a descriptive concept. *Progress in Human Geography, 39*(3), 249–267.

Woodruff, S. C., & Stults, M. (2016). Numerous strategies but limited implementation guidance in US local adaptation plans. *Nature Climate Change, 6*(8), 796–802.

# Part I
# From Risk Management to Urban Resilience: Towards a Theoretical Framework

# Chapter 2
# The Definition of Urban Resilience: A Transformation Path Towards Collaborative Urban Risk Governance

**Julie-Maude Normandin, Marie-Christine Therrien, Mark Pelling, and Shona Paterson**

**Abstract** Resilience as a theoretical concept and policy proposition is constantly being redefined and clarified. But when it comes to implementation, public managers and bureaucrats have to take ownership of resilience and translate it into practical forms that make sense to them intellectually and operationally. In this chapter, we first explain how resilience is presented in the literature as, variously, a paradigm change, a governance model to better manage complex issues, and a destination to reach. Second, we analyse how public managers and bureaucrats responsible for implementation in London and Montreal have interpreted and used resilience. Finally, we discuss how paradigm change, governance transformation and goal attainment perspectives end up converging into a relatively similar meaning in both cities. What these cities are lacking to take the next step toward urban resilience is the strategic endorsement of political authorities to support this important transformation.

**Keywords** Urban resilience · Conceptual development · Governance transformation · Exploratory comparative research

---

J.-M. Normandin (✉) · M.-C. Therrien
École nationale d'administration publique, Montréal, Canada
e-mail: julie-maude.normandin@enap.ca; mctherrien@enap.ca

M. Pelling
King's College London, London, UK
e-mail: mark.pelling@kcl.ac.uk

S. Paterson
Future Earth Coasts, Ringaskiddy, Ireland
e-mail: shona.paterson@ucc.ie

 
G. Brunetta et al. (eds.), *Urban Resilience for Risk and Adaptation Governance*, Resilient Cities, https://doi.org/10.1007/978-3-319-76944-8_2

## 2.1 Resilience Behind the Buzzword

International organisations and governments increasingly promote urban resilience as a promising policy to help communities face actual and future urban risks (ISDR 2011, 2005; Ministry of Public Security 2014). Beyond the discourse, public managers and bureaucrats need to take ownership of the concept and translate it into practical actions, programs, investments, and other tools. All indications are that this mandate is filled with pitfalls. Resilience has become a buzzword (Mitchell 2013): a broad concept, difficult to implement (Wagenaar and Wilkinson 2013), that is interpreted in different and often contradictory ways (Boin et al. 2010). How do municipal authorities understand the concept and make sense of the objective to enact resilience policy? We sought to answer this question through a comparative analysis of the meaning attached to resilience by public managers and bureaucrats in the cities of London (UK) and Montreal (Canada).

First, we analyse the different definitions of resilience in the scientific and grey literature (government documents, organizational reports, international initiatives, etc.) in the urban context, but also more broadly. These definitions reveal that resilience has been presented as a paradigm change, a governance model and a destination to reach. Second, we compare how municipal actors engaged in developing resilience in London (UK) and Montreal (Canada) understand and use the concept. This analysis identifies components of the resilience discourse that stakeholders recognise as helpful in achieving their mandates. We find that resilience is used to promote specific policy agendas, and to favour governance changes based on a more integrated and collaborative approach to urban management Finally, we identify gaps between definitions employed in the literature and the meanings assigned by stakeholders responsible for resilience in their cities, along with the collaborative networks involved in implementation.

### *2.1.1 Resilience as a Paradigm Change, Governance Model and a Destination*

In this section, we analyse the meaning of resilience as a solution that public authorities need to adopt and implement to respond to present and potential problems. We examine a wide range of scientific literature in organizational theory, public administration and urban studies, as well as grey literature published by governmental, international and non-governmental organizations. We demonstrate that resilience is presented, variously, as a paradigm change, a governance model that could be adopted, and a destination to reach or a desirable state of being. These three perspectives are not antithetical, and can coexist and overlap.

#### 2.1.1.1 Paradigm Change

In *Searching for safety*, Wildavsky (1988) renewed the discussion about safety by considering that maladaptive avoidance strategies may endanger society. He distinguished between avoidance strategies — that should be applied to risks where the probability of occurrence and consequences are documented and predictable — and resilience strategies that are better for managing risks where little information exists and it is difficult to predict probabilities and consequences. Wildavsky regarded resilience as "the ability to cope with unforeseen dangers after they are manifested" (p. 77). He warned against the rationalist premise that everything could be monitored, measured and predicted. This paradigm shift in the field of security called for the development of new capacities in terms of adaptation, flexibility and agility (Perelman 2007). Wildavsky's proposition was officially translated into public policy, as demonstrated by the following excerpt from the National Security Strategy of the United States: "As we do everything within our power to prevent these dangers, we also recognize that we will not be able to deter or prevent every single threat. That is why we must also enhance our resilience — the ability to adapt to changing conditions and prepare for, withstand, and rapidly recover from disruption "(p. 18). In other words, this change in paradigm spells the rejection of a purely rationalist vision of urban risk control.

The acceptance that man cannot control everything raises operational considerations about when to develop resilience. As an adaptive behaviour in response to a new situation, resilience can be accomplished *passively* in the time following a crisis, or it can be developed *proactively* before a crisis occurs by improving the system's capacity to deal with complex situations (Somers 2009). This distinction has been described as recovery resilience versus precursor resilience (Boin and Van Eeten 2013), or as *passive resilience* focused on recovery and reconstruction versus *transformational resilience* focused on reducing risks and vulnerabilities (Sudmeier-Rieux 2014: 75).

Is resilience a fundamental paradigm change? We consider that for urban risk managers trained to rationalize risk, the shift to resilience represents a major transformation. However, scholars taking a community perspective argue that resilience is inherently conservative as it aims to maintain the status quo by promoting a return to pre-crisis conditions at the expense of social transformation and challenges to unequal power relations in communities (MacKinnon and Derickson 2013; Sudmeier-Rieux 2014). These authors go back to Holling's (1973) original definition of resilience in ecological systems that denotes a system's ability to resist change, i.e. the severity of disturbance it can experience before changing state, and the pace at which it returns to equilibrium. In the same vein, Walker et al. (2004) consider that systems have the capacity of transformability, i.e. to create a fundamentally new system when external conditions (ecological, economic, social, political) make the existing system untenable.

In sum, discussions about resilience raise the possibility of a paradigm shift, but for whom, at what level and under which conditions? Some answers to these questions are provided in the next section, where we describe how the implementation of resilience implies governance change.

#### 2.1.1.2 A New Governance Model

The public administration literature emphasizes the important management changes required for the implementation of resilience policy. The core Weberian public administration principle of conformity, used to manage predictable and routine situations, is thought to be challenged by the emphasis on flexibility, network management and collaborative governance required to implement resilience (Bourgon 2009). Recent research recognizes the magnitude of transformation required (Matyas and Pelling 2015; Pelling and Manuel-Navarrete 2011; Scolobig et al. 2014; Stark 2014). Revi et al. (2014) reinforce the point, stressing that climate adaptation (sometimes used interchangeably with resilience) requires implementation reflexivity. For example, efficiency pressures and procedural rationality have been shown to frame how public administration might develop adaptability and point to the kinds of innovative approach public servants might envisage to address resilience (Stark 2014).

Resilience also entails the recognition of vulnerabilities caused by interdependencies between administrations, sectors and in society generally. Its implementation depends on collaboration in networks that bring together a wide range of public and private stakeholders (Boin and McConnell 2007; Therrien 2010), including citizen participation (Drabek and McEntire 2003). It further requires cultural and structural changes in public administration from stovepipe approaches to collaborative networks (Bourgon 2009; Perelman 2007), and the development of new adaptive capacities (Stark 2014; Matyas and Pelling 2015). In this sense, the development of resilience needs to be supported by a collaborative approach to governance (Emerson et al. 2012) in order to understand the different powers urban stakeholders can enact to call their city resilient.

In this regard, various resilience frameworks include components involving collaboration between stakeholders. For example, one of the four dimensions of the "100 Resilient Cities Framework" (Rockefeller Foundation & Arup 2015) calls for leadership and strategy driven by trust relationships between government, business and society, empowerment of a broad range of stakeholders, and long-term, integrated planning with a holistic vision. The United Nations Office for Disaster Risk Reduction (UNISDR) considers that an institutional and administrative framework to enhance the cooperation of all departments and external organizations towards urban risk reduction is one of the 10 essential elements to build resilience (UNISDR 2012). To assess inter-organizational resilience, Therrien (2010) focuses on factors such as coordination based on a common understanding of problems, resource exchanges, common language and values, and official agreements.

This discussion about the transformation of governance models needed to respond to contemporary challenges is consistent with broader propositions about the need for horizontal governance (Peters 1998) and a whole-of-government approach (Christensen and Lægreid 2007) to address inter-sectorial problems in general. It is important not to underestimate the internal challenges that accompany such a transformation, where units may see their interests threatened (loss of resources, power, and monopoly over certain issues, etc.) by organizational changes. For political leaders and managers in public organizations, there is also a need for clarification on "the resilience of what, from what? [and] for whom?" (Lebel et al. 2006. p. 15) and with regard to who is in charge.

#### 2.1.1.3 An Optimal State of Being or a Destination to Reach

Resilience carries different meanings that intermingle in the discourse and comprehension of actors involved in the development and implementation of resilience policy. Definitions of resilience in social science are influenced by different approaches to risk management, including ecological perspectives, safety science, organizational theories, technical-social system studies and disaster management (De Bruijne et al. 2010). Each of these approaches focuses on specific aspects of resilience and disregards others.

Many frameworks developed by scholars and organizations propose sets of qualities or resources that must be in place in order to be resilient. For example, various indexes assess community resilience by the aggregation of indicators such as educational level, access to water and health facilities, or GDP per capita (Normandin et al. 2009; Therrien et al. 2015). Organizational resilience indexes follow the same logic by evaluating organizational capacities in terms of sense making, innovation, positive relationship between employees, commitment to safety, etc. (Therrien 2010). These indexes are often based on empirical research that identifies factors that support the capacity of organizations to effectively cope with dangers and bounce back after crises. These factors are also based on organizations recognized as resilient despite managing high-risk products (La Porte 1996). The assumption is that if a system were to develop these characteristics and capacities before a crisis, it would be able to enact resilience when needed.

In this sense, resilience corresponds to a state of being supported by multiple characteristics and capacities. The definition of resilience highlights the notion of withstanding or recovering from a shock, but rarely addresses the underlying assumption that for such 'capacity' to occur in a system, its governance components need to be geared towards 'processes and outcomes' (Boin and Van Eeten 2013). This would imply a strategy of assessing the appropriate level of achievement of these features, followed by a development plan to address weaknesses.

However, evidence on outcomes that could help set goals and measure steps toward resilience is rarely discussed. Outcomes (or results) of resilience could be identified as reductions in urban risks and vulnerabilities before an event occurs. For example, a city that identified hot spots during heat waves could plan to diminish these over time. Outcomes could also be identified in terms of increasing the coordination capacity of municipal stakeholders to manage urban risks. A city could, for example, enable multijurisdictional teams to perform actions across departments.

This is not to be confused with indicators that measure the resilience of a system, but rather points to managerial and policy outcomes that could indicate a path to resilience.

In terms of outcomes, one of the challenges is to demonstrate what benefits accrue from being resilient. To promote investments in resilience, estimates have been developed to demonstrate the economic benefit of disaster risk reduction policies. However, these studies are based on different hazard types, levels of vulnerability and geographies (Shreve and Kelman 2014), which prevents the development of a general principle. Nevertheless, international organizations and governments use such data drawn from particular event estimates; for example, the World Bank has operated on the principle that one dollar invested in risk reduction will save seven dollars in the future. Further difficulties in measuring the outcomes of resilience arise if we consider that a resilient city living through a shock, crisis, or pandemic would be able to withstand the event with few measurable effects, rendering it difficult to prove that investment in resilience was actually beneficial.

In conclusion, we can therefore see that resilience needs to consider both. An optimal state of being with multiple governance capacities and based on measurable outcomes in order for the destination of resilience to become a clear path.

## 2.2 Methodology

This paper is part of a larger ongoing research program on urban resilience involving the cities of London and Montreal. Based on an action research approach, the objective is to understand how municipalities implement resilience based on comparative cases studies. This chapter presents our initial exploratory comparative research and the processes involved in our fieldwork (Dewey 1938), and more particularly on the definition of resilience. This first part of our work does not aim for generalization but was undertaken to gain knowledge in a specialized area and understand the context for the implementation of resilience, about which little is currently known. We use the same methodology from our paper on agenda setting and implementation (Therrien *et al.*, unpublished).

The analysis is based on data collected during 20 exploratory interviews (12 in London and eight in Montreal) and three workshops conducted with an initial convenience sample of resilience team members in both cities, municipal stakeholders working in related issues (environment, police, etc.), other municipal levels (borough), and external stakeholders. These workshops were important to understand how these actors interacted in efforts to implement resilience, and observe discussion of their different ideas about what was required of them. These interviews and workshops were conversational and the responses will require expansion at a later stage in the research program. All actors are directly or indirectly involved in the implementation of urban resilience in the two cities and usually work together in networks or consultative fora. In 2014, two workshops were conducted in London and one in Montreal, with about 15 participants in each city. Workshop participants

**Table 2.1** Questions for interviews and workshops

| What is resilience? (Intra) organizational elements of resilience |
|---|
| 1. Within your organization, how do you define resilience? |
| 2. Would you say that resilience in your organization has evolved in the last few years? If yes, how? (See if they mention past crises or events + an evaluation of how this changed conception of resilience) |
| 3. How would you describe your organization's commitment to resilience? (Note mentions of culture, training, coping and learning from past experiences, management practices and norms) |

included the same 20 people who were interviewed, along with other municipal actors involved in resilience in each city. Data were collected at these meetings in detailed notes taken by a member of the research team. In order to stimulate conversation, we developed two sets of questions. The first explores how resilience is defined, how it has evolved in their organization, and how they describe their organization's commitment to resilience. The second set of questions is designed to improve our understanding of how they cooperate with other organizations to increase city resilience. This chapter focuses only on the first set of questions. The specific questions are provided in Table 2.1. Data collection from interviews and focus groups was supplemented by a review of secondary sources such as city policies, internal documents produced by the actors and public documents produced by the cities of London and Montreal.

We use an interpretative inductive approach to better understand actors' views (Creswell 2013). Our interpretation is then analysed based on current knowledge about resilience and policy processes.

## 2.3 London and Montreal: How Networks Define Resilience

The cities of London and Montreal are both engaged in the "100 Resilient Cities" project pioneered by the Rockefeller Foundation in 2014, though their reflection on resilience predates that project. We will begin with a brief summary of the stakeholders involved in the policy discussion about resilience in each city, then present the results of our analysis.

### *2.3.1 Definition of Resilience in London*

At the beginning of 2000, following experience with a fuel crisis, outbreaks of bovine spongiform encephalopathy (BSE) and foot and mouth disease, severe flooding, and terrorist alerts following 9/11, the British government reviewed its emergency planning (Rothstein et al. 2013). This reflection at the national level led to the creation, in 2002, of the London Resilience Team (LRT), an organizational unit composed of permanent staff specialized in urban risk management, in liaison with experts from partner organizations such as emergency services, utility companies and transport organizations. The work of this team is guided by the London Resilience Partnership (LRP), a complex network composed of 170 public (emergency services, local authorities, Greater London Authority (GLA), health authorities and others government agencies) and privates partners (utility companies and various transport organizations). The Civil Contingencies Act of 2004 formalized the role of the LRT and LRP. Following the LRTs significant involvement in resilience ahead of the London summer Olympics, human resources on the Team have decreased, along with its position in London. Since 2015, this team has been hierarchically located in the Fire Brigade.

Besides the Team and Partnership, others actors are also involved in discussions about resilience, especially public organizations involved in climate change like the London Climate Change Partnership.

#### 2.3.1.1 Analysis

The term resilience appears very frequently in the titles of public servant positions and units (e.g., London Resilience Forum, Head of Governance and Resilience at GLA, Department for Communities and Local Government, Resilience and Emergencies Division) and in official documents (e.g., Business Resilience Planning, Emergency Preparedness, Resilience and Response of the National Health Services). This appropriation began in 2001 with the reflection surrounding emergency planning. Without looking at the exact meanings, we note that this widespread use of the term demonstrates that resilience is viewed positively by public authorities as an objective for which various units are responsible, and that it is relevant in many fields, including business, health and climate change.

So far, resilience appears as an attractive public policy solution, as demonstrated by its extensive use in London. During workshops, external actors in the resilience partnership stated their desire to see resilience principles included in additional priorities of the Mayor's Office. For them, the inclusion of resilience in official policy documents is an asset. As a broad concept related to various urban risks, it allows actors to add their specific issues (risk of flooding, infrastructure maintenance, climate change, etc.) to discussions. In this sense, stakeholders are working as policy entrepreneurs (Kingdon 1995) by strategically framing some political priorities in terms of resilience (Zahariadis 2003).

In addition, stakeholders mention that they have actively modified the language they use in order to influence existing definitions. This is starkly illustrated by the use of resilience as a substitute for terms that are more politically charged, such as climate change, or intimidating, like adaptation. Resilience is therefore used as a strategy for re-engineering existing definitions and creating opportunities for actors involved in climate change.

In terms of written definitions, the London Resilience Strategy (London Resilience Partnership 2013) describes resilience as "the ability of London to detect, prevent and if necessary to withstand, handle and recover from disruptive challenges"

(p. 4). This definition was directly influenced by the lexicon of UK civil protection terminology (Cabinet Office 2010), which defines resilience as the "ability of the community, services, area or infrastructure to detect, prevent, and, if necessary to withstand, handle and recover from disruptive challenges", and community resilience as "communities and individuals harnessing local resources and expertise to help themselves in an emergency, in a way that complements the response of the emergency services". First, these definitions emphasize the timing of resilience by explaining that it occurs before, during and after an event, with a preference for prevention. This broad definition allows stakeholders to find a place in the process according to their interests. Second, a resilient system is considered as able to recover from "disruptive challenges", like disorganizing and unusual events. This notion may be a winning card for those trying to attract the attention of policy-makers to events that cause extensive damage (such as computer attacks), while stakeholders working to reduce vulnerabilities must demonstrate clear evidence of the impacts these problems may pose.

During the workshops in London, three consensus points emerged. First, the actors consider that resilience requires having a broad, horizontal or systemic vision of the problems to solve. They also consider it essential to understand the interdependencies between challenges in order to act on vulnerabilities. In this sense, actors discuss resilience in terms of a new governance model to enable entities to work closely together. Workshop participants state that resilience requires changing the ways in which work is done, including breaking down silos. A second governance change frequently raised by participants is the need to adopt a long-term perspective because resilience results cannot be achieved in the short term. This is challenging as the governments that direct efforts and accord means have short-term time frames. More than just a governance discussion, this issue raises a desire to change the paradigm in which their actions take place by focusing less on resilience as a preparedness and recovery objective, and more on resilience as a proactive approach to reducing vulnerabilities. Third, despite the official definitions, stakeholders are still struggling to understand what the end-point of resilience is, what they are aiming for, and what specifically they must accomplish in order to reach a resilient state. So far, there has been no political or public discussion of the specific expectations of resilience. Workshop participants made the point that while there is a lot of writing on resilience by scholars and think thanks, these do not include concrete references to what actors are meant to accomplish. This is what we refer to as outcomes.

In summary, several stakeholders in London use the term resilience to better position problems they want to solve on the political agenda. On the one hand, resilience can become a buzzword that carries a number of divergent meanings, generating confusion when the population, stakeholders and policy makers are exposed to several definitions, and eventually making the term less attractive. This is particularly important as even the actors involved in resilience implementation have difficulty distinguishing the concrete goals they must reach. On the other hand, stakeholders use the term because it carries a sense of "transformation", even if that change is slow to operate effectively.

### 2.3.2 *Definition of Resilience in Montreal*

In Montreal, the administrative unit responsible for civil security is the "Civil Security Center". This unit is made up by bureaucrats trained in emergency management and various fields (e.g. environment, communication) and provides analysis and advice on urban risk management, prevention of major disasters and preparedness for the 19 boroughs, the departments of the central city administration, and the 14 reconstituted municipalities of the island of Montreal. It also supports coordination in case of an intervention. The coordinating body for civil security on the island of Montreal is the Civil Security Organization of Greater Montreal (Organisation de sécurité civile de l'agglomération de Montréal, OSCAM), which includes the 33 boroughs and other municipalities, central services of the city of Montreal (public works, police, water services, etc.) and external partners (public transport services, provincial Ministry of Health and Social Services, etc.). Its mandate is to prepare interventions and participate in the coordination of interventions when necessary (Centre de sécurité civile 2006). It is therefore a structure mainly oriented toward preparation and intervention, and, secondarily, to adaptation and the reduction of vulnerabilities. Following Montreal's recent nomination as one of the 100 Resilient Cities, part of the CSC unit became the Office of Resilience and Civil Security (*Bureau de la resilience et de la sécurité civile*) with an enlarged mandate.

#### 2.3.2.1 Analysis

The use of the term resilience by civil security stakeholders is relatively new in Montreal and in the province of Quebec. Resilience is slowly becoming embedded in bureaucrats' discussions due to its influence in international discussion on urban risk management. In 2008, the Ministry of Public Security of Quebec published a document on basic concepts in public security where it mentions that "the last decades have (...) been marked by the emergence of the concept of resilience" (p. 35). Inspired by the United Nations International Strategy for Disaster Reduction, resilience is defined here as "the ability of a system, community or a society potentially exposed to hazards to adapt, resist or change in order to establish and maintain structures and an acceptable level of functioning. Resilience therefore relates mainly to the ability to resist hazardous situations with minimal damage and to effectively recover thereafter" (Ministry of Public Security 2008, p. 31. our translation). On the one hand, this definition describes resilience as an inclusive process involving all phases of crisis management, and, on the other hand, it reaffirms the importance of preparation for emergency situations.

In its first Civil Protection Policy 2014–2024, the Ministry of Public Security included elements about community resilience. This definition becomes part of the stakeholder environment in which urban resilience discussions take place. "A resilient community is one that has developed, on a permanent basis, features that prevent the manifestation of a hazard where possible, limit its probability of occurrence, resist it when it occurs, and recover properly afterwards. Resilience is associated with the physical, social, economic and environmental conditions of society. Because it is based on characteristics of the community, it is related to issues and concerns that go well beyond those associated with civil security." (Government of Quebec 2014, p. 6, our translation). This official definition highlights two important aspects of resilience. First, resilience is presented as a state, with a series of conditions to develop in various fields. Second, the Ministry implicitly calls for a paradigm change by referring to issues that go beyond the traditional fields of civil security.

As in London, the term resilience is integrated, though less broadly, into policies, as seen in the Ministry of Health and Social Services of Quebec's Civil Security Policy (2014), and in Public Safety Canada's Action Plan for Critical Infrastructure (2014) and Counter-terrorism Strategy "Building Resilience against Terrorism" (2013). The term resilience has not been integrated into the names of key organizations and positions, nor does it appear in the titles of many documents.

Overall, our observations, document reviews, interviews and workshops in Montreal indicate that stakeholders involved in emergency management use the concept of resilience, though not on a routine basis. They know the term exists, often referring to it as a "buzzword", but rarely use it, partly because they are struggling to grasp what it involves in concrete terms. In the workshops, they state wanting to know more about what the outcomes of a resilient city might be, expressing, for example, that having resilience indicators would be helpful. These could be used as a tool to evaluate their own resilience and compare with other organizations or cities.

During the workshops, stakeholders explain that they view resilience as a general goal that implicitly requires network governance involving departments of the city, boroughs and external partners. This governance transformation links with their core mission of preparedness to better perform in the face of crisis. This view is associated with a reactive resilience. Fewer participants express a broader view of resilience that integrates all steps in crisis management (prevention, preparedness, intervention and recovery) to reduce vulnerabilities and promote adaptation. Moreover, some suggest that resilience involves addressing urban risks related to vulnerable populations, therefore adding a 'social resilience' dimension to the definition. Actors with this vision usually have coordination roles within the city, or manage issues that have multiple ramifications, such as the environment and health. Thus, a divergence in meanings exists, with some actors regarding resilience as proactive and others seeing it as reactive.

Strategically, the Office of Civil Security and Resilience also perceives resilience as an orientation that justifies the development of new tools and new forms of horizontal coordination, and ultimately enables them to fulfill their mission more effectively. To this end, actors plan to relate resilience to the priorities of the mayor and the director general of the city (the most senior official) by demonstrating how the implementation of resilience could be relevant to accomplishing two major priorities in the city: breaking down silos between administrative units, and developing the so-called 'Smart City'. The goal is to demonstrate how resilience can provide a solution to problems on the political agenda of the new administration and help garner political and administrative support for transformations needed to address integrative issues that affect civil security.

In conclusion, resilience is currently applied and understood by the majority of stakeholders as a logic applicable to preparation for crisis management, pointing toward a more reactive than proactive resilience. Changes implied by this conception would focus on partnership management. Nevertheless, some central actors consider that resilience must be achieved proactively by focusing more on adaptation through the reduction of vulnerabilities. Here, a more inclusive vision of resilience is proposed, which will require cultural change. The definitional ambiguity does not prevent resilience from being used as a means of "selling" governance transformation.

## 2.4 Discussion: Paradigm, Collaborative Governance, Networks and Outcomes of Resilience

### *2.4.1 A New Paradigm or Not?*

In looking into the definitions of resilience proffered by actors in London and in Montreal, it would seem that the question of paradigm shift remains unresolved. The change in paradigm expressed by Wildavsky (1988) in which a purely rationalist vision of urban risk control gives way to acceptance of the unpredictability of hazards, is not emerging. In fact, our analysis shows that rigorous action on urban risks is an important narrative in efforts to reduce vulnerabilities and promote adaptation. One observation of Wildavsky's has, however, taken root in the logic of actors, namely that urban risks are increasingly numerous, complex and, most importantly, inevitable. In this sense, avoidance as a goal has been replaced by reduction and mitigation; with this comes the realization that governance models need to change.

There is evidence of paradigmatic debate in discussions about the focus of action on resilience. In London, the change of vision to include both reactive and proactive resilience is well understood by street-level bureaucrats. Reactive resilience is what they used to do and need to do better, and proactive resilience is what they aim to accomplish by acting on adaptation and reducing vulnerabilities. They do not, however, have a clear policy directive, and this makes it difficult for them to enact the change in paradigm. They are stuck between their own systemic vision of resilience, and policy-makers who push the term resilience but do not follow through with implementation processes, even to the point of decreasing the resilience team's resources and inserting it lower down in the hierarchy of city organizations. In other words, the paradigm shift in London is being held back by the same actors who initiated the resilience movement.

Montreal is even further behind in terms of a paradigm shift. The focus on resilience is emerging mostly from operational street-level bureaucrats who are responsible for emergency preparedness and interventions, indicating that the shift towards proactive resilience is not yet fully accepted by the larger civil security community. In this case, we consider that alternative directions could be taken. First, the civil security community could integrate resilience into its basic paradigm by juxtaposing the traditional mandate of reactive resilience with a new vision of proactive resilience. Based on the desired transformation of the governance model, this option could include the integration of new actors working on long-term issues. This cultural change will need to be supported by leaders, and include a transformation of interests and powers, incentives, and knowledge dissemination. Second, the civil security community could decide to focus only on reactive resilience, leaving another community (either newly formed or existing) the opportunity to move ahead on proactive resilience.

Divergences in the views of different actors had already been identified in previous studies (Wagenaar and Wilkinson 2013). Our analysis demonstrates that this competition in positioning is not only a major challenge for the future of communities composed by stakeholders with a particular mission and professional culture, but that it takes place in a context influenced by leadership support, hierarchical position, and access to material and political resources. In both cities, resilience came to be advocated as a coherent policy concept to integrate activities in related areas in a large policy domain where participants are renegotiating shared ideas and objectives (May et al. 2006). This is challenging in a field that is not supported by integrated interests, often lacks political support, and is not driven by important interest groups or the general population (May et al. 2005).

### 2.4.2 New Network Governance?

Our analysis demonstrates that in both cities central actors consider that the objective of resilience must be accompanied by a transformation in modes of governance within the city and with external partners. Central actors favour horizontal management, integration of internal and external partners, and actions inserted within a long-term vision.

Despite the ambiguity of the concept, practitioners in both cities consider that they benefit from using the term resilience because it provides an opportunity for empowerment at the local level. Resilience is used at the operational level as a boundary object (Brand and Jax 2007; Meerow and Newell 2015) that needs to be supported by a compatible governance model. Boundary objects are conceptually plastic enough to adapt to different community visions, but also carry common vocabulary and ideas to coordinate different groups in intersecting social worlds (Star and Griesemer 1989, p. 393). In this sense, resilience may be conceptually more precise within each group and fuzzier in discussion among intersecting groups.

In this context, resilience not only crosses community barriers, but also supports the creation of an inter-sectorial governance model. A central idea in both reactive and proactive resilience is that implementation requires concerted action by a number of public and private organizations. Resilience is a solution for complex issues involving interdependent components and various fields of expertise. Instead of focussing on *ad hoc* projects, actors in both cities wish to position resilience within a strategic unit in the city to mobilize departments around shared objectives and develop strong relationships with external actors. It is for this reason that the hierarchical position of the units responsible for resilience becomes a criterion for success.

### 2.4.3 Resilience as a Destination or State of Being?

The two cities are facing similar challenges, as both are lacking a clear policy directive. In London, resilience is seen as a top-down policy with a variety of different interpretations at local government levels. The term has been commonly used in policy for a long time, but this does not prevent stakeholders from expressing difficulty in identifying the strategic aims of resilience. They question when a system can be considered resilient. In Montreal, the term resilience has been integrated only recently into the vocabulary of civil security practitioners and they have difficulty with the definitional question. In both cases, resilience is considered an important issue, but stakeholders are still trying to understand the strategic destination.

Urban stakeholders frequently attest to the difficulty of appropriating such a broad and multifaceted concept while struggling to develop a clear implementation plan. In London, actors state being aware of multiple frameworks developed to assess resilience. But for them, these frameworks do not add value to the work they already accomplish around a variety of urban risks. In Montreal, the issue of specific destination to reach has not yet been raised as the city is still at an early stage in the process. However, in both cities, actors use arguments publicized by international organizations about the economic returns on risk reduction in order to convince elected officials to give them the means to act.

Finally, neither city is, as yet, using frameworks or indexes to assess their level of resilience, track their progress or benchmark their achievements against others. The 100 Resilient Cities initiative may be able to address this issue, at least in the 100 cities selected, as the project developed a framework for cities to use in their policy planning. This affords cities the advantage of restricting resilience to its essential aspects rather than including so much in the concept that actors lose sight of purpose.

## 2.5 Conclusion

Resilience is an ambitious policy objective. First, communities of practice discuss the opportunity to change the paradigm from a mission of preparedness to one of vulnerability reduction. Second, city governance is called upon to build up horizontal management, network governance with other public and private organizations, and strong population involvement, to address complex problems, alongside other level of governments, with a comprehensive long-term vision. Third, the translation of resilience into a concrete objective where actors can understand the progress made and assess what remains to be done is challenging despite multiple attempts to elaborate frameworks.

This analysis demonstrates the existence of similar challenges around paradigm change and the appropriation of resilience in the city of London that embarked on the process several years ago, and in Montreal where the reflection was initiated more recently. The governance transformation is, in turn, influenced by specific political issues in each city and by the allocation of resources. Ultimately, however, our analysis implies that despite the important involvement of civil servants in resilience work and their support for the concept, officials and political actors will need to be pressured to act on this issue in order to provide direction and resources for this ambitious program. Otherwise, many of the defining issues involved in resilience are likely to remain on hold. This also implies that political actors, civil servants and the public in general will need to work out the definition of resilience: Who is it for? What level is needed? Resilience for what? Resilience to what?

## References


Boin, A., & McConnell, A. (2007). Preparing for critical infrastructure breakdowns: The limits of crisis management and the need for resilience. *Journal of Contingencies and Crisis Management, 15*(1), 50–59.

Boin, A., & Van Eeten, M. (2013). The resilient organization. *Public Management Review, 15*(3), 429–445.

Boin, A., Comfort, L. K., & Demchak, C. (2010). The rise of resilience. In L. K. Comfort, A. Boin, & C. C. Demchak (Eds.), *Designing resilience: Preparing for extreme events* (pp. 1–12). Pittsburgh: University of Pittsburgh Press.

Bourgon, J. (2009). New directions in public administration. *Public Policy and Administration, 24*(3), 309–329.

Brand, F. S., & Jax, K. (2007). Focusing the meaning(s) of resilience: Resilience as a descriptive concept and a boundary object. *Ecology and Society, 12*(1), 23–38.

Cabinet Office. (2010). *A strong Britain in an age of uncertainty: The national security strategy*. London. Retrieved from https://www.gov.uk/government/uploads/system/uploads/attachment_data/file/61936/national-security-strategy.pdf

Centre de sécurité civile. (2006). *Politique de sécurité civile de l'agglomération de Montréal*. Montreal. Retrieved from http://ville.montreal.qc.ca/pls/portal/docs/page/centre_secu_civile_fr/media/documents/politique_secu_civile_agglo_mtl.pdf

Christensen, T., & Lægreid, P. (2007). The whole-of-government approach to public sector reform. *Public Administration Review, 67*(6), 1059–1066.

Creswell, J. W. (2013). *Qualitative inquiry and research design: Choosing among five approaches*. Thousand Oaks, CA: Sage.

De Bruijne, M., Boin, A., & Van Eeten, M. (2010). Resilience: Exploring the concept and its meanings. In L. K. Comfort, A. Boin, & C. C. Demchak (Eds.), *Designing resilience : Preparing for extreme events* (p. 349). Pittsburgh: University of Pittsburgh Press.

Dewey, J. (1938). *Experience and education*. New York: Simon and Schuster.

Drabek, T. E., & McEntire, D. A. (2003). Emergent phenomena and the sociology of disaster: Lessons, trends and opportunities from the research literature. *Disaster Prevention and Management: An International Journal, 12*(2), 97–112.

Emerson, K., Nabatchi, T., & Balogh, S. (2012). An integrative framework for collaborative governance. *Journal of Public Administration Research and Theory, 22*(1), 1–29.

Government of Quebec. (2014). *Politique québécoise de sécurité civile 2014–2024: Vers une société québécoise plus résiliente aux catastrophes*. Québec.

Holling, C. S. (1973). Resilience and stability of ecological systems. *Annual Review of Ecological Systems, 4*, 1–23.

ISDR. (2005). *Hyogo framework for action 2005–2015: Building the resilience of nations and communities to disasters*. Paper presented at the extract from the final report of the World Conference on Disaster Reduction (A/CONF. 206/6).

ISDR. (2011). *Strategic framework 2025*. Geneva. Retrieved from http://www.unisdr.org/files/23291_1101657inteng.pdf

Kingdon, J. (1995). *Agendas, alternatives, and public policies*. New York: Longman.

La Porte, T. R. (1996). High reliability organizations: Unlikely, demanding and at risk. *Journal of Contingencies and Crisis Management, 4*(2), 60–71.

Lebel, L., Anderies, J. M., Campbell, B., Folke, C., Hatfield-Dodds, S., Hughes, T. P., & Wilson, J. (2006). Governance and the capacity to manage resilience in regional social-ecological systems. *Ecology and Society, 11*(1), 19.

London Resilience Partnership. (2013). *London resilience strategy*. London: London Resilience Forum. Retrieved from https://www.london.gov.uk/sites/default/files/gla_migrate_files_destination/London%20Resilience%20Partnership%20Strategy%20v1%20web%20version.pdf.

MacKinnon, D., & Derickson, K. (2013). From resilience to resourcefulness: A critique of resilience policy and activism. *Progress in Human Geography, 37*(2), 253–270.

Matyas, D., & Pelling, M. (2015). Positioning resilience for 2015: The role of resistance, incremental adjustment and transformation in disaster risk management policy. *Disasters, 39*(s1), s1–s18.

May, P. J., Jones, B. D., Beem, B. E., et al. (2005). Policy coherence and component-driven policy making: Arctic policy in Canada and the United States. *The Policy Studies Journal, 33*(1), 37–63.

May, P. J., Sapotichne, J., & Workman, S. (2006). Policy coherence and policy domains. *The Policy Studies Journal, 34*(3), 381–403.

Meerow, S., & Newell, J. P. (2015). Resilience and complexity: A bibliometric review and prospects for industrial ecology. *Journal of Industrial Ecology, 19*(2), 236–251.

Ministry of Public Security. (2008). *Concepts de base en sécurité civile*. Quebec. Retrieved from http://www.securitepublique.gouv.qc.ca/fileadmin/Documents/securite_civile/publications/concepts_base/concepts_base.pdf

Ministry of Public Security. (2014). *Politique québécoise de sécurité civile 2014–2024: Vers une société québécoise plus résilience aux catastrophes*. Retrieved from http://www.securitepublique.gouv.qc.ca/fileadmin/Documents/securite_civile/publications/politique_2014-2024/politique_securite_civile_2014-2024.pdf

Mitchell, A. (2013). *Risk and resilience: From good idea to good practice*. Retrieved from Paris, France: http://www.oecd.org/dac/governance-development/FINAL%20WP%2013%20Resilience%20and%20Risk.pdf

Normandin, J. M., Therrien, M.-C., & Tanguay, G. (2009). *City strength in times of turbulence: Strategic resilience indicator*. Paper presented at the Conference on City Futures, Madrid.

Pelling, M., & Manuel-Navarrete, D. (2011). From resilience to transformation: The adaptive cycle in two Mexican urban centers. *Ecology and Society, 16*(2), 11.

Perelman, L. J. (2007). *Shifting security paradigms: Toward resilience*. Retrieved from Fairfax.

Peters, B. G. (1998). Managing horizontal government: The politics of co-ordination. *Public Administration, 76*(2), 295–311.

Revi, A., Satterthwaite, D., Aragón-Durand, F., Corfee-Morlot, J., Kiunsi, R. B. R., Pelling, M., & Sverdlik, A. (2014). Towards transformative adaptation in cities: The IPCC's fifth assessment. *Environment and Urbanization, 26*(1), 11–28.

Rockefeller Foundation & Arup. (2015). *City Resilience Index: Understanding and measuring city resilience*. Retrieved from http://publications.arup.com/publications/c/city_resilience_index

Rothstein, H., Borraz, O., & Huber, M. (2013). Risk and the limits of governance: Exploring varied patterns of risk-based governance across Europe. *Regulation & Governance, 7*(2), 215–235.

Scolobig, A., Linnerooth-Bayer, J., & Pelling, M. (2014). Drivers of transformative change in the Italian landslide risk policy. *International Journal of Disaster Risk Reduction, 9*, 124–136.

Shreve, C. M., & Kelman, I. (2014). Does mitigation save? Reviewing cost-benefit analyses of disaster risk reduction. *International Journal of Disaster Risk Reduction, 10*, 213–235.

Somers, S. (2009). Measuring resilience potential: An adaptive strategy for organizational crisis planning. *Journal of Contingencies and Crisis Management, 17*(1), 12–23.

Star, S. L., & Griesemer, J. R. (1989). Institutional ecology, translations and boundary objects: Amateurs and professionals in Berkeley's Museum of Vertebrate Zoology, 1907–39. *Social Studies of Science, 19*(3), 387–420.

Stark, A. (2014). Bureaucratic values and resilience: An exploration of crisis management adaptation. *Public Administration, 92*(3), 692–706.

Sudmeier-Rieux, K. I. (2014). Resilience - an emerging paradigm of danger or of hope? *Disaster Prevention and Management, 23*(1), 67–80.

Therrien, M.-C. (2010). Stratégies de résilience et infrastructures essentielles. *Télescope, 16*(2), 154–171.

Therrien, M.-C., Tanguay, G. A., & Beauregard-Guérin, I. (2015). Fundamental determinants of urban resilience: A search for indicators applied to public health crisis. *Resilience, 3*(1), 18–39.

UNISDR. (2012). *Making Cities Resilient Report 2012. My city is getting ready! A global snapshot of how local governments reduce disaster risk*. Retrieved from http://www.unisdr.org/files/28240_rcreport.pdf

Wagenaar, H., & Wilkinson, C. (2013). Enacting resilience: A performative account of governing for urban resilience. *Urban Studies, 30*(Special Issue: Governing for Urban Resilience).

Walker, B., Holling, C. S., Carpenter, S. R., & Kinzig, A. (2004). Resilience, adaptability and transformability in social–ecological systems. *Ecology and Society, 9*(2), 5.

Wildavsky, A. (1988). *Searching for safety*. New Brunswick: Transaction Books.

Zahariadis, N. (2003). *Ambiguity and choice in public policy. Political decision making in modern democracies*. Washington, DC: Georgetown University Press.

# Chapter 3
# Putting Resilience into Practice. The Spatial Planning Response to Urban Risks

**Grazia Brunetta and Ombretta Caldarice**

**Abstract** Risks are growing dynamics of urban daily life in the twenty-first century. It mainly affects cities because their rapid and often unplanned expansion exposes a large number of people to unexpected events. For the purpose of considering urban vulnerabilites, an increasing number of cities is engaging in designing adaptation plans and strategies focused on resilience, which is acknowledged as one of the top priorities of the development agenda and a primary guiding principle of policy governance of our time. Nevertheless, there is little empirical evidence to demonstrate that adaptation planning is useful, as it is more concerned with having a broad vision rather than about specific actions. Despite this, the critical role played by spatial planning in addressing both the mitigation of climate change and adaptation to its unavoidable impacts has been theoretically acknowledged. This view stems from the admission that the spatial configuration of cities has significant effects on climate change and has become central to enacting adaptive responses within a resilient framework.

Starting from these remarks, the paper reports a qualitative comparison between the local climate adaptation strategies of Bologna (Italy) and London (UK). The analysis was conducted in order to understand why cities continue to engage adaptation planning, and the response of spatial planning in vision-oriented strategies and action-based plans for adaptation towards resilience. Basically, the paper discusses the gap between theory and practice of planning in achieving a resilient approach to urban risks through adaptation.

**Keywords** Urban resilience · Adaptive planning · Urban risks · Climate change · Spatial planning

---

G. Brunetta (✉) · O. Caldarice
Interuniversity Department of Regional and Urban Studies and Planning, Politecnico di Torino, Torino, Italy
e-mail: grazia.brunetta@polito.it; ombretta.caldarice@polito.it



G. Brunetta et al. (eds.), *Urban Resilience for Risk and Adaptation Governance*, Resilient Cities, https://doi.org/10.1007/978-3-319-76944-8_3

## 3.1 Risk, Resilience and Spatial Planning[1]

In 2008, the world reached a momentous milestone (Clapp 2014) with over 50% of the world's population living in urban areas, 75% in Europe alone. Hence, the next decades will witness the unprecedented development of urban areas driven by the extremely rapid growth of the world's urban population. In this perspective, cities have become key areas for global challenges. As outlined in the Pact of Amsterdam signed in 2016, post-industrial societies are characterised by unprecedented dynamism and a fast rate of change, since most of the world population lives in urban settings today.

This rapid and often unplanned expansion of cities is exposing a greater number of people and economic assets to the risk of disaster and the effects of climate change. Urban systems face a variety of risks. On the one hand, we find natural risks related to climate change that generate hydrogeological risk (landslides and floods), meteorological risk (urban heat island and storms), geophysical risk (volcanic activity and earthquakes) and climatological risk (drought and wildfires). On the other hand, anthropic risks are evident in the population growth and urbanisation related to soil sealing, the scarcity of available energy resources, the increasing gap between rich and poor countries and the ageing of technological products. Practically, risk occurrences present a growing trend, if we consider that the disasters have almost quadrupled worldwide during the past 30 years (UNISDR 2012). Risks are today becoming central issues of twenty-first century daily life (Fig. 3.1). In Europe, for instance, it has affected more than 70 million people from 1900 to 2017, leading to the death of more than 9 million people and causing estimated economic damages for up to 400 billion USD.

At the same time, cities have been perceived as a haven from disaster and as buffers against environmental changes (Wamsler et al. 2013). Hence, the city is not only the site where major actions are needed but also a place where feasible solutions can be developed and, later, learned and replicated (Caldarice 2018). Not surprisingly, and disproving some opinions, cities have come back as the most important area of research and experimentation (Glaeser 2011) in which innovation can be pushed (Florida 2017).

In order to provide a broad answer to climate change and socio-economic uncertainties, city authorities worldwide should increasingly place resilience at the heart of their policy-making activity. Indeed, resilience is now one of the top priorities of the sustainable development agenda for risk-related issues as it is considered the primary guiding principle of policy governance and a key political category of our time (Chandler 2014). In this perspective, resilience is deemed a positive concept and "*cities of resilience may suggest images such as staying power, or flexibility, or adaptability […]*". (Pickett et al. 2004: 370).

[1]This Chapter is the result of the combined research activity of the two authors. The final written version of the Chapter can be attributed to both the authors.

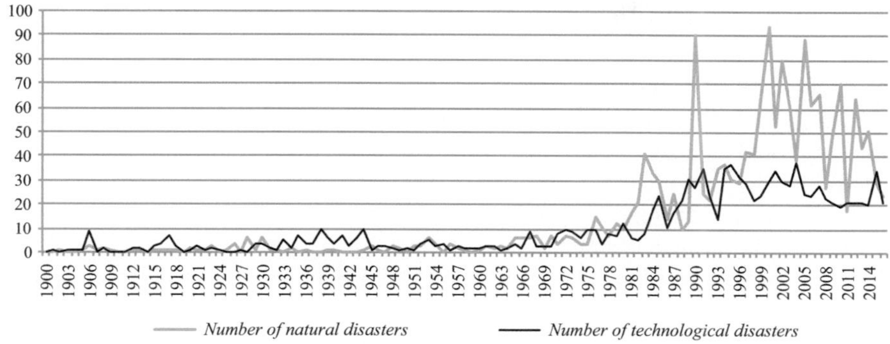


**Fig. 3.1** Number of natural and technological disasters in Europe (1900–2016) (*Source:* Authors' elaboration based on the International Disasters Database)

From a practical point of view, city authorities are increasingly facing risks by defining local adaptation plans and strategies, which underpin the concept that better urban management and governance is at the heart of reducing disasters and making cities safer. Hence, building resilience has become a major component of climate adaptation, environmental management, regional economic development and strategic planning (Davoudi et al. 2013). Nevertheless, most of the principal global cities are strongly engaged in climate adaptation planning[2] (Aylett 2014), but there is little empirical evidence to prove that urban adaptation planning is useful as it focuses more on a broad vision than on specific actions. (Siders 2017).

Intended to contribute to the theoretical knowledge and practical implications of spatial planning, the goal of this paper is to critically review and compare local climate adaptation strategies in two experimentations conducted on a metropolitan scale, the first in the UK (London) and the second in Italy (Bologna). They are internationally recognised as pioneer approaches on a local scale. On the basis of this empirical analysis, the paper discusses the gaps between theory and practice from a planning perspective, discussing the response of spatial planning to achieve a resilient risk dimension for urban risks' management. Basically, the paper supports the idea that fostering resilience needs to involve planning not only for recovery from shocks, but also to encourage preparedness and seek potential transformative opportunities that emerge from changes. This is true because spatial planning needs to acknowledge that different types of risks are inter-connected, and that adaptation plans can be useful to manage and respond to risks only if the location and nature of a risk are known, and if adaptation is mainstreamed in the planning process.

[2] As outlined by Aylett (2014), 75% of the 350 cities that are members of ICLEI - Local Government for Sustainability - are writing or have already defined a national climate adaptation strategy.

## 3.2 Defining Resilience in the Planning Domain

Resilience is a challenging concept that is interpreted both as a vacuous buzzword, considering its overuse and ambiguity (Rose 2007), and as a generative metaphor since it should emphasise innovative urban dynamics (DeVerteuil and Golubchikov 2016).

Resilience has become a central priority in international urban agendas because it sustains the principal strategies for climate change adaptation, i.e., the 2030 Agenda for Sustainable Development of the United Nations Development Programme. This latter principle has been identified by Europe as the primary approach to be used to tackle societal challenges in the actual Horizon 2020 actions. Furthermore, in its document that prepares the next Horizon 2020 Programme, Europe confirms that the preparedness of society to face crisis has to be increased, and technologies and institutions have to strengthen resilience, which becomes a prominent concern.

The meaning of resilience has been studied since the first half of the twentieth century in different scientific fields. In the course of time the concept has undergone a significant evolution. From the original idea pertaining to engineering, which considers resilience as the capacity of systems to return to an equilibrium or steady state after a disturbance (Holling 1973), to a more recent one belonging to socio-ecological systems, which defines resilience as the capacity of a system to adapt, innovate and transform certain conditions into new more desirable configurations (Folke et al. 2010). Despite the differences in these approaches, they are based on the ability of systems to spring back after being stressed or to maintain their functions by absorbing or adapting to change, despite stresses and pressures placed on them.

Mainly in the last decade, the concept of resilience has firmly established itself in planning theory and practical debates. It has emerged as an attractive perspective for cities, which are generally theorised as highly complex socio-spatial systems with an extensive and unpredictable feedback process that operate at multiple scales and timeframes (Portugali 2008). This perspective, which is defined *evolutionary resilience* (Davoudi 2012), implies that resilience is not a fixed asset but a continually changing process that is not simply the reaction to external disturbances but, conversely, a new paradigm for spatial development that depends on a certain degree of self-organisation and creativity in urban systems (Brunetta 2016). Basically, "*urban resilience refers to the ability of an urban system and all its constituent socio-ecological and socio-technical networks across temporal and spatial scales to maintain or rapidly return to desired functions in the face of a disturbance, to adapt to change, and to quickly transform systems that limit current or future adaptive capacity*". (Meerow et al. 2016: 39).

The resilient approach is extensively used in the risk domain as it refers to the importance of reducing disaster and climate change at a local level (Kim and Lim 2016). Unlike risk management, i.e., the ability of countries to manage change by maintaining their living standards in the face of shock or stress without impairing

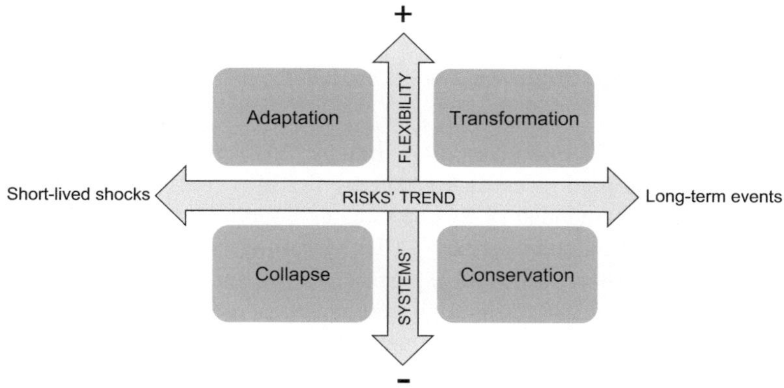


**Fig. 3.2** Elements of risk resilience approach (*Source:* Authors' elaboration)

their long-term prospects (DFID 2011), risk resilience focuses not only on risk reduction, prevention, preparedness and mitigation but also on building adaptive capacity. This approach is mainly characterised by four core elements (Fig. 3.2):

- Context, whose resilience is being built - such as social group, socio-economic or political system, environmental context or institution - related to the degree to which the urban system can organise itself;
- Disturbance, what type of shock - sudden events, like conflict or disasters and stress or long-term trends, like resource degradation, urbanisation, or climate change - the city aims to be resilient to;
- Capacity to respond, the ability of a system to deal with a shock or a stressor depends on the trend of the risk and on system flexibility, i.e., the degree to which a system responds to a stressor;
- Resilient dimension, the range of possible responses to shock and stresses, i.e., collapse, conservation, adaptation, and transformation.

Summing up, risk resilience is conceived as a process that identifies a critical element, assessing the alternatives to manage urban risks, and implementing the most sustainable actions to improve urban quality. As outlined by Davoudi et al. (2013), cities become more or less resilient depending on their preparedness, the capacity to enhance their chances of resisting disturbances (being robust), or by absorbing disturbances without crossing a threshold into an undesirable and possibly irreversible trajectory (being adaptable), or by moving towards a more desirable trajectory (being evolutionary and transformative). In this scenario, spatial planning is facing the big challenge of defining how to intervene to manage risks and increase resilience in order to establish evolutionary adaptation strategies that merge together risk trends and the socio-economic and environmental features of urban systems.

## 3.3 London Climate Change Adaptation Strategy (2011)

### 3.3.1 Institutional Framework

The London Climate Change Adaptation Strategy "Managing Risks and Increasing Resilience" was adopted in October 2011 by the Mayor Boris Johnson[3] as the adaptation strategy is mandatory for the Greater London Authority. The Mayor's office kicked off the initiative in early 2010 by calling on London's more than 8 million residents to share ideas and expertise in an online forum. London's adaptation plan is based on the national climate change projections 2002–2009 created by the UK's official centre for climate change research at the regional level. (Siders 2017).

### 3.3.2 Planning System and Tools

The United Kingdom is characterised by a performative planning system. This means that planning tools use indicative zoning and that the strategy for urban development is transferred into a non-binding land use plan, while the rights of land use are directly assigned by urban projects (Muñoz Gielen and Tasan-Kok 2010). This model refers to the indicative theory of zoning (Faludi 1987), and aspires to promote projects capable of 'performing' a collective strategy.

As stated in the National Planning Policy Framework (2012), the British planning system is based on a two-tier system, which distinguishes tasks to be implemented in the supra-local level from those which can be enacted at the local level through specific projects and programmes (Rydin 2011). In this perspective, local authorities have to identify the core strategy, i.e., a general growth and development strategy, which will subsequently be supported by more detailed programmes for specific areas. The City of London is planned through the London Plan, a spatial development strategy revised in 2016 that sets out an integrated economic, environmental, transport and social framework for London's development. In addition, after the Localism Act (2011), local communities are required to draw up neighbourhood plans for their area, which must generally comply with the policies of the London Plan. Finally, in order to implement the London Plan, supplementary planning guidances (SPGs) offer project-related details.

[3] From 2007, the Mayor of London has a "climate change duty" and is responsible for climate change adaptation, mitigation and energy strategies.

**Table 3.1** London Adaptation Strategy key actions to the manage high temperatures

| Action | Lead | Partners | Delivered by |
|---|---|---|---|
| The Mayor will work with partners to enhance 1.000 ha of green space | GLA | Boroughs, private and voluntary | Winter 2012 |
| The Mayor will work with partners to increase central green cover by 5% by 2030 and a further 5% by 2050 | GLA | Boroughs, private and voluntary | 2050 |
| The Mayor will work with partners to increase green cover across London by 5% by 2025 | GLA | Boroughs, private and voluntary | 2025 |
| The Mayor will work with partners to enable the delivery of 100.000m$^2$ of new green roofs | GLA | Boroughs, private and voluntary | Winter 2012 |

*Source:* Authors' elaboration based on Nickson et al. (2011)

### 3.3.3 Reasons to Adapt to Climate Change

London ranks among the most resilient and sustainable (Batten and Edwards 2015), yet vulnerable cities in the world (Barkham et al. 2014). The aims of the London Adaptation Strategy are to assess the consequences of climate change in order to prepare London for the impacts of climate change and extreme weather, and to enhance the quality of life of Londoners.

Davoudi et al. (2011) identify three reasons why London needs to adapt:

- London's population is projected to grow by over a million people over the next two or three decades;
- London does not adapt very well to extreme weather that can have a negative impact on the city;
- London will experience an increasing risk of floods, drought and uncomfortably hot weather.

Basically, without defining actions to manage risks, the impact of a changing climate will increasingly affect the prosperity and quality of life of the city.

### 3.3.4 Structure of the Climate Strategy

The London Climate Change Adaptation Strategy is organised in three steps:

- understanding and managing the impact in terms of flooding, drought and overheating;
- assessing the impact on issues that affect Londoners' health, London's environment, London's economy (business and finance) and London's infrastructural framework (transport, energy and waste);
- implementing the strategy with key actions by providing a *roadmap to resilience* (Table 3.1).

### *3.3.5 Integration with Planning Systems and Tools*

The adaptation principles are mainstreamed in the London Strategy and, consequently, in the 33 neighbourhood plans of the London GLA. In particular, Chapter V of the London Plan is focused on London's response to climate change. The chapter sets out a comprehensive range of policies to underpin London's response to climate change, including risk management issues. These policies cover climate change mitigation and adaptation, waste, contaminated land and hazardous substances. For instance, Policy 5.10 is oriented towards urban greening as it uses the benefit of vegetation and green spaces to cool the city, absorb and retain floodwater and support biodiversity.

## 3.4 Bologna Adaptation Strategy Plan (2015)

### *3.4.1 Institutional Framework*

The experience of the Bologna Adaptation Strategy Plan "Bologna Resilient City" started in the context of the European Mayors Adapt Initiative, a project launched in 2014 to involve European cities in developing climate change resilience and adaptation actions (Brunetta and Caldarice 2019). Practically, the Bologna Adaptation Plan to Climate Change has been defined in the project LIFE+ BLUE AP "Bologna Local Urban Environment Adaptation Plan for a Resilient City", and was created to increase Bologna's resilience in the face of climate change[4]. By taking part in the Mayors Adapt Initiative, the City of Bologna has undertaken to:

- evaluate the potential threats of climate change and vulnerability;
- detect and prioritise actions by designing a local adaptation strategy;
- enforce local adaptation actions;
- monitor progress by issuing a report every two years;
- adapt and implement the resilient strategy at the local level.

### *3.4.2 Planning System and Tools*

Unlike the UK, Italy is characterised by a conforming planning system. This means that planning tools use binding zoning and an urban development strategy, while the rights of land use and transformation are preventively inserted in the land use plan

[4]BLUE AP was coordinated by the City of Bologna and involved partners as the Kyoto Club, Ambiente Italia and ARPA Emilia Romagna. Today, Bologna and Ancona are the unique Italian cities to have a local adaptation plan. The City Padua has currently written the municipal adaptation plan.

(Muñoz Gielen and Tasan-Kok 2010). This model refers to the mandatory theory of zoning (Faludi 1987) and aspires to 'conform' single projects to a collective strategy.

From 2008, the City of Bologna is planned through the Municipal Structural Plan (PSC), which defines the strategic scenarios of urban development, the Municipal Operative Plan (POC), which defines big urban projects, and the Urban Planning Scheme (RUE) that defines the rules of urban development. Basically, the Bologna local plan takes into account the issue of improving diffused and urban environmental quality within its boundaries.

### 3.4.3 *Reasons to Adapt to Climate Change*

Bologna is located in an increasingly vulnerable area and its urban environment is characterised by three major critical elements in terms of climate change:

- drought and water shortage caused by the amount of water consumed for civilian use (43.2 million cubic metres in 2012);
- unconventional weather events and hydro-geological risk;
- heat waves in the urban area with an average temperature increase of 2 °C.

In this perspective, the main goals of the Bologna plan are, on the one hand, the intention to complement the political agenda with themes that link climate change through the involvement of all stakeholders in the decision-making process and, on the other hand, the awareness of a wider perception of effects produced by climate change among citizens (Barbi et al. 2016).

### 3.4.4 *Structure of the Climate Strategy*

The Bologna Adaptation Strategy Plan is structured in one vision, three objectives, 12 strategies and 60 actions both for the municipal and the metropolitan level (Fig. 3.3). It defines goals until 2025, and describes actions required to accomplish them. Formerly, the Bologna City Council approved the Plan in October 2015. The process was characterised by a strong participation of citizens, politicians and industry representatives. Finally, the implementation of the Plan moves along an action-monitoring phase to assess its efficacy in order to promote knowledge transfer and good practices to other Italian municipalities.

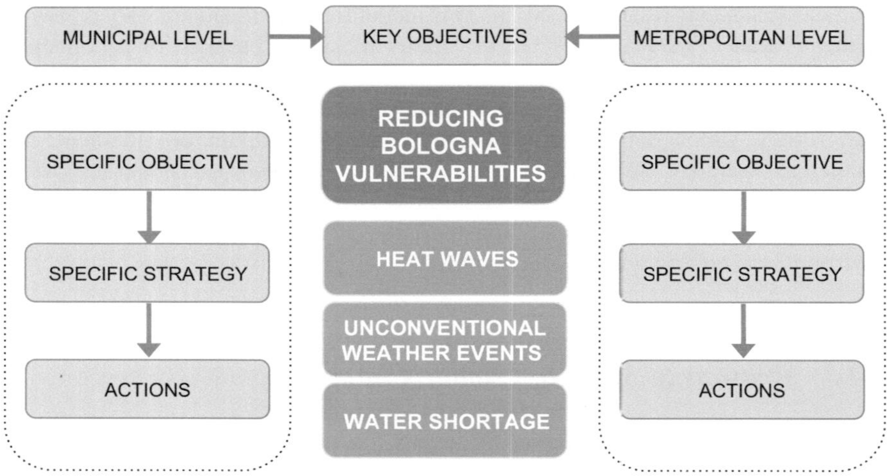


**Fig. 3.3** Logical Scheme of the BASP (*Source:* Authors' elaboration based on Barbi et al. 2016)

### *3.4.5 Integration with Planning System and Tools*

A distinctive element of the Bologna Adaptation Strategy Plan is its integration with other public policies, not only focused on the environment but also concerning urban planning and social and economic measures. From a planning perspective, some amendments were applied to urban plans adopted in 2008 and still in force, such as PSC, POC and RUE, to integrate them with the principles and actions of the Bologna Adaptation Strategy Plan. The Urban Planning Scheme (RUE) was modified in 2015 to provide stricter criteria regarding drought and water shortage, i.e., mandatory collection and recovery of rainwater for residential and rural settlements. The Municipal Operative Plan (POC) was modified in 2014 and the single urban projects were reviewed according to the guidelines of the Bologna Adaptation Strategy Plan in terms of better rainwater management in case of intense precipitation events by exploiting permeable parking, while reducing water consumption in the new urban settlements.

## 3.5 The Necessary Resilience-Planning Alliance

For more than a decade, spatial planning has been seen as playing a critical role in determining and delivering sustainable development (Howard 2009), since planning has not been able to properly deal with the uncertainty of complex urban systems (Marshall 2012). As a result of recent changes in the nature of spatial planning, land use planning is now identified as the most effective tool to reduce exposure and sensitivity to extreme events in many instances. As outlined by Campbell (2006:

206), "*there is then a growing sense that spatial planning not only has an important role in addressing the causes and impacts of climate change, but that it is increasingly required to do so*".

Spatial planning is conceived in this paper both as critical thinking about space and place (Mazza and Bianconi 2014) and as a human institution that creates a set of public rules enforced by the legal system. In this perspective, spatial planning involves not only the regulatory framework but also the institutional and social resources through which the system is implemented, challenged and innovated (North 1990). As outlined by Davoudi et al. 2009:15), this means that "*[…] spatial planning processes provide a key arena in which integrated approach to adaptation and mitigation can be designed, trade-offs between these and other social and economic goals can be negotiated, conflicts of interest can be mediated and intra- and inter-general equity concerns can be considered*". In addition, spatial planning can transform urban systems from passive to proactive in facilitating the climate change challenge. Hurlimann and March (2012) recognised six spatial planning capacities to adapt urban systems to climate change:

- planning has the ability to act on and coordinate matters of collective concern or public good;
- planning can facilitate the consideration of competing interests;
- planning is a way of thinking and acting across various spatial, temporal and governance scales, while targeting local problems;
- planning can provide a mechanism to deal with changing circumstances;
- planning has the capacity to be a repository for spatial knowledge sets;
- planning is oriented to the future and has the potential to coordinate different actions to achieve long-term benefits.

Hence, theoretically, spatial planning seems to be able to fully respond to the unavoidable challenge of climate change. But most scholars agree that the integration of climate protection with spatial planning seems to have taken place mainly at the level of rhetoric and principle, and there are actual challenges in translating these good intentions into practice (Barnes and Nel 2017; Carmin et al. 2012). The analysis of Bologna and London experiences reveals the strong tie between National Adaptation Strategies and the local framework that underlies the necessary integration of climate change into the planning system (Table 3.2). As a matter of fact, while national governments have legally binding commitments to control climate change, local authorities play a key role in designing and implementing planning policies for efficient climate adaptation. (Agyeman et al. 1998).

From this comparative analysis, it could be first said that the strategic themes and the specific adaptation measures proposed by London Strategy and Bologna Plan are similar and are, thus, relatively neutral with regard to their context. Examples of actions include issues, such as using vegetation to reduce temperature or rainwater harvesting and storage for dealing with water scarcity or drought.

Secondly, both London and Bologna have interpreted resilience in the domain of the climate change challenge as a reactive approach oriented towards the management of critical urban issues. This is the case because adaptation strategies and actions

**Table 3.2** Comparison between London Strategy (2011) and Bologna Plan (2015)

| | London climate change adaptation strategy | Bologna adaptation strategy plan |
|---|---|---|
| *Goal* | Managing risks and increasing resilience | Reduce Bologna's vulnerabilities |
| *Leader* | Mayor | Mayor |
| *Institutional level* | Greater London Authority | City of Bologna |
| *Compulsoriness* | Yes. From 2007, the Mayor of London have a 'climate change duty' | No. The Plan has been defined in the European project LIFE+ BLUE AP (2012–2015) |
| *Relationship with NAS* | Yes with the Climate Change Act (2008) | Not directly as the draft of the Italian NAS was approved in 2015 |
| *Key areas of action* | Flood Risk<br>Drought<br>Overheating | Drought and water<br>Unconventional weather events and hydro-geological risk;<br>Heat waves |
| *Relationship with planning system and tools* | The Chapter V of London Plan – revised in 2016 – is focused on the London's response to climate change. The chapter sets out a comprehensive range of policies to underpin London's response to climate change covering climate change mitigation and adaptation, waste, aggregates, contaminated land and hazardous substances | The Urban Planning Scheme (RUE) was modified in 2015 in order to provide stricter criteria related to drought and water shortage. The urban projects will be review according to the guidelines of the Bologna Adaptation Strategy Plan |
| *Implementation* | It forecasts an action plan with some key actions directly implemented | It forecasts some key actions and pilot projects that will be implement in the Municipal Operative Plan (POC) and in different Executive Urban Plans (PUA) |

*Source:* Authors' elaboration

mainly focus on the physical features of cities rather than on the construction of social, cultural and institutional change.

Thirdly, despite the different planning systems, both London Strategy and Bologna Plan are not totally mainstreamed. This is particularly evident during the implementation phase as they do not provide any guidance regarding aspects, such as urban actor's responsibilities, incorporation of risk reduction and adaptation in the municipal budget, and monitoring of the planned actions.

In a nutshell, Bologna and London experiences suggest that adaptation planning may well be able to achieve the benefits desired by city authorities in terms of motivation rather than implementation or use in decision-making. Nevertheless, spatial planning is theoretically able to respond to adaptation. The perception deriving from the above analysis is that planning practices are not able to completely understand the relationship between urban changes and adaptation.

This article is rooted in the perspective of a certain inability of planning tools to apply resilience in practice, and the need to move towards a more inclusive theoretical framework that can make sense of socio-spatial practices entailing more complex problems (De Roo et al. 2012; Innes and Booher 2010). In the light of this, the article sustains that adaptation should be inserted into planning by thinking in a mainstreamed comprehensive way. Hence, adaptation should be systematised and systematically incorporated into planning policies by shifting focus from specific responses to vulnerabilities to a resilient perspective that integrates adaptive strategies, dynamic processes and urban development. This idea entails that know-how related to adaptation should converge into local policy decision-making aimed at procuring resources in order to support the broad range of activities needed to boost the resilient transition of urban ecosystems (Gabellini 2016). Spatial planning should be rethought for action in order to reframe resilience as a process in which adaptive strategies leave the room to embrace emerging solutions relevant to the specific context. Hence, adaptation should be incorporated into spatial planning as a cross-cutting issue that is not bound either spatially or temporally. Practically, adaptation should not be planned as a desired end state but as an ongoing process in which spatial planning will be reframed within a resilient perspective for climate change. This means that resilience is correctly interpreted as an ability of urban systems to positively develop dynamic capacity in a co-evolutive dimension (Davoudi 2012) in which urban authorities, civil society and citizens are highly involved.

## References


Agyeman, J., Evans, B., & Kates, R. (1998). Greenhouse gases special: Thinking locally in science, practice and policy. *Local Environment, 3*(3), 245–246.

Aylett, A. (2014). *Progress and challenges in the urban governance of climate change: Results of a global survey*. Cambridge: MIT.

Barbi, V., Fini, G., & Gabellini, P. (Eds.). (2016). *Bologna città resiliente. Sostenibilità energetica e adattamento ai cambiamenti climatici*. Bologna: Bologna Urban Centre.

Barkham, R., Brown, K., Parpa, C., Breen, C., Carver, S., & Hooton, C. (2014). *Resilient cities: A Grosvenor research report*.

Barnes, A., & Nel, V. (2017) Putting spatial resilience into practice. *Urban Form*. https://doi.org/10.1007/s12132-017-9303-6

Batten, J., & Edwards, C. (2015) *Sustainable cities index 2015*. Arcadis.

Brunetta, G. (2016). The resilience concept and spatial planning in European protected areas development. Facing the challenges of change. In T. Hammer, I. Mose, D. Siegrist, & N. Weixlbaumer (Eds.), *Parks on the future! Protected areas in Europe challenging regional and global change* (pp. 25–36). Munchen: Oekom.

Brunetta, G., & Caldarice, O. (2019). Planning for climate change: Adaptation actions and future challenges in the Italian cities. In F. Calabrò, L. Della Spina, & C. Bevilacqua (Eds.), *New metropolitan perspectives. ISHT 2018. Smart innovation, systems and technologies* (Vol. 101, pp. 609–613). Cham: Springer.

Caldarice, O. (2018). *Reconsidering welfare policies in times of crisis. Perspectives for European cities*. Dordrecht: Springer.
Campbell, H. (2006). Is the issue of climate change too big for spatial planning? *Planning Theory & Practice, 7*(2), 201–230.
Carmin, J., Nadkarni, N., & Rhie, C. (2012). *Progress and challenges in urban climate adaptation planning: Results of a global survey*. Cambridge: MIT.
Chandler, D. (2014). Beyond neoliberalism: Resilience, the new art of governing complexity. *Resilience, 2*(1), 47–63.
Clapp, J. A. (2014). *The city: A dictionary of quotable thoughts on cities and urban life*. New Brunswick/London: Transaction.
Davoudi, S. (2012). Resilience: a bridging concept or a dead end? *Planning Theory & Practice, 13*(2), 299–307.
Davoudi, S., Crawford, J., & Mehmood, A. (Eds.). (2009). *Planning for climate change: Strategies for mitigation and adaptation for spatial planners*. London: Earthscan.
Davoudi, S., Mehmood, A., & Brooks, L. (2011). *The London climate change adaptation strategy: Gap analysis*. Newcastle: Newcastle University.
Davoudi, S., Brooks, E., & Mehmood, A. (2013). Evolutionary resilience and strategies for climate adaptation. *Planning Practice and Research, 28*(3), 307–322.
De Roo, G., Hillier, J., & van Wezemael, J. (2012). *Complexity and planning*. Farnham: Ashgate.
DeVerteuil, G., & Golubchikov, O. (2016). Can resilience be redeemed? *City, 20*(1), 143–151.
DFID. (2011). Defining disaster resilience: A DFID approach paper. *DFID*. https://www.gov.uk/government/uploads/system/uploads/attachment_data/file/186874/defining-disasterresilience-approach-paper.pdf. Accessed 12 May 2017.
Faludi, A. (1987). *A decision-centred view of environmental planning*. Oxford: Pergamon Press.
Florida, R. (2017). *The new urban crisis: How our cities are increasing inequality, deepening segregation, and failing the middle class—and what we can do about it*. New York: Basic Books.
Folke, C., Carpenter, S. R., Walker, B., Scheffer, M., Chapin, T., & Rockström, J. (2010). Resilience thinking: Integrating resilience, adaptability and transformability. *Ecology and Society, 15*(4), 1–20.
Gabellini, P. (2016). Two plan, one enviromental policies. In V. Barbi, G. Fini, & P. Gabellini (Eds.), *Bologna città resiliente. Sostenibilità energetica eadattamento ai cambiamenti climatici* (pp. 9–13). Bologna: Urban Centre.
Glaeser, E. (2011). *The triumph of the city*. London: Macmillan.
Holling, C. S. (1973). Resilience and stability of ecological systems. *Annual Review of Ecology and Systematics, 4*(1), 1–23.
Howard, J. (2009). Climate change mitigation and adaptation in developed nations: A critical perspective on the adaptation turn in urban climate planning. In S. Davoudi, J. Crawford, & A. Mehmood (Eds.), *Planning for climate change: Strategies for mitigation and adaptation for spatial planners* (pp. 19–32). London: Earthscan.
Hurlimann, A. C., & March, A. P. (2012). The role of spatial planning in adapting to climate change. *Climate Change, 3*(5), 477–488.
Kim, D., & Lim, U. (2016). Urban resilience in climate change adaptation: A conceptual framework. *Sustainability, 8*(4), 405.
Innes, J. E., & Booher, D. (2010). *Planning with complexity*. London: Routledge.
Marshall, S. (2012). Planning, design and the complexity of cities. In J. Portugali, H. Meyer, & E. Stolk (Eds.), *Complexity theories of cities Have come of age* (pp. 191–205). Berlin: Springer.
Mazza, L., & Bianconi, M. (2014). Which aims and knowledge for spatial planning? Some notes on the current state of the discipline. *Town Planning Review, 85*(4), 513–532.
Meerow, S., Newell, J. P., & Stults, M. (2016). Defining urban resilience: A review. *Landscape and Urban Planning, 147*, 38–49.
Muñoz Gielen, D., & Tasan-Kok, T. (2010). Flexibility in planning and the consequences for public-value capturing in UK, Spain and the Netherlands. *European Planning Studies, 18*(7), 1097–1131.

Nickson, A., Woolston, H., Daniels, J., Dedring, I., Reid, K., Ranger, K., Clancy, L., Street, R., & Reeder, T. (2011). *Managing risks and increasing resilience: The Mayor's climate change adaptation strategy*. London: Greater London Authority.

North, D. (1990). *Institutions, institutional change and economic performance*. Cambridge: Cambridge University.

Pickett, S. T. A., Cadenasso, M. L., & Grove, J. M. (2004). Resilient cities: Meaning, models, and metaphor for integrating the ecological, socio-economic, and planning realms. *Landscape and Urban Planning, 69*(4), 369–384.

Portugali, J. (2008). Learning from paradoxes about prediction and planning in self-organizing cities. *Planning Theory, 7*(3), 248–262.

Rose, A. (2007). Economic resilience to natural and man-made disasters: Multidisciplinary origins and contextual dimensions. *Environmental Hazards, 7*(4), 383–398.

Rydin, Y. (2011). *The purpose of planning. Creating sustainable towns and cities*. Bristol: The Policy Press.

Siders, A. R. (2017). A role for strategies in urban climate change adaptation planning: Lessons from London. *Regional Environmental Change, 17*(6), 1801–1810.

UNISDR. (2012). *Making cities resilient—My city is getting ready* (Geneva, United Nations International Strategy for Disaster Reduction). Available at http://www.unisdr.org/english/campaigns/campaign2010-2015/. Accessed 10 May 2017.

Wamsler, C., Brink, E., & Rivera, C. (2013). Planning for climate change in urban areas: from theory to practice. *Journal of Cleaner Production, 50*, 68–81.

# Chapter 4
# Adaptive Governance for Resilience of Peri-Urban Socioecological Systems

**Pablo Torres-Lima, Sandra Lee Pinel, and Kristen Conway-Gómez**

**Abstract** In light of the ongoing international debate on governance, and given the complex challenges of governing the expansion and urban development that transforms the metropolitan interstices and their agricultural and natural landscapes, this paper explores the adaptive governance approaches and some of the challenges for applying them to these types of political and spatial landscapes to achieve the transition towards strengthening urban resilience. The main features of the institutional systems and the integration of local actors and communities in the resilient processes are outlined. From a portrait of the literature review, conceptual criteria are discussed to carry out future research on the topic of adaptive governance for peri-urban socioecological systems resilience.

**Keywords** Adaptive governance · Peri-urban socio-ecological systems · Resilience

## 4.1 Introduction

This chapter is based on a recent review of the literature on theories and paradigms of adaptive management and collaboration for resilience in the context of socially complex peri-urban landscapes. Diverse definitions of adaptive governance share the idea that cultural units, social and civil states from local to national units, form a flexible governance network, nested in processes of communication and social learning; democratic participatory structures; shared authority and accountability;

P. Torres-Lima (✉)
Universidad Autónoma Metropolitana, Xochimilco, México
e-mail: ptorres@correo.xoc.uam.mx

S. L. Pinel
Antioch University, Keene, New Hampshire, USA
e-mail: spinel@antioch.edu

K. Conway-Gómez
California State Polytechnic University, Pomona, USA
e-mail: kconwaygomez@cpp.edu

 
G. Brunetta et al. (eds.), *Urban Resilience for Risk and Adaptation Governance*, Resilient Cities, https://doi.org/10.1007/978-3-319-76944-8_4

and results monitoring. However, critics call for more research that addresses different legal and cultural frameworks of the power relations between the state and social norms and institutions that govern the commons (Agrawal and Benson 2011) This chapter is used as an example of some of the challenges of applying adaptive governance for resiliency theory in peri-urban SES in order to broaden the research agenda necessary to consider the focused use of adaptive management in urban contexts. The text is structured in three parts. The first part considers adaptive governance and SES systems; the second looks at governance in peri-urban SES and the third presents some themes for research. Finally, we present brief conclusions.

The evaluation of social and economic interdependencies in rural-urban highlight different elements in territorial development dynamics that are relevant in policy formulation and decision-making (Berdegué et al. 2015). In addition the perception of policy and planning for peri-urban areas among urban populations in developing countries includes concerns with deforestation, deterioration of natural resources and change of rural land use – processes of intense conflict and risk. We identify collective property and use of natural resources by the rural poor as critical factors that are often marginalized in land management policies and practices in the peri-urban periphery.

Simplified assumptions about urban and rural geographic spaces and inattention to their social-ecological context result in the elaboration of public policy that misinterprets local realities and results in dysfunctional management paradigms being imposed on societies and their local and regional landscapes. For example, government responses to impacts of climate change are inequitably distributed in megacities and their geographies, demographic pressures, poverty levels and dependence on natural resources. A lack of integrated territorial planning results in poor coordination and collaboration between different levels of government to address vulnerability and risk and avoided the creation of local capacity, particularly for natural resource conservation and ecosystem services (Sosa-Rodriguez 2014).

The political dimension is a central aspect of megacity development. Due to the institutional complexity of decision-making and the limited capacity of territorial and sectorally organized institutions to address the diversity of socio-institutional structures among national, state and local organizations and their urban planning processes, environmental problems and socioecological system productivity in peri-urban areas are suppressed by jurisdictional and political divisions. This is especially true when different urban dynamics are occurring along the interstices and rural peripheries of these complex metropolitan systems (Aguilar 2008).

Emerging natural resource governance and territorial planning concepts could be productively applied to addressing these complex challenges of peri-urban areas. Concepts of governance and institutions have expanded well beyond government agencies and actors to explore how social systems govern common natural resources, how social institutions regulate social conflict, and how management of natural resources comprises a set of interdependent systems and subsystems between ecosystems themselves and the social, economic and political aspects governing them (Ostrom 2009; Bernal et al. 2014). Governance includes social and legal practices, institutions, knowledge, meetings, values and diverse decisions that may be best understood from the micro political as constructed by institutions in specific locations (Healey 2009), in this case peri-urban territories.

The concept of governance has been applied to different levels of government that range from global institutions (Held 2007), which seek to manage the complex process of globalization in the most democratic way, to self-governance (Parés 2009), which refers to the capacity of social entities to govern themselves autonomously. In particular, and because of the importance of global urbanization processes, the concept of urban governance (Blanco 2009) is a key stage for theoretical and empirical arguments characterizing local, territorial and metropolitan governance (Yáñez et al. 2008; Telleira 2014). In this sense, some theories that address spatial governance from urban and regional planning (Healey 2009) make up a new institutionalism and social science theories under a regulatory framework for the governance of urban and regional places. Likewise frameworks exist to explain the creation of government processes that are inclusive and promote social learning among diverse interests and government representatives (Innes and Booher 2010). In order to consider equitable and effective governance based in geographic spaces and territories, rather than legal and administrative boundaries, Healey (2009) recommends understanding urban regions as dynamic, with human relations beyond administrative and economic institutions.

Within the social and ecological system and climate adaptation literature, a broad discussion has developed on the challenges scaling up public management and local governance to address social and ecological territories and the interaction between macro and micro policies. The SES literature promotes adaptive governance, which includes self-organizing social networks capable of coping with complexity, uncertainty and change (Armitage et al. 2009), as a way to overcome conflicts among natural resource and development priorities in order improve resilience – communities' ability to respond to events that challenge their functioning (Ernstson et al. 2010) and sustainable development (Adger and Jordan 2009). Network analysis is being applied to understanding how formal and informal institutions can interact to change management practices across boundaries and within a polycentric system of formal and informal institutions (Girvan and Newman 2002; Ansell and Gash 2008). Guarneros-Meza (2009) proposes urban governance that promotes institutional change.

Very little is reported in the international literature on the development of peri-urban SES as part of metropolitan processes and trends. What improvements will be needed to the adaptive management and networked governance paradigm to apply it to the governance of peri-urban territories that are characterized by migrant and local community practices currently excluded from urban development and planning policy? The complexity and dynamics of peri-urban SES offer an opportunity to examine social, ecologic, economic and political factors of vulnerability, risk management and the ability to strengthen resilience and adaptive strategies to climate change by agents, institutions, communities and civil society at different scales and levels of government (Archer et al. 2014).

### 4.1.1 Adaptive Governance and Socioecological Systems

The approach of adaptive governance is to address ecological and social uncertainties of climate change and development in complex and large-scale environmental systems, such as urban governance, which focus on regional planning and natural resource management. However, like the focus of "good governance" and participatory natural resource management paradigms, adaptive management is a paradigm that requires contextualized interrogation of assumptions (Ostrom 2009). Adaptive management has evolved in its recognition of ecological complexity and decentralized participatory paradigms after implementation of centralized conservation approaches were seen as insufficient for the challenges of biodiversity loss, indigenous movements, human rights, livelihoods, and climate change research, especially in urbanizing areas. Peri-urban SES systems consistently reflex the intrinsic complexity not only of hydrology, climatology and ecology, but also the important relationship between the environment, land use and social impacts. In other words, different expressions of urbanization processes offer diverse scenarios in building social resilience and adaptability of natural systems under urban governance of common resources. Likewise, SES in urban contexts represents geographic and institutional possibilities juxtaposed at various scales and levels of government due to the institutional complexity, adaptive management of resources, varying socio-cultural environments and political involvement. For example, in broad urban contexts, resilient governance is framed as interaction among actors, physical and ecological infrastructure and formal and informal institutions (Archer et al. 2014).

Literature that recommends community-based approaches to adaption to climate change; integrated conservation and development, co-management (alliances between government and one or more indigenous or local groups); and collaborative governance networks commonly assumes that collaborative planning and collaborative natural resource management theories and processes for multi-party decision making that require the development of evaluation measures (Conley and Moote 2003) (Pinel et al. 2015). Other definitions of co-management include organizations in which multiple interested parties participate in decision-making by consensus (Ansell and Gash 2008); users and government agencies work together to resolve dilemmas through constructive dialogue and information sharing (Gerlak and Heikkila 2007); and actors use shared norms, rules and institutional structures operating interactively for collective decision-making (Imperial 2005).

Adaptive co-management has been defined as a system of governance involving heterogeneous actors juxtaposed through vertical and horizontal scales (Olsson et al. 2004). Reports of adaptive management in the literature range from governmental transparency and inclusion, to building social capital, reducing conflict between resource users, perceived equity in tackling social and ecological values, to environmental sustainability and resilience. In this sense, adaptive management has extended ideas about collaboration and co-management to include social learning and ongoing monitoring, and the integration of local, traditional and scientific knowledge in new decisions. This concept contains ecological and social aspects, which in turn are included in dynamic processes, continuous self-organization, developed through relevant knowledge and institutional arrangements (Folke et al. 2003). Some promising approaches to evaluation are exploring efforts to quantify environmental results and the use of science (Zollman 2012) to build relationships on the basis of long-term social capital to address emerging conflicts over time (Innes and Booher 2010).

The literature on SES governance suggests that adaptive and collaborative management should be expanded to address complex social and ecological systems through governance networks (Pahl-Wostl et al. 2007), under the premise that these systems can be more sustainable if governance is guided by participation of multiple actors and a network of multi-scaled social and institutional processes. Social variables, such as politics, laws, resources, interests and actors increase exponentially with more complex scales like peri-urban SES complexes, which makes it difficult to transfer lessons learned about collective governance to less complicated scales. Some natural resource and human ecology literature ignores references to geographic and urban studies that question scale and how individuals consider data, information and knowledge at different spatial scales (An et al. 2005). Based on a wide variety of empirical case studies and critical analyses, authors from various disciplines have discussed assumptions, case studies and, more recently, developed comparative frameworks of variables affecting the results of adaptive management (Ostrom and Cox 2010). Efforts to develop predictive governance variables found a need for analysis of different institutional or legal contexts governing common goods, such as differences between rights of use of common goods (Ostrom and Cox 2010). These differences can be divided into three categories: culture and history; social rights and rules governing common goods; and legal regulations for decision-making. First, heterogeneous resources, users and institutions may have led to the unequal allocation of resources, responsibilities and decision making (Abbot et al. 2007). First, power is a key issue in the management of natural resources, under which a community is defined as much for its cultural context (cultural capital) and relations between the social structure in which they are embedded (social capital) and the distribution of power (symbolic capital) tends to be specific for each community (Boyce 2007). Secondly, research on governance can only be transferrable with specific analysis of different institutional and legal conditions that govern common resources (Ostrom and Cox 2010). Thirdly, social and legal contexts of peri-urban SES differ widely, including trans-frontier areas subject to multiple federal and local administrative procedures and community structures of natural resource management, which prevent effective regulation of land use in order to prevent degradation and deforestation, as well as settlements of populations in high-risk areas as part of broader development processes.

In the field of urban and regional planning, the concept of "nested governance" networks suggests that when a peri-urban area is not under the authority of one regional governmental institution, it may be feasible to develop coherent territorial governance systems across spatial scales and functional or territorial boundaries and, to accommodate local actions according within a collaboratively developed regional policy. In this sense, urban governance remains an emerging political domain that includes the complexity of physical-territorial, ecological, social and institutional factors as part of the power relations that occur in an urban plan, as is the case of peri-urban SES systems.

The trajectory of a collaborative network focus may be limited by power dynamics and the time and other transaction costs for participants and thereby also require the support of formal policies. For example, as far as the geographical scale of watersheds and tributaries planning systems increases, so do the transaction costs of network governance (Jobin 2008). While planning studies contribute to define the rules of institutions and collaboration, the current proposals for adaptive governance deny the role of formal organizational structures, statutes, resources and staff participation in processes of larger scales (i.e. as peri-urban socio-ecological systems) (Gerlak and Heikkila 2007). Peri-urban areas, located outside the boundaries of conventional administrative units with multiple population dynamics and levels of economic development, present challenges in achieving the ideals of SES for adaptive governance and strengthening resilience; but at the same time they represent an opportunity to do so by their social and political dimensions, under the framework of nested units making decisions that operate across multiple scales, typically polycentric urban institutions (Olsson et al. 2004). Thus, social propositions in the literature on adaptive management can be tested in the socio-ecological complexity of megacities and peri-urban areas.

### *4.1.2 Adaptive Peri-Urban Governance*

In general, government politics for peri-urban SES systems of megacities have conceptualized these territories as subordinate to the demands and benefits of urban spaces and specificities of urbanization processes. Planning for urban and peri-urban SES ecosystems have created a false dilemma between the urban vision and valorization of environmental functions. Many of these policies take the form of regulation and as sectorial policies are implemented without government and social coordination, which often contradict or compete with each other.

The urgent need to reform current policies under a different operational dimension, as part of an urban plan, including different levels of government and territorial scales, signifies expanding the focus of the relationship between the local and national, between the productive base and planning decisions. That is, the complexity of policy relations, economics and local cultures under the regional structure and not only linear chains of production and consumption, that the peri-urban dwellers recuperate the domain of economic production and social reproduction from local adaptive capacities and the strengthening of their resilience. Local sustainable development must include not only a positive impact on the quality of life, but also a significant impact on job creation and profitability of productive activities and in some cases microenterprise, in addition to addressing the issue of security of land tenure.

The institutional capacity of local urban governments immersed in the dynamic of federal agencies and with the majority of their efforts directed at environmental planning, with the help of scientific and technical entities, are sometimes referred to as environmental governance (Arroyo and Corvera 2011). This happens without implementing a strict control over activities that damage natural resources with the application of economic instruments to strengthen peri-urban resilience. For example, these urban policies don't result in making the urban population – the principle agent of pressure on the territory – conscious of the interdependence that these people have with the natural resources, farming and forest spaces, not only in terms of recreation and food, but also in relation to the provision of environmental services and economic costs.

In developing countries, peri-urban SES spaces have been subjected to a series of policies of paternalism, clientelism with little to no participation of the social base in planning strategies and actions towards the strengthening resilience. For example, in many Latin American societies a regional social capital exists, in the form of norms of reciprocity and networks of civic engagement (Sharp and Smith 2013), which are linked to community actions of sociocultural development of the traditional type that implies certain degrees of organization and participation for the protection of natural resources or collective management of peri-urban market spaces for their peri-urban agricultural products. Often, these relationships are not been considered by public policies.

The existing yet limited capacity and regional collective regulation, based on social ownership of the land (i.e. communal property) and social and economic structures of integration and partnership between these peri-urban dwellers, has prevented the promotion of collaborative local governance directed towards regional sustainable development and community or social harmonization in light of the dilemmas and pressures of urbanization processes themselves at scales of great complexity. An example of this is the inexistence of local organizations or businesses that provide quality services in the processes of peri-urban SES landscape production (University, IAS, and IR3S/UTIAS 2016), for the better management of forestry and agricultural systems, to increase the capacity of communities to self-regulate and resolve issues of urban growth and local economic development. That is to say, until now the literature has few reported cases on institutional innovation and planning processes for urban risk reduction and vulnerability as part of an adaptive peri-urban governance system that includes social and civil institutions in accordance with formal government systems to strengthen the resilience of peri-urban SES systems and to consolidate a political culture of decision making.

Nonetheless, this dichotomy between optimal use of natural resources and making residents understand the advantage of applying economic instruments to strengthen the resilience of SES peri-urban systems allows the politics to hide part of the economic and social responsibility of federal and local governments for conservation of the peri-urban space. This obfuscation affects environmental and sociocultural function. A need exists to design comprehensive policies for sustainable development and strengthen the resilience of cities, where more than design guidelines for social and economic well-being, it is necessary to implement processes of democratization of political space with the active participation of those who live in and manage the natural resources within the peri-urban landscape.

As documented by several studies in peri-urban Mexico City and elsewhere, peri-urban residents in the informal economy rely on the availability of water and farmland and access to urban center, which in turn, rely on the products and services produced by these residents. In the absence of a policy framework generated with the peri-urban dwellers and with an understanding of the social and geographic dynamics of urbanization, unplanned local market openings and outsourcing of the economy at national and regional levels, pressure will increase on natural resource conservation, agricultural activities, and ecosystem services. Consequently, adaptive capacity and the resilience of SES systems are limited by the lack of structural mechanisms for territorial planning that would provide a framework for the recommended collaborative and networked governance approaches. Therefore, adaptive governance theory should be reviewed as the institutional principle for the use of common resources but be modified to recognize resource conflicts. One cannot assume infinite abundance or one common public good of natural and economic resources in peri-urban areas. The comparison of adaptive management ideals with peri-urban realities demonstrates the need for a severely critical view of current impacts of urbanization, human activities in regional SES, and the multi-scalar functions of urbanizing territories that surround cities.

Urban governance in these peri-urban systems needs to have implicit policies with a comprehensive multi-scalar approach that promotes a distribution of political power. Top-down, bottom-up, and spatial agreements and structures are needed that democratize local community organization while enhancing the capacity for development and reducing the risk and vulnerabilities of physical, social, and ecological systems to climate-change-related disasters and resource shortages. These considerations may allow the responsible management of natural resources and facilitating equitable access to basic public services, housing, technology and regional agricultural markets. Policies should also provide for compensation for environmental and cultural services that natural resources and peri-urban landscapes offer to the city, such as biodiversity conservation and ecotourism.

A proposed policy framework of organizational type for regional SES that allows them to confront problematic principles in peri-urban areas in developing nations should respond with greatest urgency to the development of human and institutional capacities for planning and organization. Although metropolitan planning institutions have a half-century of development in much of the world, urbanization processes extend across larger territories without strong local government institutions. In both the North and South, rural development programs and policies are distinct from urban policies, despite an urbanization process and demographic reality that links the two (Hendriks 2008). This divide further impedes the development of integrated programs that simplify financing, introduce technology packages or create environmental regulation or certification schemes that restrict land use or promote green products and environmental services in geographic areas adjacent to cities.

Integrated urban and rural programs for urbanizing areas would recognize the movement of goods services and people and promote natural resource management skills in terms of the following goals: a) create local-regional productive networks of credit, supply and commercialization; b) generate and socialize proper agroecological, agro-industrial and administrative technologies; c) create public mechanisms and economic instruments for environmental compensation for the ecosystem services offered by peri-urban areas; d) formalize regional sustainable development planning that includes ensuring adequate housing and adequate services for future generations; e) create processes for environmental education; and, f) generate multiple mechanisms for evaluation of socio-environmental and economic performance of the public policies like those for regional SES. However, as noted in the New Urban Agenda adopted by the UN Habitat Program in 2016 (https://unhabitat.org/new-urban-agenda-adopted-at-habitat-iii/), urban planning and rural development policy will remain divided in international and national development discourse unless spatial geography is fully recognized.

### *4.1.3 Future Investigation into Adaptive Governance for Resilience in Peri-Urban Areas in Developing Countries*

Currently, most of the urban geography in developing countries offers a view of fractured cities in different manifestations. For example, the perception of the Latin American city has shifted from a notion of unitary urban systems to a perception of fragmentation with contrasting inequalities and scattered spaces that exacerbate polarization and socioeconomic segregation (Rodgers et al. 2011), and increase the risks of disasters and socio-environmental vulnerability. All these circumstances result in a significant conceptual reflection on urban development, resilience and sustainability, at the same time that another type of function and intervention of the state as a societal regulator occurs (International Institute for Sustainable Development 2011). The acceleration of urban growth has created diverse challenges for theoretical and practical aspects of the urban phenomenon, for example: (1) management of disaster risks and vulnerability to climate change; (2) integrated urban development; and, (3) transparent fiscal management and governance (InterAmerican Development Bank 2006). Considering the majority of the literature on adaptive governance is concerned with natural resource conservation, national organizations and local participation, without paying attention to multiple levels of government and territorial multi-scale complex urban systems; the following lines of research into the conditions structures, and processes of governance (Plummer 2009) in order to improve planning and institutional innovation to strengthen resilience in urbanization. This direction is also supported by Sandström et al. (2014)'s framework for comparing institutional contexts and related network structures and Rist and Moen's (2013) suggested use of propositions to compare ecosystems, adaptation and approaches to resiliency.

Geographers in particular have suggested the symbolic importance of peri-urban landscapes, with social exclusion and social capital, are fundamental to understanding the social and environmental impacts of urbanization (Winton 2011) on space, which differs from one peri-urban population to another (Zimmerer 2000). The sense of place in the literature emphasizes that places are social and political constructs, and participants in collaborative natural resource management have different impressions of the significance of the management and governance of these places (Cheng 2006).

In particular, the analysis of collective action and the distinction between civic (i.e. empowerment) and social (networks) rationality may be related with debates on identity, strategies and actions that developed urban social movements to claim their role in the innovation and creativity in developing their adaptive capacity and strengthen their resilience. Under community management, one traditional form of managing natural resources that is part of the socio-cultural identity of many populations even today in peri-urban places in developing countries (Bernal et al. 2014), it could be understood that the community plans can address both distinctions – the political and economic. There are dual objectives of structuring and strengthening community management and participation in natural resource conservation and socio-ecological production landscapes. Thus, in the urban context, the definition of the commons revolves around physical and symbolic concepts of public space, but also, and increasingly, given the economic circumstances around specific rights that reduce vulnerability and urban disaster risks such as decent housing and public services.

The challenge of managing common property in urban settings resides in part on the pressure the capital exerts over cities; and, in the great diversity of urban contexts and inhabitants. Thus the existence of a variety of community dynamics driven by social fabric becomes important. The management of assets is addressed while different phases of participatory community processes are exercised with strategies and dimensions in the definition and implementation of public policies in their interaction with local bureaucratic administrative systems under the framework of current regulations (Blanco 2009). From the perspective of urban adaptive governance, it is feasible to analyze with more dynamic logic complementarity linked to urban spaces that occur in the actual processes that operate similarly in very different geographic contexts.

The above discussion raises a central question: How are we to systematically understand the structure and processes of public decision-making in peri-urban SES under the framework of multiple scales and institutional levels? Theoretically, the policy outcomes of adaptive governance include reduction of public conflicts, incorporation of traditional ecological knowledge in plans and decisions, and, the strengthening of institutional capacity, for example for the strengthening of resilience. Thus, future studies should draw from a combination of political science, anthropology, institutional economics, geography, natural resource management, and urban planning theories and case studies to support the development of adaptive urban governance policies and practices.

Therefore, recognizing the need to formulate different agendas of research with attention to both structure and decision-making and economic processes in the context of demographic, economic, and climate change (Pinel 2011). Comparative and transdisciplinary applied research projects can begin with these theories to identify common challenges and urbanization processes to understand the limitations and possibilities of adaptive governance paradigms by highlighting common dynamics and differences in the variables of process, structure and context (power and accountability from economic and conservation policy approaches) for community resource management at multiple spatial and temporal scales such as evidenced in peri-urban areas. The concept of scale-dependent comparative advantage has the potential for understanding the best level for some strategies and activities to strengthen resilience, including research on the roles and relationships of organizations operating at different scales in the networks of "nested governance" (Wyborn and Bixler 2013). However, for governance to be equitable and sustainable, we also need to understand and value the relationship between local meanings of space/place for actors and their strategic use of policy governance institutions (Chang et al. 2012) to meet their interests.

## 4.2 Final Considerations

Are the adaptive management paradigms embedded in resilience literature relevant to the functional realities and governance challenges of rural and urbanization processes to support decisions that address both the natural landscape and social needs and interests over time?

The geography of peri-urban SES involves a diverse range of natural resources, ecosystem functions, landscapes, territories, co-management structures, social dynamics, regulations and formal institutions, and decision-making processes, among others. The adaptive and collaborative management for risk and vulnerability reduction and strengthening of urban resilience should involve a number of factors. These factors include: large spatial, temporal and jurisdictional scales; governance structures to create a context of forms; more creative and participatory institution building processes are carried out, sustained in terms of culture and governance standards; and, where knowledge of co-production, mediation and translation are included and negotiated to facilitate decision making (Cash et al. 2006). That is, historically governance systems have been incompatible with peri-urban SES systems in the interplay of institutions at multiple scales. Regional planning theory has a long history of addressing these externalities, but in the context of international decentralization and collaborative governance, formal approaches to nested governance have fallen out of favor.

The spatial impact of national policies in developing countries varies across regions and environmental systems, is mediated by the characteristics of the population (Biles and Pigozzi 2000) and should be addressed in the science and policies for urbanization. From a structural perspective, governmental, managerial, and political and administrative institutions and civil service systems should be designed to be comprehensive across development and natural resource sectors, and adaptive to applying knowledge (Rogers and Weber 2010). The debate between structural adjustment and environmental approaches must go beyond sectorial and empirical evidence to expand the level of understanding on the extraction of natural resources and the level of urbanization in the case of peri-urban SES. This applies when we want to understand the direction and magnitude of the impacts of urban processes on the environment and peri-urban SES as part of adaptive processes of urban governance systems. That is, to attend to structural changes in the economy and governmental performance, in civil society, studies and policies should be based on quantitative and qualitative data, using rigorous methods based on evaluation of SES. In addition, adaptive governance cannot occur without formal policies that encourage cities to work across their boundaries and plan for social concerns and resource sustainability for the city and for the populations living in peri-urban areas. Regional planning institutions can create a context for conflict management (Pinel et al. 2015).

Future research on adaptive governance to strengthen resilience in peri-urban SES should include the perceptions of the proper functioning of regulatory frameworks and the provision of infrastructure; promotion of economic development, generation of employment and eradication of poverty; making decision-making processes transparent and accountable, including all social sectors; and, preservation of the environment, cultural and historic heritage of the city (Aguilar and Escamilla 2011). Research should also comparatively apply integrated frameworks of multiple social science disciplines to capture the complexity of cultural, institutional, and sociopolitical differences, to achieve the ideals of adaptive governance, resilience and sustainability.

The link between SES governance and institutional practices of territorial government may be defined by using the conceptual framework of territorial cohesion, which implies three different determinants: concentration; connection; and, cooperation (Pillet et al. 2013). Concentration refers to the overcoming of differences in density; connection includes overcoming distance; and, cooperation consists of surpassing division. Together, these elements serve to guide and make effective the adoption of strategies in territorial planning policies (Jouen 2008). Concentration, connection and, cooperation are intended to generate innovative models of local governance, considering the capacities and responsibilities of societies and urban and regional governments, highlighting institutional efforts, cooperation and social participation from a "multi-level" and "multiscale" vision (Jiménez 2016). As Bennett et al. (2016) profile in their Bright Spots database, focusing our attention on systems that are achieving positive human-environment relationships is a way forward with information on different spatial, institutional, social and economic contexts using participatory methods.

An agenda of working issues, in terms of adaptation strategies in spatial planning policies for peri-urban SES, may include solution to the dichotomy between optimal use of natural resources and helping the citizenry understand the advantage of using economic instruments that enable policies to conceal the federal and local government's economic and social responsibility for preserving peri-urban areas. This responsibility covers not only the environment, but also the socio-cultural context, including the need to design sustainable development policies for entire metropolitan areas. In addition to mapping out policies related to social and economic well-being, democratization, and space-usage, policies must be implemented with the active participation of the people who live in peri-urban rural areas and use the natural resources there. For instance, the public agenda of Mexico City involves not only the inclusion, equity and quality of life of its inhabitants, but also human security and the viability of the development of peri-urban areas.

Based on cultural, historical and regional differences, any research on peri-urban adaptive governance must first define the outcome variables and articulate the premises of the public interest in common property and local goods that make decisions on resource management have the greatest opportunity to influence the achievement of sustainable development and strengthening of resilience. As outlined above, the general features of adaptive peri-urban governance requires understanding of similar power dynamics across diverse contexts and giving policy makers guidance on how to ask these questions in the design of fair, equitable, regional and workable approaches to just governance in light of new processes of urbanization (Healey 2009).

**Acknowledgements** This text was part of a RCN-SEES project funded by the National Science Foundation, through the University of Idaho; Fulbright NEXUS, funded by the Department of State; the Prometheus Project of the Secretary of Higher Education, Science, Technology and Innovation of the Government of Eduador SENESCYT, Project Prometeo; and the Automous Metropolitan University, Xochimilco.

## References

Abbot, J., Campbell, L., Hay, C., Naesje, T., Ndumba, A., & Purvis, J. (2007). Rivers as resources, rivers as borders: Community and transboundary management of fisheries in the Upper Zambezi River floodplains. *The Canadian Geographer, 51*(3), 280–302.

Adger, W., & Jordan, A. (Eds.). (2009). *Governing sustainability*. Cambridge: Cambridge University Press.

Agrawal, A., & Benson, C. (2011). Common property theory and resource governance institutions: Strengthening explanations of multiple outcomes. *Environmental Conservation, 38*(2), 199–210.

Aguilar, A. (2008). Peri-urbanization, illegal settlements and environmental impact in Mexico City. *Cities, 25*, 133–145.

Aguilar, A., & Escamilla, I. (2011). Introducción. In A. Aguilar & I. Escamilla (Eds.), *Periurbanización y sustentabilidad en grandes ciudades* (pp. 5–23). México: H. Cámara de Diputados-UNAM-CONACYT-Miguel Ángel Porrúa.

An, L., Linderman, M., Qi, J., Shortridge, A., & Liu, J. (2005). Exploring complexity in a human-environment system: An agent-based spatial model for multidisciplinary and multiscale integration. *Annals of the Association of American Geographers, 95*(1), 54–79.

Ansell, C., & Gash, A. (2008). Collaborative governance in theory and practice. *Journal of Public Administration Research and Theory, 18*(4), 543–571.

Archer, D., Alamnsi, F., DiGregorio, M., Roberts, D., Sharma, D., & Syam, D. (2014). Moving towards inclusive urban adaptation: Approaches to integrating community-based adaptation to climate change at city and national scale. *Climate and Development, 6*(4), 345–356.

Armitage, D. R., Plummer, R., Berkes, F., Authur, R. I., Charles, A. T., Davidson-Hunt, I. J., & Wollenberg, E. K. (2009). Adaptive co-management for social-ecological complexity. *Frontiers in Ecology and the Environment, 7*(2), 95–102.

Arroyo, J., & Corvera, I. (2011). Gobernanza medioambiental en la zona metropolitana de Guadalajara: una comparación con la de Monterrey y Puebla-Tlaxcala. In A. Aguilar & I. Escamilla (Eds.), *Periurbanización y sustentabilidad en grandes ciudades* (pp. 119–155). México: H. Cámara de Diputados-UNAM-CONACYT-Miguel Ángel Porrúa.

Bennett, E. M., Solan, M., Biggs, R., McPhearson, T., Norström, A. V., Olsson, P., Pereira, L., Peterson, G. D., Raudsepp-Hearne, C., Biermann, F., Carpenter, S. R., Ellis, E. C., Hichert, T., Galaz, V., Lahsen, M., Milkoreit, M., López, B. M., Nicholas, K. A., Preiser, R., Vince, G., Vervoort, J. M., & Xu, J. (2016). Bright spots: Seeds of a good Anthropocene. *Frontiers in Ecology and the Environment, 14*(8), 441–448.

Berdegué, J., Carriazo, F., Jara, B., Modrego, F., & Soloaga, I. (2015). Cities, territories, and inclusive growth: Unraveling urban–rural linkages in Chile, Colombia, and Mexico. *World Development, 73*, 56–71.

Bernal, A., Rivas, L., & Peña, P. (2014). Propuesta de un modelo de co-gestión para los pequeños abastos comunitarios de agua en Colombia. *Perfiles Latinoamericanos, 22*(43), 159–184.

Biles, J., & Pigozzi, B. (2000). The interaction of economic reforms, socio-economic structure and agriculture in Mexico. *Growth and Change, 31*, 3–22.

Blanco, I. (2009). Gobernanza urbana y políticas de regeneración: el caso de Barcelona. *Revista Española de Ciencia Política, 20*, 125–146.

Boyce, J. (2007). Is inequality bad for the environment? *Research in Social Problems and Public Policy, 15*, 267–288.

Cash, D., Adger, W., Berkes, F., Garden, P., Lebel, L., & Olsson, P. (2006). Scale and cross-scale dynamics: Governance and information in a multilevel world. *Ecology and Society, 11*(2), 8.

Chang, C., Allen, J., Dawson, S., & Madsen, G. (2012). Network analysis as a method for understanding the dynamics of natural resource. *Society & Natural Resources, 25*(1–3), 203–208.

Cheng, A. S. (2006). Build it and they will come? Mandating collaboration in public lands planning and management. *Natural Resources Journal, 46*, 841–858.

Conley, A., & Moote, M. (2003). Evaluating collaborative natural resource management. *Society and Natural Resources, 16*, 371–387.

Ernstson, H., Leeuw, S., Redman, C., Meffert, D., Davis, G., Alfsen, C., & Elmqvist, T. (2010). Urban transitions: On urban resilience and human-dominated ecosystems. *Ambio, 39*(8), 531–545.

Folke, C., Colding, J., & Berkes, F. (2003). Synthesis: Building resilience and adaptive capacity in social ecological systems. In F. Berkes, J. Colding, & C. Folke (Eds.), *Navigating social ecological systems: Building resilience for complexity and change*. Cambridge: Cambridge University Press.

Gerlak, A., & Heikkila, T. P. S. P. (2007). Collaboration and institutional endurance in US water policy. *Political Science and Politics, 40*(1), 55–60.

Guarneros Meza, V. (2009). Mexican urban governance: How old and new institutions coexist and interact. *International Journal of Urban and Regional Research, 33*(2), 463–482.

Healey, P. (2009). City regions and place development. *Regional Studies, 43*(6), 831–843.

Held, D. (2007). Reframing global governance: Apocalypse soon or reform! In D. Held & A. McGrew (Eds.), *Globalization theory: Approaches and controversies*. Cambridge: Polity Press.

Imperial, M. (2005). Using collaboration as a governance strategy-lesson from six watershed management programs. *Administration and Society, 37*(3), 281–320.

Innes, J., & Booher, D. (2010). *Planning with complexity: An introduction to collaborative rationality for public policy*. New York: Routledge.

InterAmerican Development Bank. (2006). *Building a new continent, a regional approach to strengthening south American infrastructure*. Washington, DC: Iniative for the Integration of the Regional Infrastructure of South America, InterAmerican Development Bank.

International Institute for Sustainable Development. (2011). Summary of the UNCSD (Rio+20) Regional Preparatory Meeting for Latin America and the Caribbean. *Earth Negotiations Bulletin, 27*(7), 1–12.

Jiménez, L. (2016). *Hacia ciudades y territorios inteligentes, resilientes y sostenibles*. Madrid: Asociación para la sostenibilidad y el progreso de las sociedades.

Jobin, D. (2008). A transaction cost-based approach to partnership performance evaluation. *Evaluation, 14*(4), 437–465.

Jouen, M. (2008). *Territorial cohesion: From theory to practice* (Vol. 54). Paris: Notre Europe.

Olsson, P., Folke, C., & Berkes, F. (2004). Adaptive co-management for building resilience in social-economic systems. *Environmental Management, 31*(1), 75–90.

Ostrom, E. (2009). A general framework for analyzing sustainabiltiy of social-ecological systems. *Science, 325*, 419–422.

Ostrom, E., & Cox, M. (2010). Moving beyond panaceas: A multi-tiered diagnostic approach for social-ecological analysis. *Environmental Conservation, 37*(4), 451–463.

Pahl-Wostl, C., Craps, M., Dewulf, A., Mostert, E., Tabara, D., & Taillieu, T. (2007). Social learning and water resources management. *Ecology and Society, 12*(2), 1–19.

Parés, M. (Ed.). (2009). *Participación y calidad democrática: Evaluando las nuevas formas de democracia participativa*. Barcelona: Ariel.

Pillet, F., Cañizares, M., Ruíz, A., Martínez, H., Plaza, J., & Santos, J. (2013). Los indicadores de la cohesión territorial en el análisis de la escala supramunicipal o subregional: policentrismo y áreas funcionales urbanas. *Ería, 90*, 91–106.

Pinel, S. (2011). *Mountain governance draft literature review. Draft 2*. Moscow, USA: University of Idaho. Mimeo.

Pinel, S. L., Torres, P., & Lopez, F. V. (2015). *Why "Local" is not enough in climate change adaptation planning: Perspectives on regional institution building from three peri-urban and rural mountain landscapes across the Americas*. Proceedings from the mountains of our future earth conference, Perth, Scotand, October 8, 2015.

Plummer, R. (2009). The adaptive co-management process: An initial synthesis of representative models and influential variables. *Ecology & Society, 14*(2), 24.

Rist, L., & Moen, J. (2013). Sustainability in forest management and a new role for resilience thinking. *Forest Ecology and Management, 310*, 416–427.

Rodgers, D., Beall, J., & Kambur, R. W. P. N. U.-W. (2011). *Latin American urban development into the 21st century. Towards a renewed perspective on the city*. Helsinki: United Nations University World Institute for Development Economics Research.

Rogers, E., & Weber, E. (2010). Thinking harder about outcomes for collaborative governance arrangements. *The American Review of Public Administration, 40*(5), 546–567.

Sandström, A., Crona, B., & Bodin, Ö. (2014). Legitimacy in co-management: The impact of preexisting structures, social networks and governance strategies. *Environmental Policy and Governance, 24*(1), 60–76.

Sharp, J., & Smith, M. (2013). Social capital and farming at the rural-urban interface: The importance of nonfarmer and farmer relations. *Agricultural Systems, 76*, 913–927.

Sosa-Rodriguez, F. (2014). From federal to city mitigation and adaptation: Climate change policy in Mexico City. *Mitigation and Adaptation Strategies for Global Change, 19*(7), 969–996.

Telleira, I. (2014). Crisis de la gobernanza urbana y gestión de los comunes. *Revista de Investigaciones Políticas y Sociológicas, 13*(1), 33–47.

University, U. N., IAS, and IR3S/UTIAS. (2016). *Socio-ecological production landscapes and seascapes (SEPLS) in Africa*. Tokyo: United Nations University Institute for the Advanced Study of Sustainability.

Winton, A. (2011). Comunidad, Estado y periurbanización; procesos e impactos sociales de la reubicación de asentamientos irregulares en la delegación Tlalpan. In A. Aguilar & I. Escamilla (Eds.), *Periurbanización y sustentabilidad en grandes ciudades* (pp. 413–438). México: Cámara de Diputados-UNAM-CONACYT-Miguel Ángel Porrúa.

Wyborn, C., & Bixler, R. (2013). Collaboration and nested environmental governance: Scale dependency, scale framing, and cross-scale interactions in collaborative conservation. *Journal of Environmental Management, 123*, 58–67.

Yáñez, G., Orellana, A., Figueroa, O., & Arenas, F. (Eds.). (2008). *Ciudad, poder, gobernanza*. Santiago de Chile: Universidad Católica de Chile.

Zimmerer, K. S. (2000). The reworking of conservation geographies: Nonequilibrium landscapes and nature-society hybrids. *Annals of the Association of American Geographers, 90*(2), 356–369.

Zollman, K. (2012). Social network structure and the achievement of consensus. *Politics Philosophy Economics, 11*(1), 26–44.

# Chapter 5
# Ecosystem Services Based Approach for Participatory Spatial Planning and Risk Management in a Multi-Level Governance System

**Carolina Giaimo, Carlo Alberto Barbieri, and Stefano Salata**

**Abstract** The application of the Ecosystem Services (ES) concept in spatial planning can help improve the connection between land use patterns and the understanding of the functioning of settlement systems to improve risk prevention and management and contribute to enhanced human well-being.

The Project LIFE SAM4CP *Soil Administration Model 4 Community Profit* aims at delivering by June 2018 new tools to improve planning, land management and use, including mapping and assessing ecosystem services generated by the land as tangible and intangible benefits for humans.

The article aims to investigate how the government can deal with sustainable land use through new frameworks of cooperative urban and territorial planning involving all the different public actors. Then the paper will highlight how ES mapping tools and techniques can be used effectively in the decision-making processes associated with urban and regional programming and planning. The thesis is that the ecosystem-based approach adopted by the Project can be helpful in assessing the possible environmental benefits of each planning option and thus facilitate risk prevention and management decisions through the direct involvement of citizens and a multi-level governance model.

The Project consists in drawing up 4 urban plans (or their variants) based on the co-planning procedure provided for in the planning legislation of Piedmont (Italy). The pilot project concern the Municipality of Bruino followed by three subsequent Municipalities of the Metropolitan City of Turin (Settimo T.se, Chieri and None).

**Keywords** Spatial planning · Land consumption · Soil sealing · Ecosystem services · Resilience · Multilevel governance

---

C. Giaimo (✉) · C. A. Barbieri · S. Salata
Interuniversity Department of Regional and Urban Studies and Planning, Politecnico di Torino, Torino, Italy
e-mail: carolina.giaimo@polito.it; carloalberto.barbieri@polito.it; stefano.salata@polito.it

 
G. Brunetta et al. (eds.), *Urban Resilience for Risk and Adaptation Governance*, Resilient Cities, https://doi.org/10.1007/978-3-319-76944-8_5

# 5.1 Introduction

## 5.1.1 The Research Context

The theme of sustainability applied to government of urban transformations is today essentially aimed at limiting the land take and preventing disasters risks. Nowadays, therefore, it is characterized by a highly specialized discipline which makes necessary to systematize the "communicability", or, in other words, the "contamination" between those thematic disciplines and practical territorial government actions. In this perspective, it becomes essential to reframe the concepts of sustainability, development and growth into the contemporary dynamics of global climate and socio-economic changes.

The hypothesis to be tested by the project LIFE SAM4CP[1] is the development of appropriate tools for sustainable territorial governance aimed at mitigating disasters risk and increase the safety of urban systems through a multiscalar and interdisciplinary planning activity based on the integration of ecological, economic and socio-political aspects. At the same time, a model of effective practice of integrated governance, and multi-level cooperation is required, relying on a shared planning method.

In the light of the above considerations, the SAM4CP research represents a pioneering study of practical testing of ES assessing methodologies with a view to shaping planning policies and actions to spare soil and use it well. In this sense, SAM4CP is committed to ensuring a good functioning of spatial planning processes, which directly influence land use regulations. Indeed, it is at local level that the right to change land use is acquired through a permission, in accordance with the local plan; therefore, the municipal level is where a trade-off between land uses is generated, often without considering the potential – and optimal – alternative uses of a scarce and non-renewable resource such as the soil.

The project faces a two-fold challenge: on the one hand, the multi-systemic assessment of ES bio-physical and economic values requires a wider and more reliable knowledge system than the traditional one; on the other hand, the knowledge system must contribute to strengthening the Strategic Environmental Assessment (SEA) in defining the various planning scenarios.

The economic assessment, together with the mapping of the bio-physical value, supports the planning and decision-making process through an analysis of the profitability of the various land uses, highlighting the existing and potential trade-offs between alternative functions, as well as the stakeholders benefitting and those penalized by the different scenarios. Furthermore, it supplies information on the ecosystem balance generated by the various types of land use, thus identifying elements to determine the feasibility of land use changes.

---

[1] See www.lifesam4cp.eu. A first, brief account is included in: LIFE+ SAM4CP, *MIDTERM Report. Covering the project activities from 03/06/2014 to 30/06/2016*, Turin, 30/09/2016.

## 5.2 Planning for Resilience

### 5.2.1 *Government, Governance and Spatial Planning*

Territorial government is a wide concept encompassing technical and political skill and expertise that can't be traced back to a single "discipline" in the traditional sense: therefore it can't be limited to land use and building regulation.

It is an integrated system of theories and practices made by knowledge and experiences with a proper horizontal and multi-scale "functional role", which goes beyond the typical skills fragmentation of urban planning disciplines. Indeed, territorial government includes a huge amount of expertise: environment, landscape, soil conservation, ecosystems protection, enhancement of cultural and environmental heritage, socio–economic development, mobility and territorial infrastructure (Barbieri 2015).

Therefore, it is necessary to adopt rules and procedures for innovative urban and territorial planning for resilience, to support policies for the protection, enhancement, and qualification of the settlement, for urban regeneration and reduction of land take as well as climate change governance.

But resilience innovation assumes the principle of subsidiary and the recognition that territorial government has to be an integrated "horizontal" process rather than a "hierarchic system"(top-down) composed by separated plans drawn up by different and separate institutions.

Therefore, it is widely recognized that the Italian hierarchical planning model (and related procedures) with an approval system based on dirigisme and an indicator approach, is now obsolete and inadequate (Barbieri and Giaimo 2015).

In the Italian complex and diverse environment, the principle of subsidiarity for local government institutions, well-established in Europe since over 15 years, has suffered a significant setback, and perhaps a reversal of the trend in national government action and legislation.

The past positive view of subsidiarity seems inadequate to support a sustainable framework for a more democratic, multi-centred government system, responsive to the contemporary plurality of economic, social and territorial systems that express a stronger demand for innovative forms of governance and planning. This is particularly true especially for metropolitan areas like the Metropolitan City of Turin, at the core of the LIFE SAM4CP Project experimentation.

However, at the same time, the introduction of Law n. 56/2014, paved the way towards new forms of governance and representativeness: it is a framework where decision-making and regional organizational patterns appear more flexible and suitable to share strategies and produce agreements.

In Italy inertia and contradictions persist in the path towards a more process-oriented planning because of the continuing absence of national law principles for territorial governance. The inertia was aggravated by the expectation of the entry into force (which had been mistakenly, considered almost certain) of the Constitutional Reform Act drawn up in April 2016 by the government then led by the ex-Prime Minister Matteo Renzi. The law, among other things, leads to the elimination of all the so-called "competing" policy domains between the State and the Regions and their "nationalization", including the territorial government as well as the suppression of Provinces.

This refers to a process-oriented spatial planning replacing the current "planning system", mainly based on management plans and land-use regulations drafted at various separate institutional levels with vertical-hierarchical relations linking the Region (top) and the Municipalities (down). This environment is not conducive to a more horizontal profile, based on subsidiarity, on the adequacy and autonomous tasks of the institutions, substantiated by sustainable methods and procedures for institutional cooperation and real *co-planning*, including plan development and approval, concertation and public-public cooperation, as well as participation of public and private entities.

The missing element is a national territorial governance framework with the key principles (subsidiarity, joint planning, public-private relationship, land use reduction and sustainability, urban planning tax, urban regeneration, urban and land equalization, planning-urban rent-private property relations) useful to provide guidelines for innovative regional laws.

The Metropolitan City (including the Metropolitan City of Turin) has relevant tasks of territorial government: it represents the Municipalities and acts as an intermediate step with respect to the Region but it also takes up an innovative role as "supranational" body (Pizzetti 2015), with respect to sustainability and development issues, and the relations with other international metropolitan areas. It is a laboratory of potential change in the Country, with its territory, resources, community and actors.

Therefore, the sustainability challenge is a matter of institutional relations, rather than a matter of technology for planning: relations must become more horizontal and based on methods and procedures for cooperation between local authorities and consultation/participation of public and private stakeholders.

### 5.2.2 Sharing Knowledge in a Multi-Level Goverance System

The recent experiences carried out within the project LIFE SAM4CP show that the innovative transition from urban planning to local governance is intimately connected with a multi-level system of shared knowledge and methods. When decision making and public deliberation processes are characterized by a multiplicity of public subjects with relevant and differentiated tasks, it is essential to practice multi-level governance.

Since the mid-2000s, multi-level territorial governance has entered European spatial planning cohesion discourse (Faludi 2012).

As evidenced by the document entitled "On The Charter For Multilevel Governance In Europe", "given that many competences and responsibilities are shared between the various levels of governance in the European Union, [we have to be aware of] the need to work together in partnership to achieve greater economic, social and territorial cohesion in Europe. No single level can deal with the challenges we face alone. We can solve citizens' problems on the ground by cooperating better and running joint projects to tackle the common challenges ahead of us" (EU- Committee of the Regions 2014).

According to Faludi (2012), the most accepted concept refers to the interaction between layers of government, each responsible for a given territory within a hierarchy of nested units.

Multi-level governance is of particular relevance in the field of urban and territorial transformations because it deals with the uniqueness of the physical space and the natural interdependence between the various environmental components (air, water, topsoil, subsoil, biotic communities, etc.) which characterize any human activity and the conditions for its operability. Such environmental organic unity and integration contradicts the separation and segmentation of tasks and functions that characterize administrative action.

Practicing multi-level governance means to implement actions, conducts, and attitudes that favour a process of decision-making avoiding an authority driven top down approach. This implies that decision maker, primarily public, may not adopt behaviors that determine the stop to the implementation of the planning process.

In the Italian public administration, a hierarchical top-down approach is still prevailing, where public bodies express their own formal "binding power" to "influence" local policy.

The experience gathered within the Project LIFE SAM4CP points to a planning system whereby public administrations, at all levels - and private operators too - work together for sustainability, with a co-planning approach to regional government, through the instrument of the Co-planning Conference.

#### 5.2.2.1 The Co-planning Conference

In the Piedmont Region, the introduction of this new approach to urban planning dates back to Regional Law n. 1 of 2007 which was a partial modification of the regional planning law in force at the time. Then it was confirmed by further partial modification introduced by the Regional Law n. 3 of 2013.

Among co-planning activities, the relations between public institutions and between institutions and users, are based on collaboration and participation in the definition of planning contents. At that stage, the institutions have to share the definition of the knowledge framework (which is complex, multi-level and at different scales), objectives, methods, and projects. Co-planning allows every institution to share their information, knowledge, skills and specificities, through the planning process. Co-planning is a time-dependent path, open and constantly updated: the local authorities share their knowledge and share outcomes with other institutions, pursuing a gradual agreement on the general objectives and guidelines.

Some innovative aspects entailed by the co-planning approach include (i) the Conference is convened and chaired by the Mayor of the Municipality putting forward the structural urban planning variant and (ii) the planning procedure is integrated with that of SEA which takes place in the framework of the Conference.

The Municipality takes part in the Conference, with voting rights, together with the Metropolitan City of Turin and the Region. Competent environmental authorities participate too. Depending on the content of the structural variant, the Municipality may invite - without voting rights - other entities or authorities, either entrusted with planning tasks or simply interested in territorial planning.

The deliberations of the Conference are valid when shared by the majority of participants with voting rights and the planning process is officially closed by the City Council, based on the Conference results.

The Municipalities that have joined the LIFE SAM4CP project will take part to the Co-planning Conference with the ES analysis and evaluations developed during the project. This analytical framework is developed prior to the final plan content definition and jointly with the SEA. As evidenced by many authors, the availability of a shared knowledge base that informs planning actors on the links between land use patterns, landscape functioning and benefits, is an essential prerequisite (Fürst et al. 2014; Gruber 2010; Ostrom 2009). An ecosystems-based approach also helps in reducing social vulnerability and defining options for sustainable and efficient adaptation to climate change (Andrade et al. 2012) and reduce disaster risk (Monty et al. 2016).

The Co-planning Conference favours feedbacks among different stakeholders positions, offering the opportunity to share communicable targets. However, the shared decision-making process prompts each actor to state what they really want to do and how.

The complexity of Co-planning Conferences is given both by a joint decision making process based on a shared cognitive framework, and by the "formal" expression of power and competences by the Authority representatives (Presidents of the Region and Metropolitan City of Turin, Mayors, or their delegates).

### *5.2.3 The Challenge for LIFE SAM4CP: Using ES to Evaluate Plan Options*

The complexity of governing urban, territorial and environmental phenomena through the plans require a great amount of analysis, interpretation and also graphic representation. Therefore these three activities are a sine qua non in the planning process.

The LIFE SAM4CP project demonstrates that an interdisciplinary framework is necessary to pursue sustainable and resilient spatial planning. Therefore, the ES analysis requires an integration of different perspectives, because the context-based assessment of biophysical and economic values assumes the integration of theories of environmental economy, geographic information systems, the mapping and representation of local data.

ES assessment for planning purposes is one of the challenges that both academic and public administration actors will have to deal in the next years. Indeed, in areas where good quality of ecosystems is maintained, the area itself and the local community have become more resilient and less vulnerable (EEA 2010). In particular, 2014 and 2015 have seen the adoption of major global agreements and decisions that recognise the importance of ecosystems in disaster risk reduction or provide entry points to upscale such approaches (Monty et al. 2016)[2].

But the incorporation of ES assessment for planning purposes requires a rethinking of the local governance system and in particular the planning development as a tool for both regulatory and strategy management.

The role of the ES analysis should enforce an integrated planning approach, especially in conjunction with SEA to evaluate planning scenarios and a shared assessment framework, with public and private actors operating at different levels and in different sectors, promoting spatial trade-offs among different land use functions. The role of ES analysis is to define the fixed and flexible elements of negotiations for land use regulation. It is, therefore, an integration of plan and SEA, essential to define broad strategies of sustainable development beyond a simple technological enhancement of environmental issues.

Generally, ES analysis helps institutional and local stakeholders to make their choices against a background strategy for sustainable development, which sets the rules for improving or recovering the identity of the region. Unfortunately however, it does not have adequate regulatory support in the planning process. For this reason, LIFE SAM4CP experiments the drawing up of a structural variant of the local plan in 4 Municipalities.

The purpose of the variants is to reduce land take and involve the Region, Metropolitan city of Turin, and the Municipality in the co-planning practice, using ES assessment. Thus, SAM4CP provides the involved Municipalities (through a consensus building process), with tools to support planning decisions and review their urban plans.

The quality assessment methodology adopted is multicriteria. It associates biophysical and economic values for the construction of the cognitive framework of the new urban plan. It also provides an integrated, quantitative/qualitative assessment that can consider the ecosystemic multifunctionality of the soils. The quality of the

[2]For example, at the CBD COP12 in 2014, a decision XII/20 titled "Biodiversity and Climate Change and Disaster Risk Reduction" was adopted. The decision encourages Parties to promote and implement ecosystem-based approaches to climate change and disaster risk reduction.

In March 2015, the Sendai Framework for Disaster Risk Reduction 2015–2030 was adopted as the successor to the Hyogo Framework for Action 2005–2015. This new framework places a stronger emphasis on the importance of ecosystems, biodiversity and proposes a more rigorous monitoring framework, which strongly advocates for capacity development and knowledge transfers for risk management.

In June 2015, the Ramsar Convention Decision XXII.13 was adopted in order to recognize the role of wetlands in disaster risk reduction.

In September 2015, the UN General Assembly adopted the Sustainable Development Goals (SDGs).

In December 2015, the Paris Agreement was adopted by 195 countries.

soils does not derive from the measurement of a single ecosystemic function but is assess in the light of the sum of different biophysical values referring to the 7 ESs considered: Habitat Quality, Carbon Sequestration, Water Yield, Nutrient Retention, Sediment Retention, Crop Production and Crop Pollination. This action allows obtaining a map of the overall soil quality in each involved Municipality.

The overall evaluation of ES quality has been achieved through GIS work sessions to map cartographic output generated by InVEST[3] software.

#### 5.2.3.1 Defining Shared Criteria for Good Land Use Planning

Starting from the mapping of the 7 ESs, one of the first results of consensus building has been the harmonization of the criteria to draw up the plan variants with the LIFE SAM4CP project.

The criteria have been synthesized in the following operational proposals:

1. Analyzing and quantifying the residential provisions of the current general town plan still not carried out (i.e. the "plan residue" or "booked soil") by legal categories
2. Assessing plan residues compared to *de facto* (LULC) and *de iure* (current general town plan) conditions, as defined by the bio-physical and economic ES mapping outputs, and the possibility that they be included in the next plan.
3. Assessing the role /potential of plan residues to improve urban ecological qualities and support the development of environmentally-friendly networks at local level compared to *de facto*(LULC) and *de iure* (current local plan) conditions, as defined by bio-physical and economic ES mapping outputs. Identifying abandoned and/or underutilized areas (if any) and comparing them with *de facto* (LULC), *de iure* (current local plan) and possible project conditions, as defined by *bio-phy*sical and e*conomic* ES mapping outputs.
4. Assessing the potential for reusing and/or transforming areas already built on and/or urbanized and comparing them with *de facto* (LULC), *de iure* (current local plan) and possible project conditions, as defined by bio-physical and economic ES mapping outputs.
5. Assessing the areas for which a request of "retrocession" has been issued and comparing them with the urban structure of the current general town plan and the ES mapping outputs, in order to improve urban ecological qualities and support the development of environmentally-friendly networks at local level

Within this operational perspective, the local plan can play a key role using the analytical-assessment dimension to test innovative solutions into the planning process. It is in this sense the scope of planning may be effectively broadened and support real choices towards the sustainability of environmental resources.

The assessment of ES - which is not only biophysical but also economical -focuses mainly on those soils where a land use change occurs in order to verify in advance the

[3] Integrated Evaluation of Ecosystem Services and Tradeoffs.

changes to the "ecosystemic balance of the soils" and the relative cost of restoration. This action allows local administrations to implement compatible urban transformations with the quality aspects of soils, analyzing the trade-offs[4] of ecosystemic values.

Briefly, biophysical mapping and assessment provide information on the ecosystem balance, based on different land uses and their transformations. The associated economic evaluation enables the decision-making and planning process to be supported by comparison of the profitability derived from the different land use options, more accurately identifying the existing and potential trade-offs between alternative functions, as well as the advantageous and disadvantaged entities from the different scenarios.

It is, therefore, necessary to raise awareness not only of citizenship but also of local actors and decision-makers, towards the "value of the services provided by nature". Paying attention to the effects - including economic - that a choice of land use transformation can have on society and the environment.

## 5.3 The Role of Ecosystem Services in Participatory Land Use Planning: Experiences from LIFE SAM4CP

### *5.3.1 Ecosystem Services and Integrated Participatory Urban Planning Processes*

Using the pioneer case study of Bruino, the other Municipalities involved in testing the methodology for multisystemic ES mapping are: Settimo Torinese (47,576 inhabitants and 31.4 km$^2$), None (7,995 inhabitants and 24.7 sq. km) and Chieri (36,782 inhabitants and 54.3 sq. km). These Municipalities were selected (from June to September 2015) through a public tender with a call addressed to all the Municipalities (315) of the Metropolitan City of Turin. The selected Municipalities were deemed representative of the different features of the Turin metropolitan area - densely urbanized and infrastructured, flat-rural, hilly/mountainous (Fig. 5.1).

The methodological assumption of the LIFE SAM4CP Project is to define a shared cognitive framework based on ES mapping to assess the planning options. It is designed to inform consensus among decision makers at all levels of government and also a shared understanding between local decision-makers and citizens on "land use-sensitive" planning measures and land management policies.

The contribution from new experimental ES assessments allows both to effectively integrate environmental, economic, social and cultural aspects into planning, and to assess the desirability of maintaining or enhancing specific services under different policy and management options (UNEP 2014). The overarching purpose is

[4]Trade-offs indicate the change in the ecosystemic value of the soil subject to urban transformation; is a situation that implies a choice between two or more possibilities, where the loss of value of one constitutes an increase in value in another.

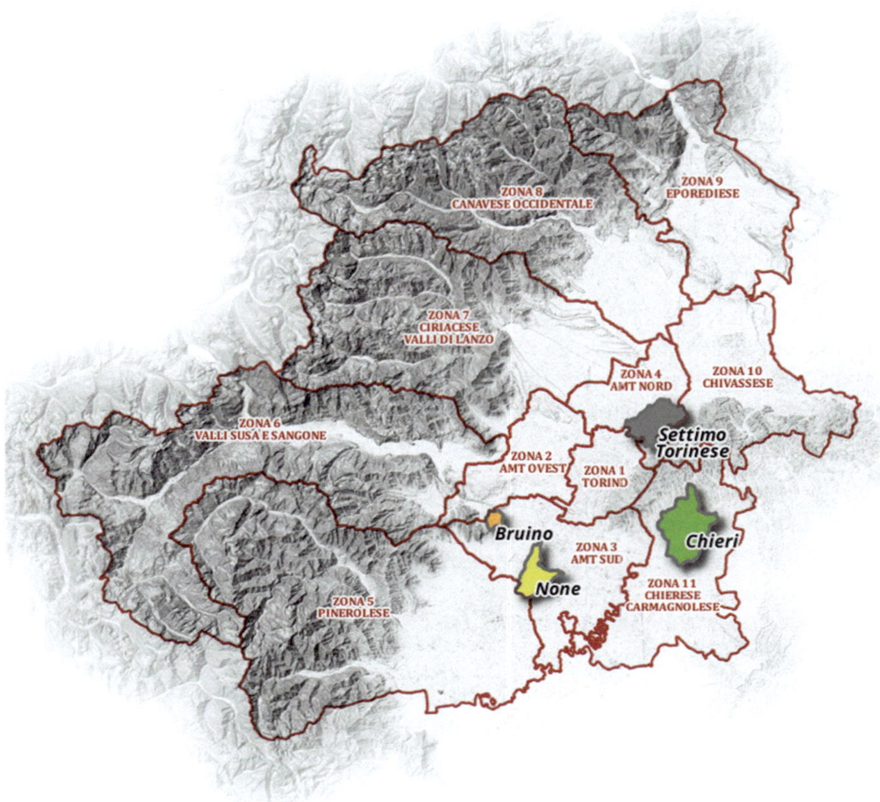


**Fig. 5.1** Metropolitan City of Turin: divided into homogeneous areas with the 4 Municipalities covered by the Project LIFE SAM4CP (Source: CSI Piemonte)

to provide adequate support and expert input to guide technical-political decisions by authorities with qualitative and quantitative ES information.

The Municipalities of Settimo Torinese, Chieri and None have adopted this approach and are re-thinking, together with the more or less organized social groupings (communities, citizens, businesses, associations, etc.) past development models to build new shared approaches to sustainable local governance.

Co-institutional planning is thus combined with consensus building aimed to manage potential conflicts arising from the various planning options, to transform and help the different stakeholders to share their perspective and harmonise the environmental, economic and social issues carried by all authority (Bobbio and Pomatto 2007).

Stakeholder consultation and involvement in the training covering the four structural variants of the SAM4CP Project aim to:

- obtain knowledge related to local specificities before making decisions;
- better Identify the different options or any mitigation measures and trade-offs;
- ensure that significant impacts are not overlooked and benefits maximized;
- create a sense of ownership around the plan actions;
- improve the transparency and accountability of decision-making;
- increase public confidence in the plan-making process and SEA.

## 5.3.2 *Citizens' Involvement in Public Choices*

Over the past decades, the involvement of citizens in public decisions has become an increasingly common practice and is now recommended by many international organizations[5]. It has been promoted by various European programmes (for ex. Urban and Leader) and has been adopted in Italian legislation, especially regional laws, in the field of urban regeneration, social policies and local development.

An extensive and well-established international scientific literature is available to help public administrations extricate themselves in the growing supply of participatory methodologies (Susskind et al. 1999; Wates 2006; Bobbio 2004; Gastil and Levine 2005).

The LIFE SAM4CP Project is testing an approach to participation in terms of a relation of society (its various components) with the public institutions (especially city councils) implying a direct intervention of the former in the decision-processes of the latter.

### 5.3.2.1 The Actors of Participatory Processes

Who are the potential actors of participation? The obvious answer is "all citizens", but it is as generic an answer as inaccurate: only some citizens will participate (a large minority) and this could lead to all sorts of imbalances in the decisions made. Realistically, we must acknowledge the impossibility of engaging the entire population and therefore essentially focus on the participation of those citizens who are already, by inclination or experience, active, competent, thoughtful or who belong to an association or any social movement.

[5]For example: *Citizens for Europe*: www.citizensforeurope.eu; *Center for Deliberative Democracy*, Stanford University: www.cdd.stanford.edu; *Center for Democracy and Citizenship*: www.publicwork.org; *Civic Evolution*: www.civic.evolution.org; *Community Planning*: www.communityplanning.net/index.php; *DDC-Deliberative Democracy Consortium*: http://participedia.net/de/organizations/deliberative-democracy-consortium; *EIPP-European institute for Public Participation*: www.participationinstitute.org/index.php?id=3&L=2; *European Commission – The European Citizens' Initiative*: http://ec.europa.eu/citizens-initiative/public/welcome?lg=en; *Iap2 - International Association for Public Participation*: www.iap2.org; *IAF – International Association of Facilitators*: *www.iaf-world.org/site/; International Observatory on Participatory Democracy*: www.oidp.net/en/, *PEP-NET Pan European eParticipation Network*: www.citizensforeurope.eu/organisation/pep-net

The inherent risk is excluding the weakest citizens, those who cannot made their voice heard. However, precisely the most vulnerable citizens, unlike active, competent or thoughtful ones, are also the most difficult to engage: according to the accounts of participation experiences in social housing neighbourhoods in Italy, several people even refuse to open the door or to participate in any citizens' meeting (Sclavi 2002; Bricocoli 2002).

Conversely, there is often the opposite challenge, i.e. to be able to involve the most powerfull citizens. Real estate operators, land owners, developers, entrepreneurs have all the means to put pressure on the authorities through more or less opaque and custom channels and they are therefore not willing to expose themselves in the public arena. Contrariwise, participatory processes imply a vital need to include the powerful actors and ensure that their reasons are addressed in a collective discussion.

Coming back to the opening questions, who participates in the experiences of participation: all or only a few?

For the LIFE SAM4CP project, participation is by definition equally directed to all the users, in various ways, in the municipality drafting a variation to the Local Plan or having some interest in the subject being discussed (land take reduction). However, those who actually take part in the process are inevitably a tiny fraction of the universe. Here is the paradox of participation: it is supposed to involve and include all the citizens but it will only engage a few in real life. There is always some form of explicit or implicit selection. The question is how it happens and by whom. The literature describes three different mechanisms: self-selection, targeted selection and random selection (Fung 2003).

The first mechanism of participation operates by self-selection, any citizen can decide whether getting involved into the process or stay out. This simple method to involve the citizens was used for example in the None Public Assembly (July 2016) and Chieri (May 2017).

Albeit this method marks a sharp reversal of the practice of "closed door" public decision making, two major difficulties occurs. The first concerns the number of participants. Experience shows that only a tiny fraction of citizens is willing to participate: in Italy, for instance, participation rates hover around 1–2 percent. This cannot but have an impact on the legitimacy of the choices made or the decisions reached by the Citizens' assemblies. Then, there is the question of the lack of diversity among participating citizens. It is likely that only certain groups of citizens will be involved in the meetings: militants, citizens involved in specific friendship networks, political groups or associations, regulars of participation practices. On the other hand it is likely that those who have more family commitments (mothers), business (self-employed) or who otherwise prefer to use their free time (young people) in a different way tend to self-exclude themselves. It is therefore unlikely that a forum based on self selection can fully reflect the different points of view in a target population. But self exclusion does not only affect passive citizens, there is also the risk of a political self-exclusion, i.e. discouraging the participation of persons belonging to political networks other than the ruling ones in the City Council.

With a view to mitigating the drawbacks of self-selection the local authorities of the LIFE Project did not confine themselves to open the meeting to all the citizens, but decided to identify potential interest groups or people where they already gather and discuss, explaining to them the participatory project and making them aware of the possible benefits of participation: there are emblematic the cases of Bruino, where they chose to hold the meeting for the presentation of the master plan Variant (July 2016) at the headquarters of Novomeccanicasrl (a productive enterprise of the town), and the cases of Settimo T.se and Chieri, where they held the public assembly at the municipal library.

Another mechanism, the so-called "targeted selection" has also been applied in Settimo Torinese with the Cycle of 7 meetings held in March and April 2016 entitled *"The local government discusses the revision of the Master Plan with the citizenship":* instead of a fully open participation, the arena of participants was carefully selected so as to mirror the interests and points of view of the community, inviting representatives of the Trade Associations, Professionals and Contractors, Crafts and Industry, Agriculture and Environmental Associations. The aim was to collect different interests, even minority interests, which could help address the revision of the Master Plan with a view to the reduction of land consumption.

The method followed in None (14 July 2016) for the Presentation of the LIFE SAM4CP Project based on both approaches, with an open discussion with the citizens and Association representatives.

Actually, since all the selected methods have their own shortcomings, experience often shows that their combination can help build participatory processes more aligned with the specific issues and circumstances (Bobbio and Pomatto 2007).

## 5.4 Conclusions

Using an ecosystem-based approach in urban planning helps to define disaster risk reduction strategies thanks to the ecological resilience framework (Monty et al. 2016). How much healthier and more resilient an ecosystem is, less damage it will experience from natural hazards and it will recover from impacts easier and faster (Adger et al. 2005) (Fig. 5.2).

Urban and regional planning must face the challenge of creating standards able to generate greater resilience of human settlements that face to potential disasters. In this direction, a significant contribution may come from the analysis and evaluation of soil ecosystem services.

Risk management in planning should be, therefore, approached in innovative ways with a series of measures to make information available and share it, develop specific rules geared to regional sustainability, monitor threats and gradually reduce vulnerability, education for prevention, and so on.

The point is not to predict precisely what, when and where it will be, but instead to explore what might threaten the system, what makes it fragile, what increases its

**Fig. 5.2** Nature-based solution as an umbrella term for ecosystem-related approaches (Source: Cohen-Shacham et al. 2016)

exposure to the adverse effects of the so-called "black swans"[6]. That is, it is planning that will enable the city to strengthen its ability to be anti-fragile, i.e. to survive and benefit from a change, even from abrupt changes such as catastrophes (Blecic and Cecchini 2016).

In this sense, the SAM4CP approach highlights a number of perspectives and points out: (i) the need to integrate a new definition of quality standards in the Plan implementation, which foresees the environmental measurement of the Plan through ES; and (ii) the need to review local authorities land management and decision making, moving towards multilevel forms of governance overcoming traditional *top-down* systems involving the various actors in land management decisions.

---

[6] Black swans are unexpected events with low probability but with enormous impacts and consequences (Taleb 2007).

Disaster management outlines the necessity of a multilevel governance system able to enhance the capacity to cope with uncertainty and surprise by mobilizing diverse sources of resilience (Adger et al. 2005).

Certainly, the Strategic Environmental Assessment (SEA) as a technical tool will have to be flexibly adapted to the various planning contexts and decision-making processes, ensuring coherence of method and substance, i.e. of assessment methods and the legal definition of the various actions (restrictive, mitigating or compensatory). The SEA must also be adapted to the specificity of the actions and contexts that need assessing. This will also influence the choice of the most adequate ES to be used.

In this direction, the participation of citizens in the planning process is crucial. It may stimulate the ability to develop a vision for the future, comparing the current state of things with possible future scenarios, and consequently identify the appropriate actions for the region.

Citizens' participation in land-use planning processes and the management of disaster risk implies the recognition of the importance of sustainability in facing uncertain contexts (Barberis Rami 2017).

The analysis of ecosystem services addresses relevant issues with a view to testing new mechanisms to manage land use and move towards the innovation of more conventional approaches to land value and land use in planning processes.

In fact, knowledge of biophysical soil quality (and the economic values associated with them) can be of great support to the government for both conscious risk prevention for disaster mitigation, reconstruction and correct integrated risk management.

## References

Adger, W. N., Hughes, T. P., Folke, C., Carpenter, S. R., & Rockström, J. (2005). Social-ecological resilience to coastal disasters. *Science, 309*, 1036–1039. https://doi.org/10.1126/science.1112122

Andrade, A., et al. (2012). *Principles and guidelines for integrating ecosystem-based approaches to adaptation in project and policy design*. Turrialba: IUCN- CEM, CATIE. https://portals.iucn.org/library/sites/library/files/documents/2011-064-Rev.pdf

Barberis Rami M. (2017). Come affrontare l'incertezza? Ovvero la gestione del rischio di disastro. *EyesReg, 7*(1), Gennaio 2017.

Barbieri, C. A. (2015). Dall'istituzione all'azione della Città metropolitana di Torino: il ruolo di una nuova pianificazione. *Il Piemonte delle Autonomie, 2*, 8–15.

Barbieri, C. A., & Giaimo, C. (2015). A new model of institutional governance for new planning policies in Italy. *Urbanistica, 153*, 92–96.

Blecic, I., & Cecchini, A. (2016). *Verso una pianificazione antifragile. Come pensare al futuro senza prevederlo*. Milano: Franco Angeli.

Bobbio, L. (Ed.). (2004). *A più voci. Amministrazioni pubbliche, imprese, associazioni e cittadini nei processi decisionali inclusivi*. Napoli: ESI.

Bobbio, L., & Pomatto, G. (2007). Il coinvolgimento dei cittadini nelle scelte pubbliche. *Meridian, 58*, 45–67.

Bricocoli, M. (2002). Uno sporco lavoro di quartiere. Il Contratto di Quartiere a Cinisello Balsamo. *Animazione sociale, 3*, 54–63.
Cohen-Shacham, E., Walters, G., Janzen, C., & Maginnis, S. (Eds.). (2016). *Nature-based solutions to address global societal challenges*. Gland: IUCN. https://doi.org/10.2305/IUCN.CH.2016.13.en
European Environmental Agency. (2010). *European environment state and outlook report 2010 - SOER 2010*. Copenhagen: European Environment Agency.
EU- Committee of the Regions. (2014). *On the charter for multilevel governance in Europe*. RESOL-V-012.
Faludi, A. (2012). Multi-level (territorial) governance: Three criticisms. *Planning Theory & Practice, 13*(2), 197–211.
Fürst, C., Opdam, P., & Inostroza, L. (2014). Evaluating the role of ecosystem services in participatory land use planning: Proposing a balanced score car. *Landscape Ecology, 29*, 1435–1446.
Fung, A. (2003). Survey article: Recipes for public spheres: Eight institutional design choices and their consequences. *The Journal of Political Philosophy, 11*, 338–367.
Gastil, J., & Levine, P. (Eds.). (2005). *The deliberative democracy handbook. Strategies for effective civic engagement in the twenty-first century*. San Francisco: Jossey-Bass.
Gruber, J. S. (2010). Key principles of community-basednatural resource management: A synthesis and interpretation of identified effective approach for managing the commons. *Environmental Management, 45*, 52–66.
Monty, F., Murti, R., & Furuta, N. (2016). *Helping nature help us: Transforming disaster risk reduction through ecosystem management*. Gland: IUCN. https://doi.org/10.2305/IUCN.CH.2016.15.en
Ostrom, E. (2009). A general framework for anaelyzing sustainability of social-ecological systems. *Science, 325*, 419–422.
Pizzetti, F. (2015). La legge Delrio: una grande riforma in un cantiere aperto. Il diverso ruolo e l'opposto destino delle città metropolitane e delle province. *Rivista on line dell'Associazione Italiana dei Costituzionalisti, 3*, 1–11.
Sclavi, M. (2002). *Avventure urbane. Progettare la città con gli abitanti*. Milano: Eleuthera.
Susskind, L., McKearnen, S., & Thomas-Larmer, J. (Eds.). (1999). *The consensus building handbook. A comprehensive guide to reaching agreement*. Thousand Oakes/London: Sage.
Taleb, N. N. (2007). *The black swan*. London: Allen Lane.
UNEP. (2014). *Integrating ecosystem services in strategic environmental assessment: A guide for practitioners*. A report of ProEcoServ. Geneletti, D. Nairobi: UNEP
Wates, N. (2006). *Community planning handbook* (2nd ed.). London: Earthscan.

# Chapter 6
# Sustainable Urban Expansion to Make Climate-Resilient Cities: The 21st Century Challenge

**Lurdes Barrico and Paula Castro**

**Abstract** The urbanization process is increasing globally, with over half of the world's population living in cities. Anthropogenic climate change is one of greatest environmental, social and economic threats. Climate change will have inevitable impacts on urban systems and populations, particularly in developing countries, where cities are growing rapidly and a high proportion of urban populations are vulnerable to climate-related disturbances. Therefore, sustainable urbanization and planning are crucial preventative tools and provide solutions to avoid or mitigate potential environmental dangers. With the increasing frequency and severity of environmental hazards such as heat, drought, and precipitation, urban design strategies, preventing sprawl patterns of cities' expansion will play an important role in reducing vulnerability, promoting health, and building resilience. Adaptation and mitigation measures can diminish the risks associated with climate change. Cities of tomorrow must become resilient to a wider range of dangers and pressures as to effectively cope with climate change. Therefore, the efforts to strengthen resilience and to promote urban sustainability must be included in the planning process. This will require a long-term collaborative effort to increase knowledge and awareness about resilient city planning and design.

**Keywords** Climate change · Urban resilience · Sustainable urban development · Mitigation and adaptation

## 6.1 Cities as Drivers of Development

Cities as areas with spatial concentrations of human activity and composed of complex inter-dependent systems are increasingly important in understanding the climate change issue (IPCC 2014a). In this more and more global and interconnected

L. Barrico (✉) · P. Castro
Centre for Functional Ecology, Department of Life Sciences, University of Coimbra, Coimbra, Portugal
e-mail: pcastro@ci.uc.pt

 
G. Brunetta et al. (eds.), *Urban Resilience for Risk and Adaptation Governance*, Resilient Cities, https://doi.org/10.1007/978-3-319-76944-8_6

world, about 54.5% of the world's population lived in urban settlements in 2016 (UN 2016a). The size and spatial distribution of the urban population are projected to continue to increase and change dramatically in the coming decades (UN 2015a). The world's population in 2030 is projected to be 60% urban and one in every three people will live in cities with at least half a million inhabitants (UN 2016a). This value will reach a level of 66% in 2050, roughly the reverse of the global rural-urban population distribution of the 1950s (UN 2015a). According to UN ( 2015a), levels of urbanization reached different values across regions. North America, Latin America and the Caribbean, and Europe have the highest urbanization level, with 81.5%, 79.5% and 73.4% of their respective populations living in urban areas. In contrast, Africa and Asia are the least urbanized with less of half of their populations living in cities (40% and 47.5%, respectively) and although they are urbanizing much faster than the other regions are still expected to be the least urbanized regions of the world. (UN 2015a). The challenges related to this unprecedented rapid urbanization include satisfying the enormous need for urban infrastructures and providing effective services while protecting the urban environment (Zhang 2016). Understanding the key trends in urbanization and the challenges and opportunities that they present for achieving sustainable development, is important for designing and implementing the 2030 Agenda for Sustainable Development (UN 2015b).

Cities are important drivers of development as they concentrate much of the economic activity, government, commerce and transportation, and provide crucial links with rural areas, between cities, and across international borders (UN 2014). Cities offer important opportunities for economic and social development as they are often associated with higher levels of literacy and education, better health, and greater access to social and cultural services, (UN 2014; Cohen 2006). However, rapid and unplanned urban growth threatens sustainable development when measures and policies are not implemented to ensure that the benefits of city life are equitably shared by dwellers. Unplanned or inadequately managed urban expansion may lead to negative impacts such as rapid urban sprawl, pollution, and environmental degradation, as well as unsustainable production and consumption patterns (UN 2014).

Cities around the world have faced economic, social, and environmental challenges, as well as fast urban changes, such as rapid population growth and social-spatial changes, and the concept of sustainable cities was advanced to help cities cope with these challenges (Rasoolimanesh et al. 2011). Therefore, improving the ability of urban planners and policy-makers to better plan their cities and achieve sustainable development goals are urgent needs of this century that will allow tomorrow's cities to be better prepared for a climate-uncertain future (Childers et al. 2015; UN 2015c; Rasoolimanesh et al. 2011). The major goal of sustainable development need to integrate the economic, social and environmental dimensions of sustainable development (UN 2015c). In order to achieve a sustainable urban development, cities need to develop social and economic strategies but without damaging the environment thus, achieving a balance between dwellers and natural resources (Rasoolimanesh et al. 2011).

The declaration on sustainable development by the United Nations Conference on Environment and Development, in Rio de Janeiro in 1992, includes a series of principles defining the rights and responsibilities of the states when it comes to sustainable development. The Rio Declaration, as it is commonly called, addressed the pressing problems of the moment but also aimed to prepare the world for the challenges of the twenty-first century (UN 1992). In this conference was adopted the definition of sustainable development that was put forward by the World Commission on Environment and Development (WCED) in 1987. The concept of sustainable development was officially defined as "not a fixed state of harmony, but rather a process of change in which the exploitation of resources, the direction of investments, the orientation of technological development, and institutional changes are made consistent with future as well as present needs" (WCED 1987, pp. 25). This concept has encouraged policy-makers to formulate new strategies for achieving a balanced economic and technological pathway to safeguard the environment now and into the future i.e., that the actual generation needs must be satisfied without compromising the capability of future generations satisfying their needs (Hadorn et al. 2006; Nijkamp and Vreeker 2000). Sustainable development cannot be successful without enabling cities to be resilient to natural hazards and ensuring that future development does not increase vulnerability (UN 2001).

## 6.2 Climate Change–a Threat for Cities

Climate change has become an important issue of the twenty-first century. The climate is changing across the globe, and changes in global and regional temperatures are already modifying weather patterns, causing a number of impacts and increasing the vulnerability of regions, economic sectors and communities (IPCC 2014a; EEA 2013). The global warming is unequivocal. Temperatures have warmed roughly 0.74 °C from 1906 to 2005, and each of the three decades from 1983 to 2012 has been successively warmer at the Earth's surface than any preceding decade since 1850 (IPCC 2014a; IPCC 2007a). In Europe, the global average annual temperature in the decade 2006–2015 was 0.83 to 0.89 °C higher than the pre-industrial average (mid-to the end of the nineteenth century), and 2015 was the warmest year on record, namely about 1 °C warmer than the pre-industrial temperature (EEA 2017a). This warming has been sufficient to disturb many of the planet's ecosystems with likely catastrophic effect on the environment, habitats, economies and human well-being (WEF 2016; IPCC 2014a). The Paris Agreement on climate change adopted in the Climate Conference, held in 12 December 2015, is a major turning point in the global fight against climate change. In Paris Climate Conference the world's nations agreed to hold global average temperature to "well below 2 °C above pre-industrial levels and pursue efforts to limit the temperature increase to 1.5 °C" (UN 2015d). However, it is expected that Earth will continue to heat, and even if each country meets the plans agreed at the Paris Climate Conference, warming is projected to reach 2.7 °C by 2100 (WEF 2016).

The evidence for human influence on the climate system has grown since the IPCC Fourth Assessment Report (IPCC 2007b) and most likely human influence has been the dominant cause of the observed warming since the mid-twentieth century (IPCC 2013). It is likely that more than half of the observed increase in global average surface temperature since the mid-twentieth century was caused by the anthropogenic increase in greenhouse gas concentrations (IPCC 2014b). Greenhouse gases contributed to a global mean surface warming which is expected to be between 0.5 °C to 1.3 °C over the period 1951 to 2010 (IPCC 2013). Anthropogenic greenhouse gas emissions have increased since the pre-industrial era, driven largely by economic and population growth, and are now higher than ever. Between 1750 and 2011, cumulative anthropogenic CO2 emissions to the atmosphere were 2040 ± 310 GtCO2, and about half of these CO2 emissions have occurred in the last 40 years (IPCC 2014b).

The climate is changing globally and in Europe. In Europe, the extent and speed of change is becoming ever more evident and as described in the EEA's report entitled "Climate change, impacts and vulnerability in Europe 2016", climate change is already causing a wide range of impacts on society and the environment (EEA 2017a). These impacts and vulnerabilities vary along main biogeographical regions in Europe (Table 6.1). The Mediterranean region is a hotspot of climate change impacts as it is facing decreasing precipitation and increasing temperatures, and these impacts are expected to worse as the climate continues to change (EEA 2017a; EEA 2012; Ulbrich et al. 2012). There is a clear relationship between cities and climate change (Meerow and Stults 2016). Although urban areas cover only a tiny fraction (less than 3%) of the Earth's surface (Schneider et al. 2010), they are major contributors to climate change, being responsible for the majority of global energy consumption and greenhouse gas emissions (Meerow and Stults 2016; IPCC 2014a). On the one hand, many global risks of climate change are concentrated in urban areas. Heat stress, extreme precipitation, inland and coastal flooding, landslides, air pollution, drought, and water scarcity pose risks in urban areas for people, assets, economies, and ecosystems (EEA 2016a; WEF 2016; IPCC 2014a). In the last decades, the growth of urban area and urban population, as well as economic wealth and human activities have increased the exposure of European cities to different climate impacts principally in hazard-prone areas. It should be emphasized that hydro-meteorological events (storms, floods, and landslides) account for 64% of the damage costs due to natural disasters in Europe since 1980, and climatological events (extreme temperatures, droughts, and forest fires) account for 20% (EEA 2012). In the future, urban area expansion, growth and aging of population in cities, will contribute to increase further the vulnerability these systems to climate change. Urban design and management, building regulations, and enhancing green infrastructure may help to mitigate these effects (IPCC 2014a; EEA 2017a; EEA 2016a).

**Table 6.1** Key observed and projected climate change and impacts for the main biogeographical regions in Europe (data from EEA 2017a)

| Arctic region (northern Europe) | Atlantic region (north-western Europe) |
|---|---|
| 1. Temperature rise much larger than global average. | 1. Increase in heavy precipitation events. |
| 2. Decrease in Arctic sea ice coverage. | 2. Increase in river flow. |
| 3. Decrease in Greenland ice sheet. | 3. Increasing risk of river and coastal flooding. |
| 4. Decrease in permafrost areas. | 4. Increasing damage risk from winter storms. |
| 5. Increasing risk of biodiversity loss. | 5. Decrease in energy demand for heating. |
| 6. Some new opportunities for the exploitation of natural resources and for sea transportation. | 6. Increase in multiple climatic hazards. |
| 7. Risks to the livelihoods of indigenous peoples. | |
| **Mountain regions** | **Coastal zones and regional seas** |
| 1. Temperature rise larger than European average. | 1. Sea level rise. |
| 2. Decrease in glacier extent and volume. | 2. Increase in sea surface temperatures. |
| 3. Upward shift of plant and animal species. | 3. Increase in ocean acidity. |
| 4. High risk of species extinctions. | 4. Northward migration of marine species. |
| 5. Increasing risk of forest pests. | 5. Risks and some opportunities for fisheries. |
| 6. Increasing risk from rock falls and landslides. | 6. Changes in phytoplankton communities. |
| 7. Changes in hydropower potential. | 7. Increasing number of marine dead zones. |
| 8. Decrease in ski tourism. | 8. Increasing risk of water-borne diseases. |
| **Boreal region (northern Europe)** | **Continental region (central and eastern Europe)** |
| 1. Increase in heavy precipitation events. | 1. Increase in heat extremes. |
| 2. Decrease in snow, lake and river ice cover. | 2. Decrease in summer precipitation. |
| 3. Increase in precipitation and river flows. | 3. Increasing risk of river floods. |
| 4. Increasing potential for forest growth and increasing risk of forest pests. | 4. Increasing risk of forest fires. |
| 5. Increasing damage risk from winter storms. | 5. Decrease in economic value of forests. |
| 6. Increase in crop yields. | 6. Increase in energy demand for cooling. |
| 7. Decrease in energy demand for heating. | |
| 8. Increase in hydropower potential. | |
| 9. Increase in summer tourism. | |

(continued)

Table 6.1 (continued)

| Arctic region (northern Europe) | Atlantic region (north-western Europe) |
|---|---|
| **Mediterranean region (southern Europe)** | |
| 1. Large increase in heat extremes. | |
| 2. Decrease in precipitation and river flow. | |
| 3. Increasing risk of droughts. | |
| 4. Increasing risk of biodiversity loss. | |
| 5. Increasing risk of forest fires. | |
| 6. Increased competition between different water users. | |
| 7. Increasing water demand for agriculture. | |
| 8. Decrease in crop yields. | |
| 9. Increasing risks for livestock production. | |
| 10. Increase in mortality from heat waves. | |
| 11. Expansion of habitats for southern disease vectors. | |
| 12. Decreasing potential for energy production. | |
| 13. Increase in energy demand for cooling. | |
| 14. Decrease in summer tourism and potential increase in other seasons. | |
| 15. Increase in multiple climatic hazards. | |
| 16. Most economic sectors negatively affected. | |
| 17. High vulnerability to spillover effects of climate change from outside Europe. | |

## 6.3 Spatial Dynamics of the Urban Change–Urban Sprawl

Is increasingly the number of disturbances that exert significant pressure on urban systems. As urban systems become more open to global pressures, urban ecological systems are affected more by global growth dynamics, increasing their exposure to ecological pressures, as well as hinders the sustainability of economic and social development (Eraydin and Taşan-Kok 2013). Urban development may affect the capacity of urban areas to continue to generate the ecosystem services needed to sustain urban quality of life (Elmqvist et al. 2013; Bolund and Hunhammar 1999). Two main factors are responsible to this: rapid changes that occur due to land use changes, and the inevitable fragmentation of ecosystems as a result of urbanization (Eraydin and Taşan-Kok 2013).

Urban sprawl (dispersed urban development) is a serious concern worldwide for a number of environmental and socio-economic reasons (EEA 2016b). It presents a major challenge with regard to making land use more sustainable (EEA 2016b) as highlighted by the International Year of Soils declared by FAO-UN in 2015 (http://www.fao.org/soils-2015/en/). In 2006, the European Environment Agency (EEA) published its first report about urban sprawl entitled: "Urban sprawl in Europe - The ignored challenge" where it described sprawl "as the physical pattern of low-density expansion of large urban areas, under market conditions, mainly into the surrounding agricultural areas" (EEA 2006, pp. 6). Urban sprawl is generally believed to result from an uncontrolled and inefficient urban dispersion accompanied by low building and population density, and occurs when urban planning is not well managed and turns open spaces into built spaces (Altieri et al. 2014; Zhang 2001).

The industrial revolution provided major changes in the geographical distribution of populations across Europe. Since the mid-twentieth century, most of Europe has been characterized by spreading cities and increased population numbers, with people choosing to move out of inner cities to suburban and peri-urban areas. This has increasingly blurred the clear division between the urban and the rural environment, as cities are being planned as more dispersed spatial units, resulting in urban sprawl (EU 2016). The EEA has warned for the scattered expansion of urban areas into the part of Europe's countryside existing on their edge–the "urban fringe". Between 2000 and 2006 around 1000 km2 of land were covered every year by artificial surfaces (EEA 2006). There is no sign that this trend is slowing down and, as a result, the demand for land around cities is becoming a critical issue in many areas. There is a high probability that approximately 77,000 km2 of the European continent will be or have been converted to urban areas between 2000 and 2030 (Seto et al. 2012). For this, and 10 years after the EEA's report published in 2006 (EEA 2006), is urgently needed to create effective measures to control urban sprawl (EEA 2016b).

Many examples of urban sprawl can be found in the literature, for example, in Milan (Camagni et al. 2002), Madrid (López de Lucio 2003), Porto (EEA 2006), Barcelona (Catalán et al. 2008), Rome (Frondoni et al. 2011), or Coimbra (Barrico 2015). The population density of the municipality of Coimbra decreased 3.4%. during the decade of 2001 to 2011. During this decade, the growth of the city area (main urban centre) of Coimbra and its population density showed opposite trends: the city area grew significantly (ca. 42%) while the population density decreased about 26% (Fig. 6.1). This physical pattern of low-density expansion in built-up areas contributed to the evolution of a less compact city, clearly an indicator of urban sprawl (Barrico and Castro 2016; Barrico 2015), causing particularly severe problems to croplands once used in those areas.

### 6.3.1 Urban Sprawl and Climate Change

Sprawl is a result of population growth and lifestyles that require more space. The development of commercial areas which require more buildings and more space, together with the preferences of single houses over blocks of flats, can explain these faster rates of building development than population growth on the peri-urban areas. Furthermore, the progressive construction of commercial areas and road networks outside the city centre also are a tendency shown by European cities (EEA 2016b). In Europe, the wish of humans to live in green environments and in single-family homes, as well as the desire for lifestyles with higher demands regarding dwelling and the search for inexpensive building lots contributed to a highly dispersed urban development (EEA 2016b; Oueslati et al. 2015).

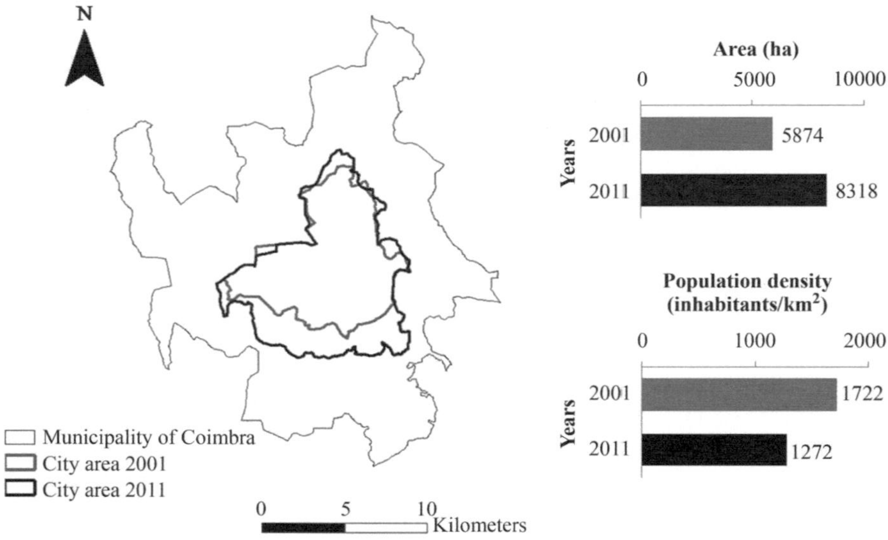


**Fig. 6.1** Coimbra city area and population density in 2001 and 2011 (data from INE–https://www.ine.pt)

The impacts of urban sprawl are evident in the extent of compaction of soil leading to impairment of soil functions such as loss of water permeability (soil sealing) (EEA 2016b; Sung et al. 2013; EEA 2006). When permeable surfaces are sealed through residential, commercial and industrial development, there is a significant impact on natural water systems. The increase in impervious surfaces causes higher and faster flows of surface water run-off, especially during high precipitation (e.g. rainfall) events. Stream flow volumes and stream peak flow rates are increased by concentrated surface water run-off that enters streams and rivers. Therefore, as a result of climate change, strong precipitation events are expected more frequently, and the likelihood of floods will also increase (EC 2012).

Positive correlation between land surface temperature and impervious surface clearly indicates a temperature increase in the sprawled area (Deilami et al. 2016; Myint et al. 2015; Xu et al. 2013). Urban areas exhibit higher temperatures than their non-urban surroundings, which can exceed 10 °C, an effect known as an urban heat island (Stone et al. 2010). The heat island effect may be caused by two factors: first, dark surfaces such as roadways and rooftops efficiently absorb heat from sunlight and reradiate it as thermal infrared radiation; second, urban areas are fairly devoid of vegetation, especially trees, which provide shade and cool the air through evapotranspiration. Therefore, as cities sprawl, the heat island effect also expands, both in extent and intensity (Bhatta 2010).

Dispersed urban expansion involves a positive feedback loop that may aggravate the heat island effect (Bhatta 2010). Urban sprawl requires the increased use of cars and, therefore, leads to increased emissions of carbon dioxide, which also contribute to climate warming (Bart 2010). Global climate change, in turn, may intensify the heat island effect in urban areas. Thus, the morphology of urban areas and greenhouse gas production contribute to warming (Bhatta 2010). With increasing global temperature, urban sprawl is more often associated with heat waves, which will increase incidences of heat-related disasters and the use of energy-demanding cooling systems (Stone et al. 2010).

Urban sprawl has increased even in regions with a declining human population. (EEA 2016b). Both the impact of urbanization (through an increase in impervious surfaces) and climate change (through higher temperatures and precipitation values) are expected to affect future watershed run-off and stream flow. Therefore, understanding the different effects of increased impervious surfaces and climate change on surface run-off can help urban planners to design suitable policies in response to the challenges of greater sealing of urban areas and climate change impacts (EC 2012).

## 6.4 "Resilience Thinking" in Urban Planning

There is a growing emphasis on enhancing the resilience of cities in the face of unprecedented urbanization and climate change (Meerow et al. 2016; Carmin et al. 2012; Tyler and Moench 2012; Leichenko 2011). According to Meerow et al. (2016, pp. 39) urban resilience refers to "the ability of an urban system - and all its constituent socio-ecological and socio-technical networks across temporal and spatial scales - to maintain or rapidly return to desired functions in the face of a disturbance, to adapt to change, and to quickly transform systems that limit current or future adaptive capacity". Therefore, urban resilience is dynamic and offers multiple pathways to resilience (e.g., persistence, transition, and transformation) and recognizes the importance of temporal scale, and advocates general adaptability (Meerow et al. 2016).

Increasing economic, social and spatial vulnerabilities in cities, the rapid depletion of natural resources, and the increasing frequency of ecological events and other causes of environmental degradation indicate that it is time to adopt resilience thinking in planning (Eraydin and Taşan-Kok 2013). During the World Conference on Disaster Risk Reduction held in Sendai, Japan, in 2015 was adopted the "Sendai Framework for Disaster Risk Reduction 2015-2030" which "address the urgency to anticipate, plan and reduce disaster risk in order to more effectively protect persons, communities and countries, their livelihoods, health, cultural heritage, socioeconomic assets and ecosystems, and thus strengthen their resilience" (UN 2015e, pp. 3–4). Therefore, well managed urban development could give rise to cities more sustainable and resilient to climate change and natural disasters (Terakado and Williams 2014).

A resilient city is a sustainable network of physical systems (built roads, buildings, infrastructure, waterways, soils, topography, geology) and human communities (schools, neighbourhoods, agencies, organizations, enterprises, task forces), and is constructed to be strong and flexible, rather than brittle and fragile. A city without resilient physical systems and resilient communities will be extremely vulnerable to disasters (Godschalk 2003). Therefore, planning for resilience in the face of urban disaster requires designing cities that combine seemingly opposite characteristics, including redundancy and efficiency, diversity and interdependence, strength and flexibility, autonomy and collaboration, and planning and adaptability (Meerow et al. 2016; Eraydin and Taşan-Kok 2013).

European policies have affect the economic competitiveness, social conditions and environmental viability, thus the key challenges that Europe faces therefore demand closer cooperation between states, regions and cities. Spatial planning is the most effective means of facilitating this cooperation, and achieving smart, inclusive and sustainable growth and territorial cohesion with more resilient structures throughout Europe, especially for urban areas (ECTP 2013). The action plan defined during United Nations Climate Change Conference in Paris (COP21), highlights the importance of climate-resilient cities (UN 2015d). The European Commission stressed the crucial importance of reaching a fair, ambitious and legally binding global climate deal at the Paris Climate Conference. Therefore, European Union will be working to achieve a comprehensive, durable and dynamic agreement that will accelerate the global shift to low-carbon and climate-resilient economies. This will help to contain climate change and increase societies' efforts to adapt to its impacts, as well as underpin long-term economic growth and sustainable development in the European and global levels (EC 2015). In order to address the increasingly complex challenges in urban areas, European Union Ministers responsible for Urban Matters have reached, at their informal meeting in Amsterdam, agreement on the establishment of the Urban Agenda for the European Union as set out in the "Pact of Amsterdam" (EC 2016). This Urban Agenda will contribute to the implementation of the United Nations 2030 Agenda for Sustainable Development, particularly the Goal 11 "Make cities inclusive, safe, resilient and sustainable" (UN 2015b) and the "New Urban Agenda" adopted in the Habitat III Conference (UN 2016b). This New Urban Agenda shares a vision of "cities for all, referring to the equal use and enjoyment of cities and human settlements, seeking to promote inclusivity and ensure that all inhabitants, of present and future generations, without discrimination of any kind, are able to inhabit and produce just, safe, healthy, accessible, affordable, resilient, and sustainable cities and human settlements, to foster prosperity and quality of life for all" (UN 2016b, pp. 4).

## 6.5 Adapt and Transform Cities into Climate-Resilient and Sustainable Places

Worldwide, cities are increasingly recognizing the need to prepare for the impacts of climate change. It is clear that new weather and climate patterns are emerging and that these changes are putting urban dwellers and assets at risk (Carmin et al. 2012).

Managing the risks of climate change involves mitigation and adaptation decisions with implications for future generations, economies, and environments. Mitigation relates to initiatives designed to reduce the causes of climate change, while adaptation relates to human interventions to address the effects of climate change (IPCC 2014a,c,d). Many municipalities across Europe are already working to mitigate the effects of climate change, decreasing energy use and reducing greenhouse gas emissions. However, they increasingly acknowledge the need to adapt to climate change and have begun to adopt various measures (EEA 2017a,b; EEA 2016a).

Total anthropogenic greenhouse gas emissions have continued to increase over 1970 to 2010 with larger absolute increases between 2000 and 2010. Despite a growing number of climate change mitigation policies, anthropogenic greenhouse gas emissions in 2010 have reached 49 ± 4.5 GtCO2-eq/yr. (IPCC 2014b). Mitigation scenarios reaching atmospheric concentration levels of about 450 ppm CO2 eq by 2100, consistent with a likely chance to keep temperature change below 2 °C relative to pre-industrial levels, include substantial cuts in anthropogenic greenhouse gas emissions by mid-century. These scenarios describe a wide range of changes in energy systems and land use, reflecting different assumptions about the scale of bioenergy production, forestation, and reduced deforestation (IPCC 2014d). Strong mitigation efforts are needed to keep climate change impacts down to a level that still allows the major services obtained from nature and society to function properly (EEA 2016a).

In Europe, current energy and climate policies are delivering substantial progress. Greenhouse gas emissions in 2012 decreased by 18% relative to emissions in 1990 and the share of renewable energy has increased to 13% in the same year as a proportion of final energy consumed. The European Union also had installed about 44% of the world's renewable electricity (excluding hydro) at the end of 2012. The energy intensity of the European Union economy has reduced by 24% between 1995 and 2011 whilst the improvement by industry was about 30%, as well as the carbon intensity of the European Union economy fell by 28% between 1995 and 2010 (EC 2014). The European Union climate change mitigation policy has targets for reducing its greenhouse gas emissions progressively up to 2050. Therefore, the 2030 climate and energy framework also proposes three key targets for 2030 (meeting at least these levels) (EEA 2017a; EC 2014):

1. 40% cut in greenhouse gas emissions (from 1990 levels);
2. 27% of European Union final energy consumption from renewables;
3. 27% improvement in energy efficiency compared with baseline.

These targets are defined to help the European Union become a competitive low carbon economy by 2050, which aims to reduce greenhouse gas emissions by 80–95% by 2050 compared to 1990 (EC 2011a).

With the Paris Agreement on climate change, the European Union thus appears to have consolidated its role as a "leadiator" in international climate policy. However, while its policy objectives on mitigation were the most ambitious among the main players, they were more moderate than what would be required to achieve the 2 °C target (Oberthür and Groen 2017a). Paris Agreement is insufficient to avoid dangerous climate change (Oberthür and Groen 2017b), since even if global greenhouse gas emissions were to stop today, climate change would continue as a result of past emissions and the inertia of the climate system (EEA 2016a). Therefore, adaptation strategies to future climate change in urban areas should pay close attention to reducing the vulnerability of people and communities (EEA 2016a; Carter et al. 2015). These strategies may increase resilience across possible future climates while helping to improve human health, livelihoods, social and economic well-being, and environmental quality. Create resilience and enable sustainable development can accelerate successful climate change adaptation globally. Therefore, integration of adaptation into planning and decision making can promote synergies with development and disaster risk reduction (IPCC 2014a,b,c). There are three different approaches to adaptation: (1) Coping–opts to cope with the immediate impacts of extreme events once they appear or when stresses become obvious, running high risks in terms of human and economic losses and requires rebuilding after each disaster; (2) Incremental–builds on existing adaptation measures and knowledge gained, for example in disaster risk management, by incrementally improving them and increasing their efficiency; and (3) Transformational–opts to fundamentally change the way to approach the challenges, by establishing new and innovative solutions that aim to develop opportunities to transform the city to be resilient and sustainable space. Transformational adaptation enables cities to find more sustainable solutions to long-term change. They can realize many joint benefits and thus turn challenges into opportunities for attractive, climate-resilient and sustainable cities (EEA 2016a).

European Commission concluded that cities of tomorrow will provide a high quality of life and welfare, will be places of advanced social progress, platforms for democracy, cultural dialogue and diversity, and be green places where environmental regeneration takes place (EC 2011b). In the transformed climate-resilient city of tomorrow, people live in houses that are secure and pleasant to live in, even when outdoor temperatures are high, rivers flood and other extreme events take place. They are cities that share knowledge about risks and opportunities, helping prepare for natural hazards. These cities are places of green, ecological or environmental regeneration, as well as places of attraction and engines of economic growth. They also finds novel ways to add green space while limiting urban sprawl and making more compact and energy-efficient cities. Smart spatial and infrastructure designs minimize the urban heat island effect, air pollution and flooding of streets and houses (EEA 2016a; EC 2011b). European cities are undertaking more and more adaptation actions and have found various innovative ways to overcome the challenge of financing their adaptation measures (Table 6.2) (EEA 2017b; EEA 2016a).

**Table 6.2** Seven cases of financing urban adaptation to climate change (data from EEA 2017b)

| City | Adaptation measures financed |
|---|---|
| Amsterdam | 1. Climate-proof collective garden. |
| | 2. Rainwater storage tanks. |
| | 3. Water-sensitive urban and building design. |
| Bilbao | 1. Opening of a water canal. |
| | 2. Elevation of ground level of buildings. |
| | 3. Establishment of green open spaces. |
| | 4. Provision of stormwater tanks. |
| Bologna | 1. Green spaces in urban areas. |
| | 2. Tree planting. |
| Copenhagen | 1. Stormwater runoff management measures. |
| | 2. Detention areas to store large volume of waters. |
| Hamburg | 1. Awareness campaigns for behavioural change. |
| | 2. Economic incentives to build green roofs. |
| | 3. Water-sensitive urban and building design. |
| | 4. Green spaces in urban areas. |
| Lisbon | 1. Leakage detection system. |
| | 2. Adjustment of water infrastructure to reduce water leakage. |
| | 3. Adaptation of drought and water conservation plans. |
| | 4. Water restrictions and consumption cuts. |
| Paris | 1. Green spaces in urban areas. |
| | 2. Planting 20,000 trees. |
| | 3. Establishing 30 hectares of new parks by 2020. |

## 6.6 Conclusions

Nowadays, climate change is already a reality, and the uncertainty and unpredictability have become key characteristics worldwide. Although risks and disasters are inherent to human existence, their speed, frequency and scale are growing.

These socio-ecological and socio-technical network systems (Meerow et al. 2016) are particularly vulnerable to climate change, due to growing urban population worldwide and the complex patterns of economic assets, infrastructures and services that characterize them. Minimizing the climate change impacts on urban areas requires that scientists, local governments, managers and other stakeholders, take steps to protect natural systems as well as human health and well-being, across different temporal and spatial scales. Therefore, promoting relevant mitigation and adaptation strategies, including novel responses to climate change and territorial resilience will be of crucial importance as to attain sustainable development.

## References

Altieri, L., Cocchi, D., Pezzi, G., Scott, E. M., & Ventrucci, M. (2014). Urban sprawl scatterplots for urban morphological zones data. *Ecological Indicators, 36*, 315–323.

Barrico, L. (2015). *Ecological processes related to urban development and land use change in the municipality of Coimbra*. Doctoral dissertation. Retrieved from http://hdl.handle.net/10316/26494

Barrico, L., & Castro, P. (2016). Urban biodiversity and cities' sustainable development. In P. Castro, U. M. Azeiteiro, P. Bacelar-Nicolau, W. Leal Filho, & A. M. Azul (Eds.), *Biodiversity and education for sustainable development, World Sustainability Series* (pp. 29–42). Cham: Springer International Publishing Switzerland.

Bart, I. L. (2010). Urban sprawl and climate change: A statistical exploration of cause and effect, with policy options for the EU. *Land Use Policy, 27*, 283–292.

Bhatta, B. (2010). Causes and consequences of urban growth and sprawl. In *Analysis of urban growth and sprawl from remote sensing data* (pp. 17–36). Berlin/Heidelberg: Springer-Verlag Berlin Heidelberg.

Bolund, P., & Hunhammar, S. (1999). Ecosystem services in urban areas. *Ecological Economics, 29*, 293–301.

Camagni, R., Gibelli, M. C., & Rigamonti, P. (2002). Urban mobility and urban form: The social and environmental costs of different patterns of urban expansion. *Ecological Economics, 40*, 199–216.

Carmin, J., Nadkarni, N., & Rhie, C. (2012). *Progress and challenges in urban climate adaptation planning: Results of a global survey*. Cambridge, MA: MIT.

Carter, J. G., Cavan, G., Connelly, A., Guy, S., Handley, J., & Kazmierczak, A. (2015). Climate change and the city: Building capacity for urban adaptation. *Progress in Planning, 95*, 1–66.

Catalán, B., Saurí, D., & Serra, P. (2008). Urban sprawl in the Mediterranean? Patterns of growth and change in the Barcelona metropolitan region 1993–2000. *Landscape and Urban Planning, 85*, 174–184.

Childers, D. L., Cadenasso, M. L., Grove, J. M., Marshall, V., McGrath, B., & Pickett, S. T. A. (2015). An ecology for cities: A transformational nexus of design and ecology to advance climate change resilience and urban sustainability. *Sustainability, 7*, 3774–3791.

Cohen, B. (2006). Urbanization in developing countries: Current trends, future projections, and key challenges for sustainability. *Technology in Science, 28*, 63–80.

Deilami, K., Kamruzzaman, M., & Hayes, J. F. (2016). Correlation or causality between land cover patterns and the urban heat island effect? Evidence from Brisbane, Australia. *Remote Sensing, 8*, 716. https://doi.org/10.3390/rs8090716

EC (European Commission). (2011a). *Communication from the Commission to the European Parliament, the Council, the European economic and social committee and the committee of the regions–A roadmap for moving to a competitive low carbon economy in 2050*. Brussels, COM(2011) 112 final.

EC (European Commission). (2011b). *Cities of tomorrow: Challenges, visions, ways forward*. Luxembourg: European Commission Directorate General for Regional Policy, Publication Office of the European Union.

EC (European Commission). (2012). *Science for environment policy*. In-depth report: Soil sealing.

EC (European Commission). (2014). *Communication from the Commission to the European Parliament, the Council, the European economic and social committee and the committee of the regions–A policy framework for climate and energy in the period from 2020 to 2030*. Brussels, COM(2014) 15 final.

EC (European Commission). (2015). *Paris climate conference: Historic opportunity to avoid dangerous climate*. Brussels: EC.

EC (European Commission). (2016). *Urban agenda for the EU. Pact of Amsterdam*. Agreed at the informal meeting of EU ministers responsible for urban matters on 30 May 2016 in Amsterdam, The Netherlands.

ECTP (European Council of Spatial Planners). (2013). *The charter of European planning: Barcelona 2013*. The Vision for Cities and Regions–Territoires of Europe in the 21st Century.

EEA (European Environment Agency). (2006). *Urban sprawl in Europe: The ignored challenge*. Luxembourg: Office for Official Publications of the European Communities. EEA Report No 10/2006.

EEA (European Environment Agency). (2012). *Climate change, impacts and vulnerability in Europe 2012: An indicator-based report*. Luxembourg: Office for Official Publications of the European Union. EEA Report No 12/2012.

EEA (European Environment Agency). (2013). *Adaptation in Europe. Addressing risks and opportunities from climate change in the context of socio-economic developments*. Luxembourg: Publications Office of the European Union. EEA Report No 3/2013.

EEA (European Environment Agency). (2016a). *Urban adaptation to climate change in Europe 2016: Transforming cities in a changing climate*. Luxembourg: Publications Office of the European Union. EEA Report No 12/2016.

EEA (European Environment Agency). (2016b). *Urban Sprawl in Europe. Joint EEA-FOEN report*. Luxembourg: Publications Office of the European Union. EEA Report No 11/2016.

EEA (European Environment Agency). (2017a). *Climate change, impacts and vulnerability in Europe 2016. An indicator-based report*. Luxembourg: Publications Office of the European Union. EEA Report No 1/2017.

EEA (European Environment Agency). (2017b). *Financing urban adaptation to climate change*. Luxembourg: Publications Office of the European Union. EEA Report No 2/2017.

Elmqvist, T., Fragkias, M., Goodness, J., Güneralp, B., Marcotullio, P. J., McDonald, R. I., Parnell, S., Schewenius, M., Sendstad, M., Seto, K. C., & Wilkinson, C. (2013). *Urbanization , biodiversity and ecosystem services: Challenges and opportunities. A global assessment*. A Part of the Cities and Biodiversity Outlook Project. Dordrecht: Springer Science+Business Media Dordrecht.

Eraydin, A., & Taşan-Kok, T. (2013). *Resilience thinking in urban planning*. Dordrecht: Springer Science+Business Media Dordrecht.

EU (European Union). (2016). Urban Europe: Statistic on cities. In *Towns and suburbs*. Luxembourg: Publications office of the European Union.

Frondoni, R., Mollo, B., & Capotorti, G. (2011). A landscape analysis of land cover change in the Municipality of Rome (Italy): Spatio-temporal characteristics and ecological implications of land cover transitions from 1954 to 2001. *Landscape and Urban Planning, 100*, 117–128.

Godschalk, D. R. (2003). Urban hazard migration: Creating resilient cities. *Natural Hazards Review, 4*, 136–143.

Hadorn, G. H., Bradley, D., Pohl, C., Rist, S., & Wiesmann, U. (2006). Implications of transdisciplinarity for sustainability research. *Ecological Economics, 60*, 119–128.

IPCC (Intergovernmental Panel on Climate Change). (2007a). *Climate change 2007: Synthesis report. Summary for policymakers*. Geneva: IPCC.

IPCC (Intergovernmental Panel on Climate Change). (2007b). In M. L. Parry, O. F. Canziani, J. P. Palutikof, P. J. van der Linden, & C. E. Hanson (Eds.), *Climate change 2007: Impacts, adaptation and vulnerability. Contribution of Working Group II to the fourth assessment report of the intergovernmental panel on climate change*. Cambridge: Cambridge University press.

IPCC (Intergovernmental Panel on Climate Change). (2013). In T. F. Stocker, D. Qin, G.-K. Plattner, M. Tignor, S. K. Allen, J. Boschung, A. Nauels, Y. Xia, V. Bex, & P. M. Midgley

(Eds.), *Climate change 2013. The physical science basis. Contribution of Working Group I to the fifth assessment report of the intergovernmental panel on climate change*. Cambridge/New York: Cambridge University Press.

IPCC (Intergovernmental Panel on Climate Change). (2014a). In C. B. Field, V. R. Barros, D. J. Dokken, K. J. Mach, M. D. Mastrandrea, T. E. Bilir, M. Chatterjee, K. L. Ebi, Y. O. Estrada, R. C. Genova, B. Girma, E. S. Kissel, A. N. Levy, S. MacCracken, P. R. Mastrandrea, & L. L. White (Eds.), *Climate Change 2014. Impacts, adaptation, and vulnerability. Part A: Global and sectoral aspects. Contribution of Working Group II to the fifth assessment report of the intergovernmental panel on climate change*. Cambridge/New York: Cambridge University Press.

IPCC (Intergovernmental Panel on Climate Change). (2014b). Climate change 2014. Synthesis report. In R. K. Pachauri & L. A. Meyer (Eds.), *Summary for policymakers. Contribution of Working Groups I, II and III to the fifth assessment report of the intergovernmental panel on climate change. Core writing team*. Geneva: IPCC.

IPCC (Intergovernmental Panel on Climate Change). (2014c). In V. R. Barros, C. B. Field, D. J. Dokken, M. D. Mastrandrea, K. J. Mach, T. E. Bilir, M. Chatterjee, K. L. Ebi, Y. O. Estrada, R. C. Genova, B. Girma, E. S. Kissel, A. N. Levy, S. MacCracken, P. R. Mastrandrea, & L. L. White (Eds.), *Climate Change 2014: Impacts, adaptation, and vulnerability. Part B: Regional aspects. Contribution of Working Group II to the fifth assessment report of the intergovernmental panel on climate change*. Cambridge/New York: Cambridge University Press.

IPCC (Intergovernmental Panel on Climate Change). (2014d). In O. Edenhofer, R. Pichs-Madruga, Y. Sokona, E. Farahani, S. Kadner, K. Seyboth, A. Adler, I. Baum, S. Brunner, P. Eickemeier, B. Kriemann, J. Savolainen, S. Schlömer, C. von Stechow, T. Zwickel, & J. C. Minx (Eds.), *Climate change 2014: Mitigation of climate change. Contribution of Working Group III to the fifth assessment report of the intergovernmental panel on climate change*. Cambridge/New York: Cambridge University Press.

Leichenko, R. (2011). Climate change and urban resilience. *Current Opinion in Environmental Sustainability, 3*, 164–168.

López de Lucio, R. (2003). Transformaciones territoriales recientes en la región urbana de Madrid. *Urban, 8*, 124–161. [in Spanish].

Meerow, S., & Stults, M. (2016). Comparing conceptualizations of urban climate resilience in theory and practice. *Sustainability, 8*, 701. https://doi.org/10.3390/su8070701

Meerow, S., Newell, J. P., & Stults, M. (2016). Defining urban resilience: A review. *Landscape and Urban Planning, 147*, 38–49.

Myint, S. W., Zheng, B., Talen, E., Fan, C., Kaplan, S., Middel, A., Smith, M., Huang, H.-P., & Brazel, A. (2015). Does the spatial arrangement of urban landscape matter? Examples of urban warming and cooling in Phoenix and Las Vegas. *Ecosystem Health and Sustainability, 1*(4), 15. https://doi.org/10.1890/EHS14-0028.1

Nijkamp, P., & Vreeker, R. (2000). Sustainability assessment of development scenarios: Methodology and application to Thailand. *Ecological Economics, 33*, 7–27.

Oberthür, S., & Groen, L. (2017a). The European Union and the Paris agreement: Leader, mediator or bystander? *WIREs Climate Change, 8*, e445. https://doi.org/10.1002/wcc.445

Oberthür, S., & Groen, L. (2017b). Explaining goal achievement in international negotiations: The EU and the Paris agreement on climate change. *Journal of European Public Policy*. https://doi.org/10.1080/13501763.2017.1291708

Oueslati, W., Alvanides, S., & Garrod, G. (2015). Determinants of urban sprawl in European cities. *Urban Studies, 52*, 1594–1614.

Rasoolimanesh, S. M., Badarulzaman, N., & Jaafar, M. (2011). Achievement to sustainable urban development using city development strategies: A comparison between Cities Alliance and the World Bank definitions. *Journal of Sustainable Development, 4*, 151–166.

Schneider, A., Friedl, M. A., & Potere, D. (2010). Mapping global urban areas using MODIS 500-m data: New methods and datasets based on 'urban ecoregions'. *Remote Sensing of Environment, 114*, 1733–1746.

Seto, K. C., Güneralp, B., & Hutyra, L. R. (2012). Global forecasts of urban expansion to 2030 and direct impacts on biodiversity and carbon pools. *Proceedings of the National Academy of Sciences of the United States of America, 109*, 16083–16088.

Stone, B., Hess, J. J., & Frumkin, H. (2010). Urban form and extreme heat events: Are sprawling cities more vulnerable to climate change than compact cities? *Environmental Health Perspectives, 118*, 1425–1428.

Sung, C. Y., Yi, Y.-j., & Li, M.-H. (2013). Impervious surface regulation and urban sprawl as its unintended consequence. *Land Use Policy, 32*, 317–323.

Terakado, M., & Williams, H. K. (2014). *Investing in sustainable cities: Challenges and opportunities*. International Development Finance Club. Special Interest Group on Sustainable Urban Development.

Tyler, S., & Moench, M. (2012). A framework for urban climate resilience. *Climate and Development, 4*, 311–326.

Ulbrich, U., Lionello, P., Belušic, D., Jacobeit, J., Knippertz, P., Kuglitsch, F. G., Leckebusch, G. C., Luterbacher, J., Maugeri, M., Maheras, P., Nissen, K. M., Pavan, V., Pinto, J. G., Saaroni, H., Seubert, S., Toreti, A., Xoplaki, E., & Ziv, B. (2012). Climate of the Mediterranean: Synoptic patterns, temperature, precipitation, winds, and their extremes. In P. Lionello (Ed.), *The climate of the Mediterranean region: From the past to the future* (pp. 301–346). London/ Waltham: Elsevier.

UN (United Nations). (1992). *Agenda 21–United Nations conference on environment & development*.

UN (United Nations). (2001). *Natural disasters and sustainable development: Understanding the links between development, environment, and natural disasters*. Department of Economic and Social Affairs. Background document for the World Summit on Sustainable Development. DESA/DSD/PC2/BP5.

UN (United Nations). (2014). *World urbanization prospects: The 2014 revision*. Highlights. Department of Economic and Social Affairs, population division. (ST/ESA/SER.A/352).

UN (United Nations). (2015a). *World urbanization prospects: The 2015 revision*. Department of Economic and Social Affairs, Population Division. (ST/ESA/SER.A/366).

UN (United Nations). (2015b). *Transforming our world: The 2030 agenda for sustainable development*. A/RES/70/1.

UN (United Nations). (2015c). *The millennium development goals report 2014*. New York: UN.

UN (United Nations). (2015d). *Adoption of the Paris agreement*. Conference of the Parties, twenty-first session Paris, 12 December 2015. FCCC/CP/2015/L.9.

UN (United Nations). (2015e). *Resolution adopted by the General Assembly on 3 June 2015. Sendai framework for disaster risk reduction 2015–2030*. A/RES/69/283.

UN (United Nations). (2016a). *The World's cities in 2016*. Department of Economic and Social Affairs, Population Division. Data Booklet (ST/ESA/ SER.A/392).

UN (United Nations). (2016b). *Draft outcome document of the United Nations conference on housing and sustainable urban development (Habitat III)*. Quito, 17–20 October 2016.

WCED (World Commission on Environment and Development). (1987). *Our common future. Report of the world commission on environment and development*. New York: Oxford University Press.

WEF (World Economic Forum). (2016). *The global risks report 2016* (11th ed.). Geneva: World Economic Forum.

Xu, H., Lin, D., & Tang, F. (2013). The impact of impervious surface development on land surface temperature in a subtropical city: Xiamen, China. *International Journal of Climatology, 33*, 1873–1883.

Zhang, T. (2001). Community features and urban sprawl: The case of the Chicago metropolitan region. *Land Use Policy, 18*, 221–232.

Zhang, X. (2016). Sustainable urbanization: A bi-dimensional matrix model. *Journal of Cleaner Production, 134*, 425–433.

# Chapter 7
# Improving Resilience through Cross-Scale Knowledge Sharing

**Elena Pede**

**Abstract** The rapid urbanization, often accompanied by uncontrolled use of land, occupation of unsafe environments as well as the increased rate of occurrence of climate events are introducing new elements of uncertainty. The idea of certainty or security that was fundamental to risk management in the past collapsed. In this context, the notion of 'risk society', introduced by Ulrich Beck in 1992, is considered as a shifting paradigm in world security, where our modern society becomes ever more interdependent and more complex, and consequently more vulnerable to threats and risks. New paths for planning and risk management practices are required in order to be able to enlarge the numbers of actors, including citizens, and able to deploy all form of expert and not expert knowledge, in innovative ways.

Planning in the context of risk society requires flexibility in knowledge systems and learning process that are the main tools for the cross-scale interaction among actors.

The present work aims to explore the potential of sharing data, information and knowledge among different stakeholders and with citizens in order to create a more efficient governance and flexibility in risk prevention, improving the resilience of territorial systems.

**Keywords** Cross-scale · Knowledge transfer · Responsibility · Complex system · Governance

## 7.1 Introduction

In the incoming decades, there will be an increase of uncertainty in the context of world 'risk society' (Beck 1992, 1996, 2009) where society becomes more interdependent and complex (in term of rapid urbanisation, uncontrolled use of land and exploitation of the resources), and, as result, more vulnerable to new threats and risks.

---

E. Pede (✉)
Interuniversity Department of Regional and Urban Studies and Planning,
Politecnico di Torino, Torino, Italy
e-mail: elena.pede@polito.it

 
G. Brunetta et al. (eds.), *Urban Resilience for Risk and Adaptation Governance*,
Resilient Cities, https://doi.org/10.1007/978-3-319-76944-8_7

In such a framework the core value of risk management is a constant search for knowledge to provide certainty toward a predictable tomorrow. Managing an emergency means identifying and framing the crisis, taking decisions under pressure, organizing individuals and sources, but it also concerns the improvement of risk management skills from one crisis to the next, studing and working in order to avoid or mitigate new hazards (Koraeus 2008). It means to be able to manage the data, information and knowledge among the myriad of actors involved in each step, from prevention and prediction to response and rocovery (PPRR chain)[1].

Despite the factor of 'uncertainty', decisions still have to be made, and planning knowledge is a decision-making tool considered as essential and directly relevant for risk society. Planning in risk society context requires leaning on all forms of knowledge e.g. expert and non-expert, as well as identifying and sharing them, in terms of having risks perceptions shared with a wide range of actors.

The resilience concept, even if concerning the question of responsibilities, tries not to see the action as a bureaucratic process but is more in favor of a shared vision. In order to do so, stakeholders must build a common conceptual space for different domains, even if nowadays there are different perspectives, due to the pluralism of professions involved and the scale chosen for the observation of the phenomena. Nevertheless, the challenge of flexibility and dynamism in planning sets also a challenge of dynamism in the knowledge systems and learning process that are the main tools for the interaction among actors and scales.

This work aims to show cross-scale[2] knowledge sharing as a key characteristic of resilience in risk management. The work want to contribute to strengthen the systemic prospective in resilience theories and practices through the exploration of risk society implications and the existing problems in knowledge transfer in risk management.

The first chapter introduces the consequences of risk society in planning practices: from 'quantificable' risks to 'no-quantificable' insecurities, the new risk perception and the demand of citizens participation. The second and the third parts investigate the interaction among actors in risk management in the new context of uncertaintly that requires flexibility and inclusion. Finally, the work argues how cross-scale knowledge can be considered a key factor in building resilience.

## 7.2 The Uncertainty of Risk Society

From 1990s a new perspective of risk as unknowable or unpredictable has come to light. The contemporary society have to face risks as far ranging as degradation to the global ecology, international terrorism, global health pandemics, or the health

[1] Prevision, prevention, response and recovery are the stages that compose the PPRR chain, the disaster management cycle (Smith 2009)

[2] The term "cross-scale" refers to interactions across different scales, for example, between spatial domains and jurisdictions. Changes in cross-scale may arise from consequences of interactions or be caused by other variable. (Cash et al. 2006).

consequences feared as a result of exposure to a myriad of technologies, chemicals, industrial toxins and pollutants and financial crisis (Jarvis 2007).

In this context, Beck's thesis on 'risk society' - introduced in 1992 - has initiated some of the most interesting debates that challenge the social science tradition. These challenges regard the constructed nature of risk, which needs to go beyond the disciplinary boundaries to understand in which way particular risks are experienced, perceived, defined, mediated, legitimated and/or ignored. In his work, Beck (1996) argued that the speeding up of modernization has produced a "watershed" between the world of quantifiable risk, in which decision-makers can think and act based on calculating risks, and the world of non-quantifiable insecurities that we are creating. The modern societies are confronted with the principles and limits of their own model because modernization and industrialization has increased complexity of risk. The concept of risk society takes this as its starting point, in order to articulate systemic and epochal transformation. In particular, there is a new challenge in the relationship of society to hazards and problems produced by it, with in turn exceed the basis of societal conceptions of security.

This turns traditional assumptions, about planning and managing the future. In the past, risk management contains the concept of control, which presumes a decision-making process based on the idea that risk can be mapped and measured, so it could be generally controlled. Nowadays there are new multiple uncertainties like "boundary", "voice" and the "reflexivity" matters (Cantelli et al. 2010). The boundary matters argue that the traditional categories of nation-state are not longer valid and the governance of uncertainty and risk involve various sectors and policy fields. The voice and the reflexivity matters, indeed, concern the risk perception and the citizen participation and inclusion. In this framework characterized by uncertainty, the challenge is to find new ways of relating risks to decisions and practices. The emerging of new types of risk requires an alternative mode of planning practice that acknowledges the new conception of security to identify, enable and mobilize the physical, intellectual, social and cultural elements of society. There is the need of bringing a new dimension to the long-standing role of planning in risk management, by shifting the discussion from managing "calculated risks" to "uncalculated risks". New roles and identities for citizens, experts and bureaucrats have evolved and multiple types of knowledge beyond scientific expert knowledge have emerged and serve as on ongoing resonance body for risk perceptions and governance requirements.

In this framework, the emerging of the 'resilience' concept, signals a turn in disaster risk reduction because it takes into account the myriad of variables that affect complex systems and introduce the question of responsibilities, including the citizens engagement. Its use has been increased with the growing of uncertainties and complexity of urban system. In particular, building resilience has become a "pillar" of disaster management, risk reduction, climate adaptation, regional economic development and strategic planning (Davoudi et al. 2013).

## 7.3 A Myriad of Actors in the PPRR Chain

The impacts of extreme natural events concern multiple stressors, temporal, spatial and jurisdictional scales with the involvement of an enormous number of divers actors with different values, levels of knowledge and practice (Adger 2006; McEvoy et al. 2013).

The activities included in risk management are several and each of them could be considered as different step of a unique cycle that covers from the prevision phase to recovery activities (Smith 2009). All the steps are consequential and each of them is crucial for the best resolution. There are three potential distinct times: before, during and after the catastrophic event. Each of these could be divided in different phases that together constituted the PPRR (prevision-prevention- response – recovery) chain (Fig. 7.1).

The number and type of actors vary according to the type of disaster (e.g. forest fire, earthquake, floods, landslide, etc.), the spatial extension and the type of damages on people and goods. Some actors work during all stages of PPRR chain, others are involved only in few of them. In the same way, some institutions are directly involved in the crisis (e.g. emergency management activities) others indirectly effected the pressure and consequently the vulnerability of the territory (e.g. territorial strategic development).

During the ordinary time, the different actors are involved in activities aimed to avoid or mitigate the possibility of an emergency situation, or however, to be prepared to deal with it (prevision and prevention phases). During these steps, actors work on vulnerability and exposure modification of the territory as well as on community preparedness.

But most of the actors are required during the "golden" hours of the emergency. It concerns the rescue of survivors and the distribution of basic supplies (food, water, medical care). The emergency response depends on the magnitude of disaster and the degree of cascade effect, but it also relates on the capacity flow of resources, equipment information and knowledge among the actors involved (Comfort et al. 2004). The main actors during this step usually are the fire fighters, civil protection, specific police forces (e.g. forest service in case of forest fire events), volunteer services and/or even military defense are required in case of wider disaster. The coordination during this phase should not be limited to the horizontal interaction among the forces involved, but it should also affect the competent authorities of previous prevision and prevention steps.

Finally the activities for returning to normality take place after the catastrophic event. The recovery includes from the rehab - first activities destined to re-enable the fundamental functions (e.g. removal of debris, reactivation of power and water supplies as well as infrastructure) - to the reconstruction that can be long-term actions to return an area to "normality" after devastation. This phase does not concern only physical activities but include also psychological counseling to community.

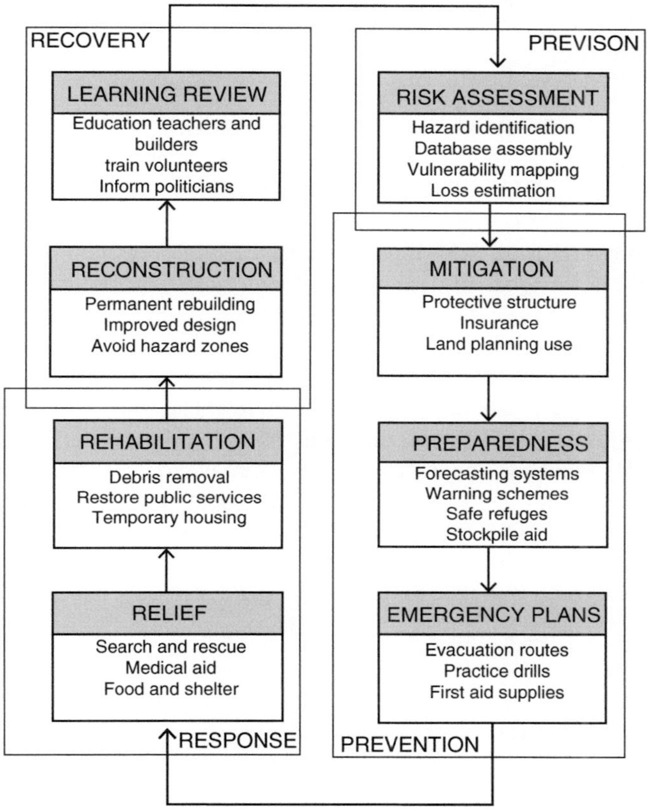


**Fig. 7.1** The PPRR chain and its activities (personal elaboration from Smith 2009)

The PPRR chain shows how the management of risks is usually entrusted to a variety of institutions that operate on different spatial scales (Bignami 2010; Sapountzaki et al. 2011; Wilbanks 2006). As Smith argues (2009) *"(…) effective risk resolution depends on the implementation of a sequential series of actions. The individual stages often overlap but it is crucial that they operate as a closed loop in order to draw benefits from experience and feedback."* (ibid, pp. 67). The same overlapping problems affect the responsibilities and competences of the different institutions involved.

Nowadays, even if there is a progressive awareness of the need of feedback and collaboration with the different actors in all the phases, a fully integrated approach to PPRR chain is rarely achieved. The result is a situation where the respective information, knowledge and policy actions run in parallel without any linkages, feedback, and mutual interaction (Sapountzaki et al. 2011).

## 7.4 The Enlargement of Responsibilities in Socio-Ecological Resilience

The concept of resilience was borrowed from ecological studies, deriving it from the manner in which ecological systems cope with stresses and disturbances caused by external factors. Then, the distinction between engineering and ecological resilience based on the concept of alternative stable states has opened up the "resilience thinking" to all fields linked to complex system and non-linear dynamics.

Traditionally, the goal of risk management is the protection of people, of properties and of the environment from the destructive force of catastrophic events. Nevertheless, building resilience does not only concern the use of land and structural engineering. It must also provide the capability to anticipate and respond to disasters. Static engineering-based conceptualization of resilience has been embraced for long time by disaster studies and the focus has been on recovery in quantitative terms. On the contrary, the metaphor of social-ecological resilience helps to synthesize the integration between ecology and social sciences due to the central role of human ecosystem framework in multiple scale (Teigão dos Santos and Partidário 2011). According to the main definitions, it could identify three characteristics that allow the use of social-ecological resilience concept for territorial systems:

- Humans are part of the ecosystems. "human ecosystem framework are not a matter of humans versus nature, but humans and ecological processes combined into a reciprocally interactive network" (Pickett et al. 2004)
- Complex system. Urban system is the sum of several subsystems in interaction.
- Adaptive cycles. The ecological, social and economic processes permit the continued adjustment and self-organization of urban systems.

Thus, in the planning process, resilience highlights the need to be more flexible. It considers transformation as normal, and dynamism as an intrinsic factor of how systems act. For this reason, it appears a deep affinity with governance theory and social learning, co-management and participation.

As Godschalk (2003) argued *"a resilient city is a sustainable network of physical systems and human communities"* (ibid., p.137). In his metaphor, the physical systems are the constructed and natural environmental components of a city, whereas human communities concern all the actors that live, work and act in that space. The physical systems include the structural engineering like roads network, energy facilities or infrastructure and building, as well as its natural systems as topography, geology and soils. For Godschalk, the physical system is the body of the city, its bones, arteries, and muscles. This means that during a disaster the physical systems must work in order to guarantee a proper and effective functioning under extreme stresses, otherwise without a persistent and resilient physical system a city will be extremely vulnerable to disasters.

In the same way, a body without a brain directing its activities, responding to its needs, and learning from its experiences is likewise fragile and vulnerable. The metaphor of the brain refers to the human communities with the social and institutional components of the city. "*They include the formal and informal, stable and ad hoc human associations, that operate in an urban area: schools, neighbourhoods, agencies, organizations, enterprises, task forces, and the like. (...) During a disaster, the community networks must be able to survive and function under extreme and unique conditions.*" (ibid., p.137).

Thus, social and institutional networks have varying degrees of organization, identity and cohesion. During a disaster they must be able to handle the emergency. Their absence or inefficiency undermines the capability to find solutions.

If we take on Beck's notion of public reflexivity in risk society it could be very close to the meaning of resilience (Sapountzaki 2007). In Beck's view the concept of "public reflexivity" shows also the enlargement of responsibility for risks from institutions to all individuals. This assumption is also promoted by the evolutionary resilience defined by Davoudi et al. (2013) as a way of enhancing preparedness to future transformations. The people-centred approach was one of the priorities for reducing vulnerability and building resilience to present and the future disasters of Hyogo and in Sandai Frameworks. Building resilience plays an increasingly important role in all international agreements, conventions and debates. Despite the growing interest among the politicians, and practitioners to use community resilience as a mean of confronting the response to local disasters (McAslan 2010), resilience remains a difficult concept to practice, thus achieving a tangible outcome is challenging compared, for instance, with the risk management concept (Mitchell and Harris 2012).

Nowadays most of actions in this context still are focused mainly on making the system physical resistant to disaster force, with less attention to the programs for community preparedness, forecasting and warning. If community's resilience will be improved, it would be more responsible for building ultimate urban resilience.

The key challenge should be built in knowledge, learning ability and adaptation of institutions that manage ecosystems at the different levels, in order to improve the resilience of natural and human systems and to contrast their vulnerability in the present context of uncertainty.

The adaptive cycle – as defined as in Holling's work – concerns the characteristics of being a systemic process of complex environmental systems, characterized by high levels of uncertainty, and affected by potential ecological social and economic impacts, due to different management options (Holling 2004). This process is continually improving its management skills, learning and adapting constantly[3]. The ability to learn is related to the cyclical approach that manages the effects of policies and/or actions and includes the results in subsequent decisions with the integration of different knowledge. One useful outcome of resilience method is exactly the ability to link together phenomena that in mainstream planning approaches still remain firmly separated.

[3] Pickett et al. (2004) use the concept *learning loop* to define the need of long term dialog among different institutions and community in order to ensure the monitoring and implementation of knowledge.

In this view, it draws an alternative theoretical perspective of planning more dynamic, fluid and interpretive, that requires a shift of methodology from "command-and-control" to "learn-and-adapt" (Davoudi et al. 2013). For this purpose planning as learning is fundamental to align ideas and goals, to shape leaderships and achieve a shared vision instead of acting as a bureaucratic process (Teigão dos Santos and Partidário 2011).

In this framework, stakeholders must create new relationships to enhance multidirectional information flows to learn from each other and to develop together flexible ways of managing their environments. In the same way, planning also as communication is crucial to enhance information flows and to build relationships. Planning should be a process of communication with its regulations, orientations and actions, but in reality the information display at the end of the planning process, often disregards the different capacities and needs of the potential end-users.

## 7.5 The Cross –Scale Challenge

What emerges is a cross-scale claim characterized by the interaction of scale and level in the knowledge system. According to Gibson et al. (2000) scales are "*the spatial, temporal, quantitative, or analytical dimensions used to measure and phenomenon*", and levels *"the units of analysis that are located at the same position on a scale"* (ibid, pp.218).

Commonly, levels have a spatial feature - e.g. small, medium or large-size phenomena – but they could be also related to time (durations, frequencies or effects). Closely related to spatial scale are jurisdictional scale defined as clearly bounded and organized political units. In the same way, jurisdictional characteristics are linked to institutional arrangements that also have fitting inner hierarchy (Cash et al. 2006).

It comes to light how the scaling problems can be related to issues of scale and/or level as well as in hierarchies and players involved. Most attention given to scale in studies of human-environment interactions has focused on spatial, temporal and jurisdictional issues, but there are also some issues not conventionally framed as a scale problem, but that have to do with the "scale" of management response and change (Cash et al. 2006; Kok and Veldkamp 2011). Following Cash et al. (2006) one of the issues in relation with scale could be the problem of "*knowledge as a scale*".

The term knowledge refers to the process of data and interpreting and understanding that should support actors in taking the right decisions (Koraeus 2008). It means that data that are created at different scales and levels are linked by a cross-level interaction in order to support decisions. In fact, cooperation is not only a problem of governance: a cooperation without information is not sufficient to increase response effectiveness (Comfort et al. 2004).

For example, many environmental management plans and actions can be grouped into hierarchical sets ranging. Thus, each phase of the risk management chain has to combine data and models at different spatial and temporal scales, or also to extrapolate information between scales and levels.

It is difficult to identify the scales of knowledge but it is undeniable that knowledge has to do with scale. This is the reason why a decision should be sensitive to the knowledge of multiple scales rather than focused on a single scale because "*(..) a single can frame an investigation too narrowly because questions and research approaches characteristic of that scale tend to dominate and because upscaling and downscaling information from other scales requires compromises that often lose information or introduce biases*" (Wilbanks 2006, pp. 24).

In building knowledge, there could be different types of interaction between the owner of the data or information and the last user. In reality, these interactions are characterized by a lack of data, limits in gathering data at multiple levels and lack of cross-level interaction (Gibson et al. 2000) rather than by a "progressive embedding".

In its history, planning theory has always made own terms as multiscale and multilevel to indicate the presence of more than one level or scale. The planning theory has often referred to interdisciplinary but with low attention in implying that there are important cross level or cross-scale interaction. The cross-level and scale interaction emerges especially from planning practice in which different policies overlap on the same territory in different times. The organizational attitude of the system is crucial for resilience and this attitude is directly linked to the capacity of system's multi-stakeholders to interact across scale. Governing ecological-social problems means coming to terms with cross-scale and cross-level dynamics (Buizer et al. 2011; Cash et al. 2006). A hot spot of the social ecological system literature is the management of them, in particular referring to adaptive management.

Over the years, several methods have been studied to understand stakeholders' behavior in the management system (e.g. multiagent simulation; social network analyses; system dynamic models) as well as a range of participatory tools and methods have been employed. Following Buizer et al. (2011) all these works acknowledge the idea that scales and levels must be considered as co-produced in processes in with institutions work together. In this field planning theories and practices have a long history in including stakeholders linked to various mode of governance across temporal-spatial scale: multilevel or multiscale governance of problems and the corresponding need to address these problems to multiple administrative levels as well as public-private relationship (Kok and Veldkamp 2011; Termeer et al. 2010). Thus, scale is not a new subject in planning, actually it has always had a key role in literature on the governance of social-ecological systems (Termeer et al. 2010).

However, the growth of sustainability issues and the increase of claims like climate change, natural disasters, pollution and biodiversity ask solutions, now more than ever, that stretch across traditional jurisdictions.

The scaling problem should be focused more on the knowledge issue not interpreted as the recognition and discussion of knowledge claims (Buizer et al. 2011) but more relate to the cross-scale knowledge sharing and interaction. This approach to cross-level claim does not entail only attention to typical spatial and jurisdictional issues of risk management but also to the knowledge and network scales.

One of the main problems is the different knowledge systems that actors use that could compromise or create incongruences in cross-level interactions (Young 2006). In particularly important times of significant change as nowadays, the acquisition and use of integrated knowledge system has a crucial role in effectively responding to the challenge of managing complex social-ecological systems. The alignment and cooperation of capabilities to create, verify, absorb, share and, apply new knowledge can be one of the crucial issue to improve the resilience of complex system to risk (Roux et al. 2006).

## 7.6 The Central Role of Knowledge

The notions of risk society and socio-ecological resilience add necessity to the well-established knowledge. All forms of knowledge seem fundamental for understanding transformations of complex social-ecological systems and play a vital role in communicating the ideas of risks and resilience elements among wider social networks as fundamental condition of the society to face threats. But in the risk society, the knowledge systems and learning processes are characterized by new dynamism that requires new solutions.

Generally, a distinction is made between three main components: data, information and knowledge. The terms data and information are often used interchangeably with the term knowledge but the three concepts are linked by a system of progressive embedding with increasing level of cognitive complexity (Koraeus 2008).

In this framework, it is important to underline that information technology play an important role in supporting knowledge creation, codification, retrieval, transfer, integration and application but it is not the solution.

Knowledge was originally defined by Polanyi (1966, 2009) as the process of interpreting and understanding, the act of making use of information and data. Knowledge is what making possible to take the right decision and to implement the right actions in order to move forward.

Thus, data can be considered as raw facts. If data are organized in a given context we obtain information and when informations are interpreted we obtain knowledge. The relationship between data, information and knowledge depends on the degree of "organisation" and "interpretation" (Cong and Pandya 2003). Data and information can be seem in terms of "measurements" and "observations" indispensable for the cognitive process of knowledge.

Information refers to organized and interpreted data, but even if information includes human participation in the organization of raw data, the end product is explicit and can be readily transferred to another entity (Roux et al. 2006). Indeed, knowledge is a mix of experiences, contextual information and it gives to decision-makers the capacity for effective action.

The plurality of views in a multidisciplinary field entails cultural differences between different domains which renders the knowledge transfer and sharing complex and overflowing with obstacles.

The professional pluralism poses a predicament for the practitioners who must devise their own way of combing the different knowledge (Schön 1983). The importance of practical experience shows also the role of tacit knowledge. Especially in professional practice, problems do not present themselves to practitioner as givens but they must be constructed from the materials of the problematic situations, than they are solved through the selection of the best solutions suited to established aims (Schön 1983). The perspective used to recognize and solve problems is related to the field of inquiry and it has a strong influence on the potential solution. The adopted worldview and the aims frame could shape very different conclusions from the same information. Also the traditional tacit knowledge can play a important role in building knowledge and find solutions. In Hyogo and Sandai Framework there is an explicit reference to the use of traditional, indigenous and local knowledge and practices to complement scientific knowledge in disaster assessment (de la Poterie and Baudoin 2015).

This is the reason why it is important to find a common conceptual space for the different domains to meet in order to collectively create knowledge (Roux et al. 2006). The elements built up the knowledge are influenced by the actors made themselves available to cooperate during the action. Usually, sectors and levels are linked by rules, procedures, organizational forms and technologies, but that does not necessarily imply a fully interaction. The different actors consult data and information generated from different sources, whatever degree of understanding they have of them. Focusing on disaster operations is not easy to describe the dynamic relationship among actors because different environments generate different types of demands that lead to the formation of different types of response patterns based on different levels of capacity in the system (Comfort et al. 2004).

In planning, we can assume that the process of the plan put the bases for interaction between actors that anyway not always implies interactivity. Otherwise, the stakeholders' values, norms, their perception of problems and the possibility to building a shared vision through interactive and communicative processes are the core element of the success of an "interactive process" (Crosta 1998)[4]. The challenge is the interaction system of multiple actors in a framework which is partially structured by shared procedures and norms. For this reason there is the need to harmonize the divergent perspectives that characterized specialists of various sectors in order to realize a joint and coordinated action. If planning is an interactive process it must include a common significance framework and the willingness of learning of the practitioners.

Donal Schön in his studies on learning from practice had developed interesting conceptualizations of the nature of learning systems and the significance of 'learning society' that had become more and more necessary. Schön argued that social

[4]The notion of interactive knowledge emerged not like a positivist paradigm but more like a strategic paradigm oriented to consensus building (Palermo and Ponzini 2014).

systems must learn to become capable of transforming themselves in a changing world.

*"The loss of the stable state means that our society and all of its institutions are in continuous processes of transformation. (…) We must learn to understand, guide, influence and manage these transformations. (…) We must, in other words, become adept at learning. We must become able not only to transform our institutions, in response to changing situations and requirements; we must invent and develop institutions which are 'learning systems', that it to say, systems capable of bringing about their own continuing transformation."* (Schön 1971, p. 28–29). The capacity of the institutions to transform themselves, to adapt or evolve it also assumes in social-ecologic resilience and adaptive cycle literature.

Each professional can bring his practice of knowledge into his work. The core competence of practitioners does not consist in the procedural knowledge, but this is just part of the usable knowledge that must be jointed to the tacit and interactive knowledge for starting the planning process *"Actions and interactions will improve the cognitive background since they generally induce further contextual understanding of strategic problems and of the relevance of spatial visions"* (Palermo and Ponzini 2014, p. 127).

Thus, interactive knowledge seems crucial for understanding transformations of complex social-ecological systems. The planning practice needs more than studying statistics and data, it requires interaction between stakeholders in order of building an incrementalist model of practice and learning.

The hardest challenge in this field concerns the sharing of tacit knowledge, but at the same time, the consideration of practical application highlights the potential and hidden role of local communities and the broader topic of public participation.

Once again, knowledge and resilience show the importance of the human dimension and the necessity of a wider cooperation among a wide range of stakeholders and institutions in order to evolve toward adaptive co-management (F. Berkes and Folke 1998). They emphasize the role of networks, leadership, diversity, collective memory and trust. But as Pahl-Wostl et al. (2007b) explained the difficult question is how these characteristics are developed and sustained.

It must be pursued processes of learning (usually called social learning) based on the development of shared meanings and practices and the relation between the individual and the organization and between different entities.

The framework is provided by multiparty collaboration embedded in a specific context and leading to specific outcomes. It should be based on feedback loop between outcomes and context. In the social-ecological systems the context of social learning includes the governance structure (all the actors, institutions and stakeholders) and the natural environment. The governance structure has a strong influence on the nature of multiparty cooperation and social learning processes because it influences the access to information that can impede social learning. The multiparty interactions regard two different aspects: the processing of factual information about a problem and the problem solving (Pahl-Wostl et al. 2007a). The issue of management is one of the core aspects of participatory processes because it makes explicit and shared the problem definition, the direction setting, the implementation, the type of ground rules and negotiation strategies chosen and the role of leadership in the process. Likewise, the development of shared meaning for a joint action is facilitated by relational practices (Pahl-Wostl et al. 2007a). Relational practices may take different forms, such as joint field visit or common training sessions. Benefits of social learning refer both to the measures implemented to deal with problems and to the capacity of the stakeholder group to face problems as well as to enhance the relationships involved.

## 7.7 Conclusions

Social and institutional networks have varying degrees of organization, identity and cohesion. During a disaster they must be able to handle the emergency. Their absence or inefficiency undermines the capability to find solutions.

In its history, planning theory has always made own terms as multi-scale and multi-level to indicate the presence of more than one level or scale. As well as it has often referred to interdisciplinary but with low attention in implying that there are important cross level or cross-scale interaction. By contrast, the cross-level and scale interaction emerges especially from planning practice.

In this work it surveys the role of cross-scale knowledge ability of institutions that work at the different levels. The ability to learn is related to the cyclical approach that manages the effects of policies and/or actions and includes the results in subsequent decisions with the integration of different knowledge.

In order to ensure an effective collaboration among competent bodies, it is necessary a co-planning system capable to integrate the different knowledge in a shared system. In this framework, stakeholders must create new relationships to enhance multidirectional information flows to learn from each other and to develop together flexible ways of managing their environments. To this aim governance is not enough. It must be supported by a rapid and continuously update system for sharing data and information. In emergency management this need is clear even more, considering the time constraints inherent to crises.

## References


Adger, W. N. (2006). Vulnerability. *Global Environmental Change, 16*(3), 268–281.

Beck, U. (1992). *Risk society. Towards a new modernity*. London: SAGE Pubblications Ltd.

Beck, U. (1996). Risk Society and the Provident State. In S. Lash, B. Szerszynski, & B. Wynne (Eds.), *Risk, environment and modernity: towards a new ecology*. London: SAGE Pubblications.

Beck, U. (2009). Critical theory of world risk society: A cosmopolitan vision. *Constellations, 16*(1), 3–22.

Berkes, F., & Folke, C. (1998). Linking social and ecological systems for resilience and sustainability. Linking social and ecological systems: management practices and social mechanisms for building resilience, 1–25.

Bignami, D. F. (2010). *Protezione civile e riduzione del rischio disastri. Metodi e strumenti di governo della sicurezza territoriale e ambientale*. Santarcangelo di Romagna: Maggioli Editore.
Buizer, M., Arts, B., & Kok, K. (2011). Governance, Scale and the Environment: The Importance of Recognizing Knowledge Claims in Transdisciplinary Arenas. *Ecology and Society, 16*(1.) [online] Available at http://www.ecologyandsociety.org/vol16/iss1/art21/ [Accessed 10 May 2014].
Cantelli, F., Kodate, N., & Krieger, K. (2010). Questioning world risk society: Three challenges for research on the governance of uncertainty. Global policy. Retrieved from http://www.globalpolicyjournal.com/articles/health-and-social-policy/questioning-world-risk-society-three-challenges-research-governanc
Cash, D. W., Adger, W. N., Berkes, F., Garden, P., Lebel, L., Olsson, P., et al. (2006). Scale and cross-scale dynamics: Governance and information in a multilevel world. *Ecology and Society, 11*(2), 8.
Comfort, L., Ko, K., & Zagorecki, A. (2004). Coordination in rapidly evolving systems: The role of information. *American Behavioral Scientist, 48*(3), 295–313.
Cong, X., & Pandya, K. V. (2003). Issues of knowledge management in the public sector. *Electronic Journal of Knowledge Management, 1*(2), 25–33.
Crosta, P. (1998). *Politiche. Quale conoscenza per l'azione territoriale*. Milano: FrancoAngeli.
Davoudi, S., Brooks, E., & Mehmood, A. (2013). Evolutionary resilience and strategies for climate adaptation. *Planning Practice and Research, 28*(3), 307–322.
de la Poterie, A. T., & Baudoin, M. A. (2015). From Yokohama to Sendai: Approaches to participation in international disaster risk reduction frameworks. *International Journal of Disaster Risk Science, 6*(2), 128–139.
Gibson, C. C., Ostrom, E., & Ahn, T. K. (2000). The concept of scale and the human dimensions of global change: A survey. *Ecological Economics, 32*(2), 217–239.
Godschalk, D. R. (2003). Urban hazard mitigation: Creating resilient cities. *Natural Hazards Review, 4*(3), 136–143.
Holling, C. S. (2004). From complex regions to complex worlds. *Ecology and Society, 9*(1.) Retrieved from http://www.ecologyandsociety.org/vol9/iss1/art11/.
Jarvis, D. S. (2007). Risk, globalisation and the state: A critical appraisal of Ulrich Beck and the world risk society thesis. *Global Society, 21*(1), 23–46.
Kok, K., & Veldkamp, T. (A).. (2011). Scale and governance: Conceptual considerations and practical implications. *Ecology and Society*, *16*(2).
Koraeus, M. (2008). *Who knows?: The use of knowledge Management in Crisis*. Stockholm: CRISMART, National Defence College.
McAslan, A. (2010). *Community resilience. Understanding the concept and its application*. Adelaide: Torrens Resilience Institute.
McEvoy, D., Fünfgeld, H., & Bosomworth, K. (2013). Resilience and climate change adaptation: The importance of framing. *Planning Practice and Research, 28*(3), 280–293.
Mitchell, T., & Harris, K. (2012). *Resilience: A risk management approach. ODI Background Note, January*. London: Overseas Development Institute. Retrieved from http://www.odi.org/sites/odi.org.uk/files/odi-assets/publications-opinion-files/7552.pdf.
Pahl-Wostl, C., Craps, M., Dewulf, A., Mostert, E., Tabara, D., & Taillieu, T. (2007a). Social learning and water resources management. Ecology and Society, 12(2). Retrieved from http://www.ecologyandsociety.org/vol12/iss2/art5/
Pahl-Wostl, C., Sendzimir, J., Jeffrey, P., Aerts, J., Berkamp, G., & Cross, K. (2007b). Managing change toward adaptive water management through social learning. Ecology and Society, 12(2). Retrieved from http://www.ecologyandsociety.org/vol12/iss2/art30/
Palermo, P. C., & Ponzini, D. (2014). Inquiry and Design for Spatial Planning. In E. A. Silva, P. Healey, N. Harris, & P. Van den Broeck (Eds.), *Three approaches to planning research in late modern cities, Research Methods in Spatial and Regional Planning*. Routledge.
Pickett, S. T. A., Cadenasso, M. L., & Grove, J. M. (2004). Resilient cities: Meaning, models, and metaphor for integrating the ecological, socio-economic, and planning realms. *Landscape and Urban Planning, 69*, 369–384.

Polanyi, M. (1966). The logic of tacit inference. *Philosophy, 41*(155), 1–18.
Polanyi, M. (2009). The tacit dimension. University of Chicago Press.
Roux, D. J., Rogers, K. H., Biggs, H. C., Ashton, P. J., & Sergeant, A. (2006). Bridging the Science–Management Divide: Moving from Unidirectional Knowledge Transfer to Knowledge Interfacing and Sharing. *Ecology and Society, 11*(1). [online] Available at http://www.ecologyandsociety.org/vol11/iss1/art4/ [Accessed 10 May 2014].
Sapountzaki, K. (2007). Social resilience to environmental risks: A mechanism of vulnerability transfer? *Management of Environmental Quality: An International Journal, 18*(3), 274–297.
Sapountzaki, K., Wanczura, S., Casertano, G., Greiving, S., Xanthopoulos, G., & Ferrara, F. F. (2011). Disconnected policies and actors and the missing role of spatial planning throughout the risk management cycle. *Natural Hazards, 59*(3), 1445–1474.
Schön, D. A. (1971). *Beyond the stable state: Public and private learning in a changing society*. London: Maurice Temple Smith Ltd.
Schön, D. A. (1983). *The reflective practitioner: How professionals think in action*. New York: Basic Books.
Smith, K. (2009). *Environmental hazards: Assesing risk and reducing disaster 5°ed*. New York: Routledge.
Teigão dos Santos, F., & Partidário, M. R. (2011). SPARK: Strategic planning approach for resilience keeping. *European Planning Studies, 19*(8), 1517–1536.
Termeer, C. J. A. M., Dewulf, A., & Lieshout, M. v. (2010). Disentangling scale approaches in governance research: Comparing monocentric, multilevel, and adaptive governance. *Ecology and Society, 15*(4).
Wilbanks, T. J. (2006). Chapter 2 How scale matters: some concepts and findings. In W. V. Reid, F. Berkes, T. J. Wilbanks, & D. Capistrano (Eds.), *Bridging scales and knowledge systems* (pp. 22–35).
Young, O. (2006). Vertical interplay among Scale-dependent environmental and resource regimes. *Ecology and Society*, *11*(1). [online] Available at http://www.ecologyandsociety.org/vol11/iss1/art27/ [Accessed 10 May 2014].

# Part II
# From Risk Management to Urban Resilience: Experiences and Best Practices

# Chapter 8
# Urban Resilience and Informality: Effects of Formalisation in Golestan, Iran

**Solmaz Hosseinioon**

**Abstract** In times of rapid changes and transformations which new paradigms and challenges are arising fast, it is felt more than ever that we require more adaptive viewpoints in urban decision making and planning. The importance of resilience thinking framework is ever increasingly felt in various aspects of built environment and human settlements whether formal or informal.

Informal settlements are an important part of urbanity due to rapid urbanization, lack of access to affordable housing, disasters, civil wars and climate change. Socio-political events as well the modernization processes have increased the people's tendency to live in these areas due to the need for affordable shelter and job opportunities for survival. These areas are very vulnerable to disasters due to their usually exposed situations. Informality as a way of life, is strongly present in informal settlements but it does not mean it is not present in all other aspects of urban areas. Informality is not a problem anymore rather a dominant type of urbanity which needs to be studied, analysed and learned from. It helps demonstrating features compatible with resilience attributes in some aspects.

This article is an inquiry into the effects of formalization of informal settlements within a resilience thinking framework. Formalization in this research includes the urban codes for regularization. Resilience is a solution for dealing with uncertainty and adaptation in complex developments. Practices such as urban planning and design codes transform the urban form. This research traces the transformations imposed by urban planning regulations for upgrading three neighbourhoods with different levels of formalization in Golestan, Tehran, Iran. It compares and studies the effects of urban upgrading processes on adaptation capacities through an urban design lens.

**Keywords** Resilience · Regulations · Formalization · Informality · Adaptation

---

S. Hosseinioon (✉)
Urban Designer, Independent Urban Scholar, Researcher, Melbourne, Australia



G. Brunetta et al. (eds.), *Urban Resilience for Risk and Adaptation Governance*, Resilient Cities, https://doi.org/10.1007/978-3-319-76944-8_8

## Case Study Fiche/Resume

| | |
|---|---|
| **Image or graphic** |  |
| **Location** | Region: Middle east, Iran, Tehran conurbation, city of Golestan |
| **Duration** | 48 months |
| **Time Frame** | Start date: 10/2011 End date: 05/2015 |
| **Consortium** | The university of Melbourne, School of Design, Professor Kim Dovey as supervisor |
| **Cost** | 108,000 US$ |
| **Resources** | Human (the researcher), spatial, socio-economic and human data, mapping and onsite data collection technologies |
| **People affected** | Directly 300,000 (population of Golestan) |
| **Donors** | The Australian government as APA scholarship, The university of Melbourne |
| **Urban Sectors** | Municipality, water, waste, building, energy, transport |
| **Resilience dimension** | Adaptation, mitigation, transformation, resistance |
| **Strengths** | 1. Conflating resilience and informality viewpoints<br>2. Studying resilience from urban design and planning point of view<br>3. Studying the effects of urban planning regulations on resilience |
| **Weaknesses** | 1-Limited resources and time for increasing the scope and depth of the case study<br>2-Limits of the study to one study instead of several cases and one country<br>3-limited data gathering which could be more effective if done in a longer time span and included longitudinal method as well |
| **Impact** | 1-Conflating two key issues and challenges that the work has: resilience and informal settlements<br>2-Taking different aspects of resilience into consideration<br>3-studying the effects of urban regulations on resilience in practice<br>4-Concertizing the resilience theoretical viewpoints into a real case study |
| **Stakeholders List** | 1. Urban resilience experts<br>2. Urban designers and planners<br>3. Municipality of Golestan and similar cities<br>4. People of informal settlements such as Golestan |
| **Challenge** | The motivation for initiating this research rises from the significance of studying resilience and adaptation in informal settlements because they are both among the main urban challenges in the world and Iran. This research conflates informality and informal settlements with resilience viewpoints. Resilience, informality and informal settlements have common features, because both are considered as solutions and survival strategies and are key challnges which need to be dealth with in todays's world. |
| **Project description** | This research has traced the transformations caused by urban regulations in three neighborhoods with different states of informality and how they affect their adaptation capacities |

| | |
|---|---|
| **Urban resilience elements** | The urban form characteristics and dynamics of urban life are related to urban resilience elements, these attributes include: Sense of place and community, social capital, self-organization and self-sufficiency, variety, and heterogeinty, redundancy. |
| **List of innovation elements.** | 1.Urban resilience measures from urban design point of view<br>2.Relationship of informality and resilience<br>3.Effects of urban regulations on adaptation capacities |
| **Lessons learned** | Informal areas have capacities which are in accordance with resilience attributes as adaptation. Yet the implementation of codes can help reduce their vulnerability. |
| **Upscaling and replication** | This principle our comes of this research including resilience attributes from urban design point of view and the adaptation capacities of informal areas can be applied to similar contexts, both nationally and internationally. |
| **Web page** | http://www.unescosost.org/en/2016/05/03/premio-resurbe-2016/<br>https://minerva-access.unimelb.edu.au/handle/11343/91435 |
| **Reference** | List any major publication speaking about the case study<br>1.Informality on the edge, an insight into the informal settlements in Tehran conurbation, Iran, ISUF 2014, conference EBook<br>2. Effects of Urban Planning and Design Regulations on Resilience of Informal Settlement. Case study, Golestan, Iran, RESURBE 3 conference, 2016 |

## 8.1 Introduction

Urban design and planning regulations are affecting built environments constantly. In the case of informal settlements, they are conducted as upgrading processes to transform them into formal areas. This research has analysed how formalization through urban upgrading processes in Iran's informal settlements have transformed them into new entities with different levels of resilience. The motivation for initiating this research rises from the significance of studying resilience and adaptation in informal settlements because they are both among the main urban challenges in the world and Iran. Informality and resilience have common features, because both are considered as solutions and survival strategies. This study conflates some of the most challenging global issues including resilience, informal settlements and informality through an urban design lens.

Resilience framework helps us deal with transformations and volatile, multidimensional challenges. The concept of resilience has been applied in many fields because of its capacity for dealing with complicated and unstable issues. Resilience thinking is a new lens for looking at the world we live in (Ward 2007). Resilience has had many definitions which refer to difference aspects covering a wide range, from being able to bounce back to the original state to rising to a higher condition with expanded capacities. Walker et al. (2004) define it as a "capacity in a system to absorb disturbance and reorganize while undergoing change so as to still retain essentially the same function, structure, identity, and feedbacks. Resilience as resistance on the hand, includes vulnerability reduction in the cities (Pelling 2002, 2003; Godschalk 2003; Wilkinson 2010, 2012; Paton and Johnston 2001; Cutter et al. 2003; Vale and Campanella 2005).

Informality is considered as one of the key challenges for most governments and urban managers. It subsists not only in the Global South, but permeates in formal areas of the Global North too. It is a concept which is much more nuanced than mere shaping of spontaneous settlements. Informality is acknowledged as an arrangement (McFarlane 2012) and "a way of life" (AlSayyad 2004). Informal settlements as areas with high saturation of informality are acknowledged as the dominant pattern of urbanity in today's world (Roy 2005; Roy and AlSayyad 2004) which is one of the complex challenges of today's urbanity. Rapid urbanization and incapability to cope with the housing needs of people in formal urban areas have led to formation of informal settlements and Iran is no exception.

Urban codes are chosen as the agents of change for upgrading process in the cases of this study. They do not solely affect the morphogenesis of informal areas; they are tools of power and control as well (Roy 2005). They lead to incremental changes by human and non-human agents (Brenner et al. 2011). As a result, other parts of the system adapt. The variables of change (slow and fast) are both external and internal forces.

This article uses urban upgrading codes as external forces and key slow variables which have affected and are transforming the informal settlements in Iran. Three neighbourhoods in the city of Golestan in Tehran conurbation, Iran are chosen as the case studies to get an in depth understanding of the socio-economic and morphological transformations of the neighbourhoods.

## 8.2 Resilience

The international urge to create 'resilient cities' is the new agenda along with sustainable development. Resilience and adaptation have become an important part of toolkits for thinking about development issues in all scales. As the world problems have become more complex, new concepts are emerging for confronting the new challenges in the world. The common consensus is that the today's challenges are too complicated and unpredictable to be led by rigid regulations. Resilience thinking is a new lens for looking at the world we live in for dealing with ever-changing unpredictable, multifaceted challenges. Resilience is 'the ability of systems to absorb changes of state variables, driving variables and parameters and still continue its functions (Holling 1973).

The concept of resilience has been applied in many fields for its capacity to deal with complicated and volatile issues. The applications of resilience are useful in different fields such as disasters (Godschalk 2003; Paton and Johnston 2006), ecological resilience (Klein 2003; Longstaff 2005) or community resilience (Pfefferbaum 2005; Coles 2004; Kimhi 2004; Ahmed 2004; Ganor 2003; Paton 2000) as well as spheres of planning (Hillier 2007; Shaw 2012; Wilkinson 2011,

2012; Porter and Davoudi 2012) and urban design, and informal settlements (Dovey 2012; Revell 2010; Mehaffy 2011, 2013). It is used for preparation of vision statements and offer solutions for the world's complex urban problems.

Resilience theory deals with dynamism, contingency and absorbing shock or disruptions in complex adaptive systems or parts of them in different scales, from global scale (climate change) to countries (sudden or chronic hazards) to local ecosystems and communities. Resilience should be considered in different scales, from the global scale (like climate change) to countries (hazards or peak oil) to local ecosystems and communities. Resilience concept has started a long journey from several disciplines such as engineering, psychology and ecology and has spread its use in urban and development debates. Resilience has several definitions referring to a wide range of close yet different aspects, "its meanings and measurements are still contested" (Adger 2003) and many scholars have tried to differentiate its several definitions and their uses (Holling 1973; Davoudi et al. 2012), hence grasping the different aspects of resilience is key in any study related to it.

One of the main definitions of resilience comes from engineering. It is about "elasticity and storing strain energy and maintaining equilibrium without breaking or being deformed" (Holling 1996; Folke et al. 2010; Norris et al. 2007; Gordon 1978). The psychological approach to the concept of resilience includes the resilience of individuals in the face of stress or threat (Norris et al. 2008; Masten 1990; Butler 2007). Resilience definition has evolved during the extensive studies on socio-ecological systems. According to Holling (1973:17) and Folke (2006:254), resilience is the capacity to remain within a specific state during a phase of change.

> "Resilience determines the persistence of relationships within a system and is a measure of the ability of these systems to absorb changes of state variables, driving variables and parameters, and persist (Holling 1973:17)."

Resilience attributes are assemblages with interrelated elements in dynamic cycles which are evolving continuously. Adaptation attributes have different implications, depending on the type of resilience, inner or outer stressors, expectations and domains of practices and the locality of cases.

The best way to cope with unexpected changes is to consider the broad basin of attraction in the systems state and study their capacities for adaptation (Walker and Salt 2006). Researchers should clearly state which aspect of resilience and what sort of drivers are being considered. Every system faces variables of change which impose sudden or chronic changes on the system. Slow changing variables are state variables with slow turnover rates or vague processes with long-term return times (Gunderson and Holling 2002; Walker et al. 2004) whereas fast variables act as shocks to the system such as abrupt disasters.

A regime shift happens in two scenarios, one in which the system internally approaches the threshold or an external "shock to the system pushes it across the threshold Resilience (Alliance 2007:19)." Understanding the drivers of the system, and the types of probable shocks and stressors is important. We must recognize what kinds of shocks we are anticipating regarding a certain system and identify the possible trajectories; they in turn, can lead us to plausible scenarios and regimes.

Identification of the stakeholder and their need and tendencies will help us understand the resilience of the specific socio-ecological systems in mind and their management (Resilience Alliance 2007:19).

Another key point which this research considers is that resilience is relative in many ways, it is neither good nor bad. Resilience can be advantageous or disadvantageous (Holling 1973). We should consider transformations rather than returning to the previous status (Vale and Campanella 2005).

## 8.3 Informality and Informal Settlements

Informality has been considered as a negative aspect by many scholars and professionals until recently, but it is one of the means of the poor to adapt their environment according to their needs (Dovey 2012; Neuwirth 2006; Roy and AlSayyad 2004; Roy 2005; Yiftachel 2009). Informality is defined by necessities, "it can change and adapt to meet the needs of the people" (Revell 2010) which reverberates with adaptation capacities. There is far more nuance to informality than mere lack of formality or as "the absence of formality" (Oxford Dictionary 2014). It is a manner of production, a mode of life which is deeply entangled with formality (Roy 2005; AlSayyad and Roy 2006; Simone 2011). Informality is represented by unorganised, unregulated labour (McFarlane 2012), inaccessible and even hidden (Revell 2010) dynamics and features. It is conducted by the people who want to survive and improve their life and challenge the structures dictated by the authorities through "the quiet encroachment of the ordinary" (Bayat 1997). Informality cannot be defined merely as something outside the control of the state since the government defines its boundaries and nature, and informality always flows within the scope of the state (Roy 2009:26; Porters et al. 1989). The informal style may include informal patterns of self-employment, negotiations, and way finding through local sources informal credit associations.

Urban informality term indicates "an organizing logic, a system of norms that governs the process of urban transformation (Roy and AlSayyad 2005:148)." Informality is a particular mode of urbanisation in contemporary human settlements, and "a critical epistemology for planning (Roy 2004:159)." The creative ways of informality can teach urban planners and designers' lessons for adaptation.

One billion people currently live in slums, and it is estimated that their population will double by 2025 and triple in the following two decades (UN-Habitat 2003; Mehta et al. 2008). Considering that by 2050, 70 percent of world population live in the cities, the significance of studying them becomes even more crucial in today's urban discourse. "The worsening state of access to shelter and security of tenure have resulted in severe overcrowding, homelessness, and environmental problems" (UN-Habitat 2003). Many global and national agendas and programmes are dedicated to analysing the informal settlements and solving their problems.

Informal settlements have versatile forms, locations and characteristics. In an urbanizing world where informal settlements are significantly growing, they are not an exceptional phenomenon, rather a prevailing state of development. The main challenge of such settlements rises from the struggle of the poor and their need for adequate housing and shelter, ownership and overcoming the insecurity of tenure. They usually lack access to rudimentary services, especially water, sanitation and amenities in the cities when they can't afford to acquire through formal channels.

## 8.4 Iran's Informal Settlements

Rapid urbanization and incapability to cope with the housing needs of people in formal urban areas have led to formation of informal settlements and Iran is no exception. Their self-grown, self-organized and self-managed nature is reflected in their complex spatial patterns and characteristics which may teach us about adaptability traits in built environment. Although Iran's informal settlements are not among the most famous in the world, but Iran is the eighths among the 20 largest slum populations (UN-Habitat 2003). Despite the high presence of formal factors in Iran's informal settlements which are more than many global cases of slums, they still have many informal characteristics. They include lack of proper sewage, low quality construction and materials, lack of safety and social problems, informal ownership and incremental formation which has occurred out of the control of the state. Studying Iran's informal settlement life cycles will add to the knowledge about their formation and adaptation capacities in this special context.

Most of the informal settlements in Iran are located around the cities not in the inner city areas and they are called "Hashieneshini" which means peripheral settlements which indicate their geographical location. However that are socially and economically pushed to the fringes as well. The city of Golestan (called Soltanabad) is one of the former informal settlements outside the city of Tehran located in Tehran conurbation area which is represented in this article as an example of these informal settlements which depicts different phases of informality and formalization.

## 8.5 Formalization Process of Informal Settlements in Iran

Urban codes for upgrading process are chosen as agents of change in the cases of this study. They do not solely affect the morphogenesis of informal areas; they are tools of power and control as well (Roy 2005). There are several organizations and institutions involved in upgrading informal settlements in Iran. The main criteria are set in 2005 by The High Commission of Urbanism and Architecture (which is now a part of Ministry of Roads and Housing).

There are four types of urban areas targeted for upgrading which are called, "distressed urban areas or decayed urban fabrics", among which are the informal settlements. They are upgraded by regulations set for redevelopment planning rules which in turn, affect their morphology and so their spatial and socio-economic characteristics. Based on these criteria, the plot size should be more than 200sqm, the width of existing access ways must be more than 6 meters wide and the buildings should be less than 20 years old. If any of the mentioned criteria are not present in 50% of an urban block, the area will be marked as obliterated fabric and subject to reformation plans. The regularization process are based on upgrading projects delivered by municipalities or assigned organizations. The first two criteria directly affect the urban form and cause morphological changes which have consequences in other aspects of urbanity as well. The formalizing regulations have changed not only the spatial but other fundamental characteristics of these areas in time.

The formalization process has transformed the morphological, economic and social structures of these areas, leading to loss of identity, segregation, social exclusion in these settlements. One of the results which is common in many international cases is pushing the poor and homeless to the fringes. The regime shift and identity change which is caused by the regularization processes is considered from resilience point of view.

## 8.6 Introduction of the Case Study: Golestan

Golestan (previously called Soltanabad) is a small city located 18 km away from Tehran, along the Saveh-Tehran highway in Baharestan prefecture. The reason for choosing this city is the presence of sequential formation phases in it from informal to formal. It includes areas formed as informal settlements which have gone through different phases of growth and formalization and informal parts are recognizable at present. Golestan is shaped along the Saveh road because it an important access way in Tehran conurbation since the 1960's, connecting several industrial zones and warehouses which are located around it to the capital and surrounding cities. A lot of its residents commute and work in Tehran and its surrounding areas.

Golestan's main spatial development constitutes of an amalgam of rural areas and farmlands overtaken by poor immigrants in time. Although most of Golestan has evolved as an informal settlement, it is now a formal city although a big part of it is still informal in many ways. Golestan is divided into three official districts by its municipality, providing basic infrastructure and maintenance which will increase the control and hence hasten the formalisation process (Fig. 8.1).

Although Golestan has become an official city with a legal boundary, it is still a mixture of formality and informality, rural and urban, industrial and residential with different official and unofficial status of tenure ownership. The economic status of people is higher in the more formalized areas, (according to the official data and field observations) such as rate of car ownership. Golestan is going through formalization as it is in the process of preparation of its structure plan in 2013 (Table 8.1).

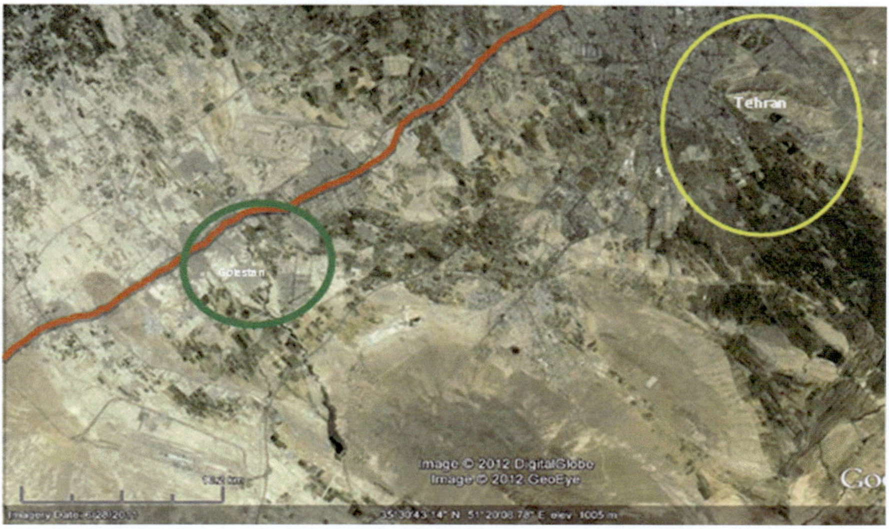


**Fig. 8.1** Golestan (Soltanabad), an informal settlement in Tehran conurbation, Iran (*Source*: Google Earth 2012)

**Table 8.1** General characteristics of Golestan (*Source*: author derived from official census and data)

| Population | Distance from Tehran(km) | Plot size ratio | Unofficial ownership | Population working in Tehran | Official urban development plan |
|---|---|---|---|---|---|
| 231,905 | 17 | 50%:15-50sq.m<br>30%:50-100sq.m | 79% | 61% | Golestan structure plan |

## 8.7 Methodology and Findings

Resilience is a relative concept and yet, at its core, there are certain capacities which are mutual for all resilient systems including natural and built environments.

In this study, resilience is emphasized upon as adaptation. In order to conduct this study, key adaptation attributes which resonate with urban form characteristics are chosen and summarized from the vast literature on the issue. However, resilience as resistance and vulnerability reduction is not neglected and is considered for studying the case studies. One of the main reasons is that urban formalization regulations in Iran are set for the purpose of risk reduction. One of the main reason for conducing this study is to examine the effects of such codes on different aspects of resilience specially its adaptation capacities which are not studied from practical point of view in actual case studies.

The three case studies' characteristics are analysed from urban design point of view. The morphological interventions in the three neighbourhoods have affected the lifestyle and socio-economic situations of the inhabitants (Lefebvre 1992). The way the residents use and perceive their public space, is changed by formalization of the spatial attributes. They have cultural and socio-economic consequences in different levels. The formalization process is considered as an element of regime shift (Carpenter et al. 2001; Walker and Salt 2006; Gunderson and Holling 2002). It changes the identity and characteristics of the areas to the point of reterritorialization and creation of similar homogenous neighbourhoods. At the moment, each area has diverse behaviours in times of stress and different adaptation capacities. The question is to find out if the process of urban formalization in the neighbourhoods makes them more or less adaptable? What are the effects of application of upgrading codes on the type of resilience?

Studying the effects of formalization process and the morphological changes, can indicate how the identity, socio-economic and adaptation capacities of these areas are transformed as a result of implementing the urban upgrading processes. Three neighbourhoods are chosen in different phases of formalization. The data gathered and analysed used for this article is gained by the researcher's on-site observations. Interviews and documentation of urban and morphological characteristics as well as people's activities and behaviour in the public realms.

Soltanabad (the first case) still has the original urban morphological characteristics, and has not gone through formalization much. It is the central core of Golestan's formation which has not changed much in the last 20 years and most of its ownership is informal. Feshargavi neighbourhood is located on the south of the Saveh road. It has a mixture of formal and informal characteristics and is going through fast change. Golestan is a formal neighbourhood located on the south with formal land subdivisions built according to formal urban regulations and shows the final product of the formalisation process although the socio-economic characteristics of the people is similar to the other areas. Most of its buildings are officially built with materials such as brick and cast iron (Fig. 8.2).

The comparison between the morphological traits and qualities of public space in the three cases shows that adaptation attributes are reduced as the cases are more formalized. Despite the fact that all of them are located in a formal city, life in Soltanabad and inner areas of Feshargavi is still informal. Traits such as self-organization, resourcefulness, robustness and social cohesion, diversity, adaptability and sense of community are reduced by formalization. The observation and conducted interviews in the field showed that as the neighbourhoods become regularized, the residents are estranged, and social bonds become weaker and top-down management takes over the areas (Fig. 8.3).

Diversity as one of the primary resilience attributes is curtailed by regularization, including plot size, height, building age, coverage, aspects of public spaces and user types, social and institutional diversity and economic mix. The types, intensity and diversity of activities in public spaces are transformed towards more zone based and less mixed-use formal functions which in turn affcet the socio-economic mix as

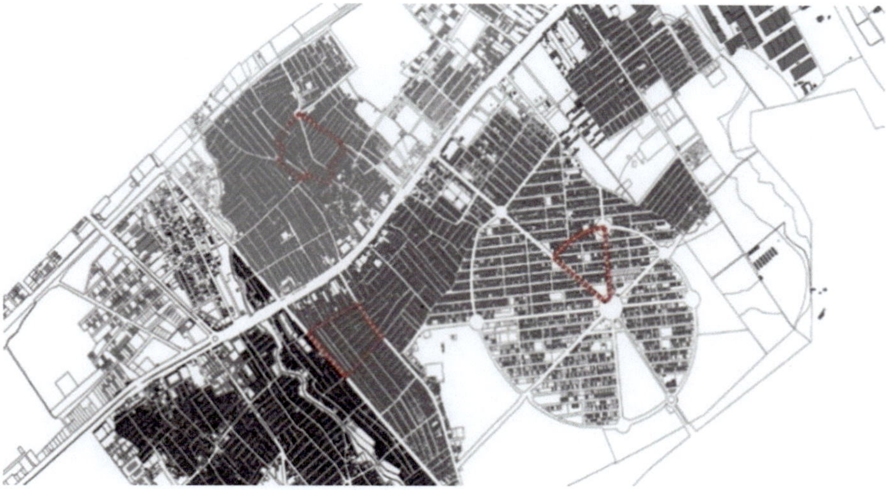

**Fig. 8.2** Location of three case studies in the city Golestan (*Source*: Parsumash 2013 marked by the autor)

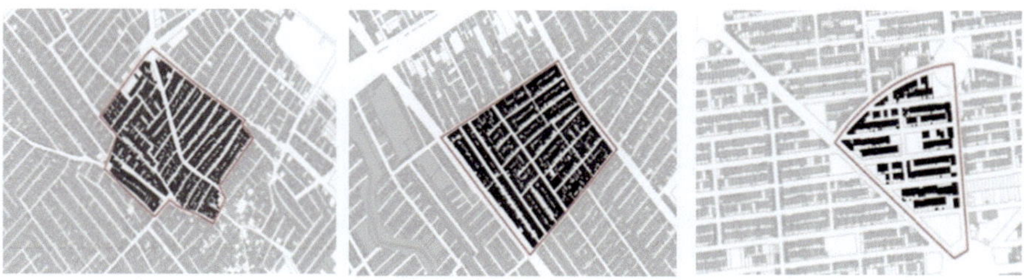

**Fig. 8.3** Figure-ground maps representing morphological characteristics of the three chosen neighbourhoods (*Source*: Author 2013)

well. The morphological transformations have caused changes in flows including dominance of car access and use, different type of permeability, reduced walkability and connectivity for pedestrians (Fig. 8.4).

Despite the fact that all the three cases are located in a formal city, Soltanabad and inner areas of Feshargavi still have dynamics of their past informal heritage. The character of upgraded informal settlements transforms them into generic areas similar to any other urban area in the country. Hence, it does not indicate any specific local identity, history, climatic condition, ethnicity and culture. This is opposed to current resilience discourse on the importance of maintaining "functions, structures, systems, and identity" (Fig. 8.5).

There are several examples of self-organization capacities through informal channels. For example, due to lack of proper civic services, electricity and water networks are not reliable. The electricity is cut off frequently and water flow is dis-

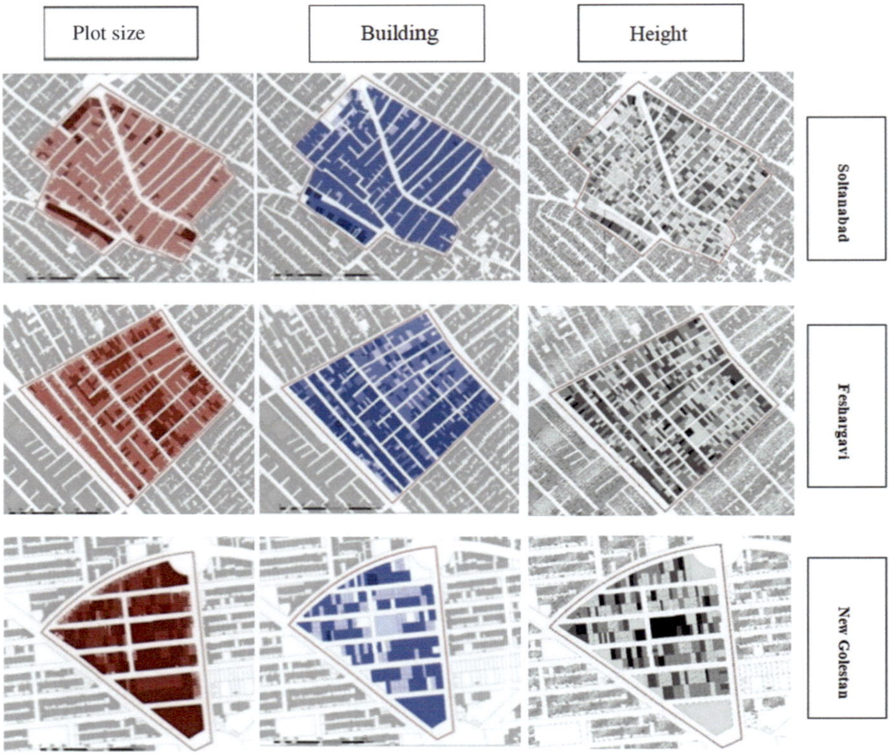


**Fig. 8.4** Diversity of plot size, building height, building age in the three cases (the darker colours indícate smaller plots, older and taller buildings) (*Source*: Author 2013)

rupted since most of the water supply comes from wells. Hence, self-thought solutions by the people for dealing with the deficiencies of infrastructure are shown as redundancy solutions in Soltanabad and Feshargavi. It is reflected in water tanks located on top of all the roofs in Soltanabad as depicted in Fig. 8.6. They are less in Feshargavi, and no signs of them can be seen in New Golestan (the formal case).

Durability and construction quality and the civic services are increased in the most formalized case. Structural vulnerability is reduced due to standard building methods and materials in formalized areas. The access ways are wider and the open space ratio are increased by regularization which assists in time of emergency and disaster for response and rescue phases.

Hence, one can conclude that the three neighbourhoods which are at different states of informality show different types of resilience according to analysis of the case studies based on fieldwork observations and interviews. Morphological changes lead to socio-economic transformations in many ways. This research argues that there should not be a generic model for approaching adaptation issues (Carmin et al. 2011). Each area must find different measures by the people based on their local characteristics and socio-political contexts and the types of stressors.

**Fig. 8.5** Comparison between the three case studies (*Source*: Author 2013)

**Fig. 8.6** Water tanks on the roofs in the first two cases as redundancy plan (*Source*: Author 2013)

## 8.8 Conclusion

This research has examined the effects of relative concepts: resilience, informality and formalization which have different levels and degrees. Studying their relations can have significant contribution to their literature and the conflation of these viewpoints can be valuable for interdisciplinary studies. This study focuses on resilience as systems' adaptation abilities while facing stressors (sudden or chronic) for cultivating more adjustable capacities to transform and evolve towards adaptation. It focuses on attributes which are related to the quality of the built form and public realm as the field of urban design. The case studies, are neighborhoods with different levels of informality. Urban upgrading codes as formalization agents are considered as the agents of change. What this study is concerned about is the consequences of the planning and design regulations on fundamental aspects of adaptability. Studying them can be a step in revising and reconsidering the causes and aims of upgrading which is done as a generic solution in all urban fabrics in Iran with the assumption of betterment. Their self-grown, self-organized and self-managed nature is reflected in their complex spatial patterns and characteristics which can teach us about adaptibility traits in built environment.

Informality and resilience have common features, because both are considered as solutions and survival strategies. Resilience is a way of dealing with unpredictable challenges, which can include adaptibilty capacities for meeting people's needs. Informality too is considered as one of the means of the poor to adapt their environment according to their needs (Dovey 2012; Neuwirth 2006; Roy and AlSayyad 2004; Roy 2005; Yiftachel 2009). Informal areas encompass traits such as self-organization, redundancy, mix and sense of place which are valuable in adaptation and urban upgrading processes can reduce these characteristics.

The three areas, encompass different states of informality even in the most formal cases. Informality permeates in all the three cases but has different representations and stages as explained in the earlier parts of this article. The upgrading regulations are considered as 'key slow variables' causing changes in the long term. They alter the public spaces, and patterns of interactions among agents/actors including leadership, political and power relations, associated organizational and institutional arrangements. The behavioral typologies and use of public space show remarkable changes among the three areas. The sense of community, place security and eyes on the street decrease as formalization prevails where official control and physical safety increases. It lessens community participation and bottom up leadership.

The analysis and comparison of the case studies shows that, Soltanabad, as the most informal case is the most adaptable followed by Feshargavi. The most formal case, New Golestan is the least adaptable. This paper depicts that resilience differs in the three cases, the informal areas are more adaptable but formalised areas are less vulnerable and resistant to risks so resilience in the sense of adaptation is reduced by formalization.

In general, formalization as the application of upgrading codes in Iran, bring about new identities which are less adaptable, and more orderly, close to generic types of urban areas which are more controllable and predictable which is desirable for the state. But this predictabilty does not mean they have better capacities to face unpredictable stressors as is the claim of resilience. The rigidity caused by urban upgrading reduces their adaptibilty.

But another argument which has risen from this study is that formalization is not entirely good or bad for resilience. Considering other definitions of resilience is significant and can help us grasp all of its aspects. This argument reiterates the importance of considering what kind of resilience is expected. Relativity of resilience is a fundamental challenge here. If we consider the resistance dimension of resilience for reducing vulnerability, we can trace positive consequences of formalization as well, because they include a significant part of the literature too. Regularization leads to better structural quality of the buildings beause one of the

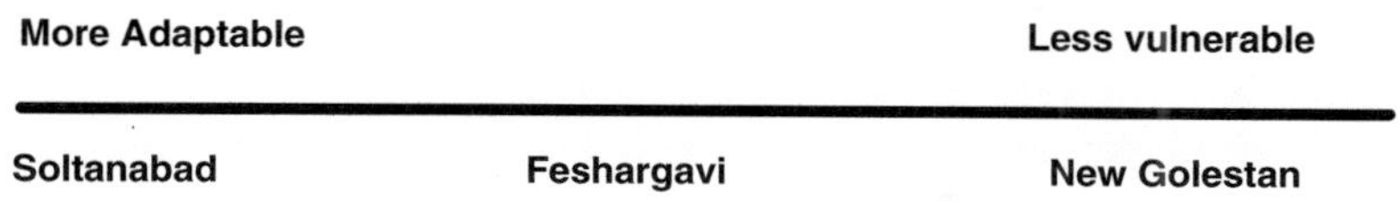


**Fig. 8.7** Different levels of resilience from adaptability to vulnerability reduction from Soltanabad as the most informal case to New Golestan the most formal case (*Source*: Author 2013)

criteria includes building retrofitting which leads to less physical vulnerability in case of earthquakes. Formal neighbourhoods are more resistant from disaster management point of view because they have wider streets and more open spaces that allow more efficiency when hazards strike and better access to emergency shelters and vehicles. A future research can be a study on effects of formalization on resistance and vulnerability reduction of informal settlements (Fig. 8.7).

If planned cities grew in the same organic fashion as informal areas, they would be more responsive to the needs of the people and the changing parameters of society in time. Cities are dynamic and have ever-changing needs. A vigorous, organically evolving, rapidly metabolizing city will be agile and flexible and able to adapt quickly. The fluid boundaries and versatile uses, increase flexibility in times at trouble.

## 8.9 Future Replication Potentials

The conflation of these viewpoints with the study of informal settlements makes the contributions and results of this research valuable for all the fields of resilience and adaptation, informal settlements and urban design.

There are many lessons we can learn from informal areas and their adaptation capacities. They are self-built self-organized areas with high levels of self- maintenance and sufficiency. Despite the fact that these attributes rise from poverty and lack of state's presence, it shows that these loose areas can adapt to changes better than formalized areas. Upgrading has its advantages too and can help these areas reach higher living standards especially regarding hygiene and access to basic infrastructure and more durable buildings. It seems that we need to consider more grey areas and seek a balance between what formal/informal, smooth and striated concepts offer to us.

We are facing unpredictable and ever-changing complex issues in a world where poverty and disasters are striking living environments more and more. We can be more adaptable in urban planning and design interventions including upgrading actions to help prepare our cities for more unstable situations and unpredictable scenarios.

Informal settlements are usually notorious for their negative aspects such as lack of hygiene, crime and socio-economic problems. One of the key aims of upgrading processes is removing and transforming them to formal areas. The state aims to bring these areas under control by formal regulations to reduce irregularity, and hence, keeping their existing identity becomes a controversial question. On the other hand, informal areas have capacities which are in accordance with adaptation attributes. The implementation of codes can help reduce their vulnerability to risks such as earthquakes. On the other hand, changeability will always be an inseparable part of urban challenges. It is time to acknowledge informality as a phenomenon which can help us adapt better and move away from "closed system and over-determined forms" (Sennett 2006) to help urban environments survive through change without losing their basic existance.

The principle outcomes of this research can be applied to similar contexts, both nationally and internationally. This research can be conducted in other scales and countries to depict how informality and adaptation viewpoints can interacat to help urban designers, planners and managers deal with today's pressing challenges.

**Acknowledgments** This article is extracted from a PhD dissertation for attaining the doctor of philosophy degree in the University of Melbourne supervised by professor Kim Dovey and supported by APA scholarship by the Australian government.

This paper was presented in RESURBE III Mexico international conference on Urban & Regional resilience conference and has won the RESURBE prize as Best Practice award in year 2016.

I want to express my deep gratitude to the honorable reviewers in the different stages of this article's acceptance from RESURBE conference, to UNESCO chair for sustainability and the award for resilience 2016 and springer publications for their valuable comments and insights.

## References

Adger, W. N. (2003). *Building resilience to promote sustainability.* IHDP Update 2: 1–3.

Adger, W. N., Barnett, J., et al. (2013). Cultural dimensions of climate change impacts and adaptation. *Nature Climate Change, 3*(2), 112–117.

Alliance, R. (2007). *Assessing resilience in social-ecological systems. A workbook for scientists*. Version 1.1. Draft for testing and evaluation, Resilience Alliance.

AlSayyad, N. (2004). Urban informality as a 'New' way of life transnational perspectives from the Middle East, Latin America, and South Asia. In A. Roy & N. AlSayyad (Eds.), *Urban informality* (pp. 7–30). New York: Lexington.

Bayat, A. (1997). *Street politics: Poor people's movements in Iran*. New York: Columbia University Press.

Brenner, N., Madden, D. J., et al. (2011). Assemblage urbanism and the challenges of critical urban theory. *City, 15*(2), 225–240.

Carmin, J., Dodman, D., et al. (2011). Urban adaptation planning and governance: Challenges to emerging wisdom. In *Resilient bities* (pp. 123–129). Bonn: Springer.

Carpenter, S., Walker, B., et al. (2001). From metaphor to measurement: Resilience of what to what? *Ecosystems, 4*(8), 765–781.

Deleuze, G., & Guattari, P. F. (1987). *A thousand plateaus: Capitalism and schizophrenia*. Minneapolis: University of Minnesota Press.

Dovey, K. (2012). Informal urbanism and complex adaptive assemblage. *International Development Planning Review, 34*(4), 349–367.

Fainstein, S. (2013). *Resilience and justice*. MSSI Research Paper Melbourne Sustainable Society Institute. No. 2.

Godschalk, D. R. (2003). Urban hazard mitigation: Creating resilient cities. *Natural Hazards Review, 4*(3), 136–143.

Gunderson, L., & Holling, C. S. (2002). In L. H. Gunderson & C. S. Holling (Eds.), *Panarchy: Understanding transformations in human and natural systems*. Washington, DC: Island Press.

Hartzog, P. B. (2005). *Panarchy: Governance in the network age*. Internet paper found at http://www.panarchy.com/Members/PaulbHartzog/Papers:2.

Hillier, J. (2007). *Stretching beyond the horizon: A multiplanar theory of spatial planning and governance*. Hampshire: Ashgate Publishing, Ltd.

Hillier, J. (2011). Strategic navigation across multiple planes: Towards a Deleuzean-inspired methodology for strategic spatial planning. *Town Planning Review, 82*(5), 503–527.

Holling, C. S. (1973). Resilience and stability of ecological systems. *Annual Review of Ecology and Systematics, 4*, 1–23.

Lefebvre, H. (1992). *The production of space*. London: Wiley.

Longstaff, P. H., et al. (2010). Building resilient communities a preliminary framework for assessment. *Homeland Security Affairs, 6*(3).

McFarlane, C. (2012). Rethinking informality: Politics, crisis, and the city. *Planning Theory & Practice, 13*(1), 89–108.

Mehaffy, M. W. (2011). A city is not a rhinoceros: On the aims and opportunities of morphogenetic urban design. *Built Environment, 37*(4), 479–496.

Mehta, B., Janus, S., et al. (2008). *Approaches to urban slums: A multimedia sourcebook on adaptive and proactive strategies*. Washington, DC: World Bank Publications.

Moroni, S., & Chiodelli, F. (2014). The complex Nexus between informality and the law: Reconsidering unauthorised settlements in light of the concept of Nomotropism. *Geoforum, 51*, 161–168.

Neuwirth, R. (2006). *Shadow cities: A billion squatters, an urban new world*. New York: Routledge.

Paton, D., & Johnston, D. (2001). Disasters and communities: Vulnerability, resilience and preparedness. *Disaster Prevention and Management, 10*(4), 270–277.

Paton, D., & Johnston, D. (2006). *Disaster resilience: An integrated approach*. Springfield: Charles C Thomas.

Pfefferbaum, B. J., Reissman, D. B., et al. (2007). Building resilience to mass trauma events. In *Handbook of injury and violence prevention* (pp. 347–358). New York: Springer.

Porter, L., & Davoudi, S. (2012). The politics of resilience for planning: A cautionary note. *Planning Theory and Practice, 13*(2), 329–333.

Revell, K. (2010). *Working with informality: Increasing resilience in cities of the Global South*. 46th ISOCARP Congress.

Roy, A. (2005). Urban informality: Toward an epistemology of planning. *Journal of the American Planning Association, 71*(2), 147–158.

Roy, A., & AlSayyad, N. (2004). Urban informality: Transnational perspectives from the Middle East, Latin America, and South Asia. In A. Roy & N. Nezar AlSayyad (Eds.), *Center for Middle Eastern Studies, University of California*. Berkeley: Lexington Books.

Sennett, R. (2006). *The open city* (pp. 1–5). Berlin: Urban Age.

UN-Habitat. (2003). The challenge of slums: Global report on human settlements, 2003. *Management of Environmental Quality: An International Journal, 15*(3), 337–338.

UN-Habitat. (2009). *Planning sustainable cities—Global report on human settlements 2009*. London: Earthscan.

Vale, L. J., & Campanella, T. J. (2005). *The resilient city: How modern cities recover from disasters*. New York: Oxford University Press.

Walker, B. H., & Salt, D. (2006). In B. Walker, D. Salt, & W. V. Reid (Eds.), *Resilience thinking: Sustaining ecosystems and people in a changing world*. Washington, DC: Island Press.

Wilkinson, C. (2011). Strategic navigation: In search of an adaptable mode of strategic spatial planning practice. *Town Planning Review, 82*(5), 595–613.

Wilkinson, C. (2012). Social-ecological resilience: Insights and issues for planning theory. *Planning Theory, 11*(2), 148–169.

Yiftachel, O. (2009). Critical theory and 'grey space': Mobilization of the colonized. *City, 13*(2–3), 246–263.

# Chapter 9
# Ecosystem Service Valuation for Forest Landscape Resilience: Managing Fire Risk

**Roberta Ingaramo, Emma Salizzoni, and Angioletta Voghera**

**Abstract** Prevention and reduction of forest fires is a crucial challenge. Forest fires are constantly increasing at a global level, particularly in areas close to settlements due to the steadily growing interface between vegetated land and settlements resulting from sub-urbanization processes. Shortcomings in fire management have also contributed to this increase. Management shortcomings are even more alarming if we consider the multifunctionality of forest landscapes, that carry out essential functions from an environmental, socio-economic and cultural point of view, strongly affecting also urban areas.

AF3 project ("Advanced Forest Fire Fighting") aims at overcoming the current operational gaps in managing forest fires through the development of new technologies and methodologies. In the context of the overall AF3 project, this research aims at highlighting forest landscape values in the Sardinia Region, and at defining the potential damages caused by fires, in order to underpin a more complex and complete definition of forest fire risk, to sensitise social actors and institutions regarding forest landscape values, and to foster the design of policy instruments that effectively address forest fire risk.

To this aim, the concept of "Ecosystem Services" (ES) has been used as an analytical and operative key, defining a method for valuing and mapping Forest Ecosystem Services (FES) from both a biophysical and economic point of view. The FES valuation method can act as a useful tool to foster an adaptive management

---

This chapter is the result of the combined research activity undertaken by the three authors and coordinated by Angioletta Voghera (DIST, Politecnico di Torino) and Roberta Ingaramo (DAD, Politecnico di Torino). The final written version of paragraph 9.1 is to be attributed to Roberta Ingaramo, that of paragraphs 9.2 and 9.4 to Angioletta Voghera and that of paragraph 9.3 to Emma Salizzoni.

R. Ingaramo
Department of Architecture and Design (DAD), Politecnico di Torino, Torino, Italy
e-mail: roberta.ingaramo@polito.it

E. Salizzoni · A. Voghera (✉)
Interuniversity Department of Regional and Urban Studies and Planning (DIST), Politecnico di Torino, Torino, Italy
e-mail: emma.salizzoni@polito.it; angioletta.voghera@polito.it

 
G. Brunetta et al. (eds.), *Urban Resilience for Risk and Adaptation Governance*, Resilient Cities, https://doi.org/10.1007/978-3-319-76944-8_9

and planning approach for resilience, since: (i) it supports community sensitisation, (ii) it promotes institutional learning, (iii) it allows a monitoring action of FES dynamics, fostering a "learn-by-doing" approach, and (iv) it strengthens compensation plans.

**Keywords** Forest landscape resilience · Fire risk · Ecosystem Service valuation · Adaptive management and planning approach

| **Image** | 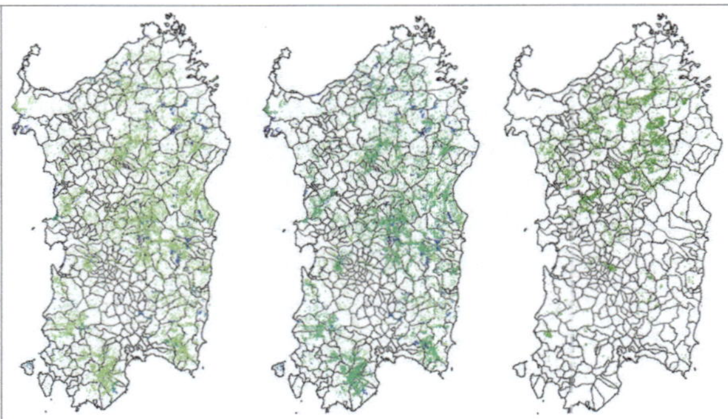 |
|---|---|
| **Location** | International project involving 19 partners from 10 different Countries |
| **Duration** | 36 months |
| **Time Frame** | 01/05/2014–30/04/2017 |
| **Consortium** | In the context of the consortium made up by 19 partners (Selex Galileo, Elbit System, EFPC, Fraunhofer-Gesellschaft Zur Foerderung Der Angewandten Forschung, Intracom Sa Telecom Solutions, Demokritos, Skytek, Politecnico di Torino, Centrum Badan Kosmicznych Polskiej Akademii Nauk, Pyro Fire Extinction, The University Of Westminster, Airbus DS, Ministry Of National Defence Greece, Universitat Politecnica De Valencia, Empresa De Transformacion Agraria, Linköpings Universitet, ARIA Technologies, Israeli Ministry of Public Security, The Italian Ministry of Interior – Department of Fire Corps, Public Rescue and Civil Defence), the research here presented is coordinated by Politecnico di Torino, Interuniversity Department of Regional and Urban Studies and Planning (DIST), and Department of Architecture and Design (DAD). |
| **Cost** | 60.000 € |
| **Resources** | 2 tenured professors, 1 research fellow |
| **People affected** | 1.658.138 (inhabitants of Sardinia 2016) |
| **Donors** | European Union (European Funds - 7th Framework Programme) |
| **Urban Sectors** | Open spaces, forest areas |
| **Resilience dimension** | Persistence, adaptation, transformation |

(continued)

| | |
|---|---|
| **Strengths** | The research addresses a current and urgent topic, namely forest fire risk prevention and reduction: forest fires are constantly increasing at a global level, especially near urban areas. |
| | The research topic – i.e. biophysical and economic valuation of Ecosystem Services (ES) – is a core issue of many current international initiatives, some of them still underway. |
| | The research does not remain a purely academic study but provides a practical tool (a Forest Ecosystem Services – FES – valuation and mapping method) for institutions that have to manage forest fire risk. |
| **Weaknesses** | The application to the case study of the FES valuation method is influenced by the availability of local data. |
| | The valuation method does not concern the whole framework of FES, but just a subset. |
| | The valuation method does not entail a trade-off analysis among different FES. |
| **Impact** | The FES valuation method proposed by the research and applied in the Sardinia Region has a potential impact on practical planning and management, driving to a deeper understanding of the forest fire risk and supporting regional adaptive policies to face forest fires. |
| | Due to the multifunctionality of forest landscapes, an improved forest fire risk management in the Sardinia Region can also underpin a better management of other types of territorial and environmental risks, such as hydrogeological ones. |
| **Stakeholders List** | Local administrations |
| | National Fire Corp |
| | National Forest service |
| | Local inhabitants |
| | Tourists |
| | Hunters |
| | Shepherds |
| | Silviculturists |
| **Challenge** | Prevention and reduction of forest fires is a crucial challenge: forest fires are constantly increasing at a global level, particularly in areas close to settlements, due to the steadily growing interface between vegetated land and settlements resulting from sub-urbanization processes. Shortcomings in fire management also contributed to this increase. Management shortcomings are even more alarming if we consider the varied spectrum of risks deriving from forest fires, connected to the multifunctionality of forest landscapes, that carry out important functions from an environmental, socio-economic and cultural point of view, strongly affecting urban areas too. |

(continued)

| **Project description** | AF3 project ("Advanced Forest Fire Fighting") aims at overcoming the current operational gaps in managing forest fires through the development of new technologies and methodologies. |
|---|---|
| | In the context of the overall AF3 project, this research defines a method for valuing FES. The method aims at highlighting forest landscape values and defining the potential damages caused by fires, in order to underpin a more complex and complete definition of forest fire risk. Moreover, the method aims at sensitising social actors and institutions with relation to forest landscape values and at supporting the design of policy tools that target FES. |
| | Considering these general objectives, the method is conceived to be: |
| | highly communicative – FES are valued not only from a biophysical point of view but also from an economic one; moreover, FES values are spatialised through Geographic Information System (GIS) procedures, to define forest landscape value maps; |
| | easily implemented by institutional actors – FES valuation is developed without using complex digital and mathematical models but only GIS-based indicators; moreover, the data sources used to calculate indicators are open access data. Indicators can therefore be easily calculated and FES dynamics simply monitored. |
| **Urban resilience elements** | The FES valuation method can be a useful tool to "operationalize" urban resilience through an adaptive management and planning approach. In this regard, the method: |
| | supports community sensitisation; |
| | promotes institutional learning, enhancing the capabilities of those who manage fire risk; |
| | allows to monitor FES dynamics by fostering a "learn-by-doing" approach; |
| | strengthens compensation actions that, if conceived as a systemic operational framework, are an important tool for adaptive management and planning. |
| | FES valuation can therefore support urban resilience, fostering more responsive and adaptive government that works in concert with its environment. |
| **List of innovation elements** | The application of the valuation method in the Sardinia Region is among the first studies that address FES values in the Region, providing useful information to integrate the current operative tools aimed at facing forest fire risk (*Piano regionale di previsione, prevenzione e lotta attiva contro gli incendi boschivi 2015–2019*). |
| | The FES valuation method is easily implemented by institutional actors, since it is developed without using complex digital and mathematical models that are often applied in ES valuation studies, but only GIS-based indicators. |
| | The project promotes effective links between theoretical valuation and actual implementation of risk management policies. |

(continued)

| | |
|---|---|
| **Lessons learned** | Since the ES concept translates landscape "values" in terms of "benefits", being moreover characterised by a strong anthropocentric perspective (i.e. the benefits provided by ecosystems to *people*), this concept acts as an effective driver of sensitisation of social actors and institutions with respect to environmental risks. |
| | To foster an adaptive approach to risk management based on institutional learning and on a "learn-by-doing" approach, simple methods must be defined to value and monitor ES. |
| | More generally, since resilience is intended as a process of co-evolution of socio-ecological systems across temporal and spatial scales, a territorial and urban planning approach oriented by ES is certainly appropriate because it explicates systemic relations among multiple spatial scales and between long-term and short/medium-term sustainability objectives. |
| **Upscaling and replication** | The general points of the FES valuation method here presented are conceived to be transferable also to other contexts, although the indicator set should be partially re-defined on the basis of local context specific features. |
| | Some barriers to the effective application of the method could be data deficiency, a common situation when it comes to forest ecosystems. |
| **Web page** | http://af3project.eu/af3/ |
| **Reference** | Ingaramo et al. (2017). |

## 9.1 Challenges for Forest Preservation

Forest fires are a widespread phenomenon across the globe (MEA 2005). In Europe, it is estimated that the most important damages caused to forests in the recent decades have been due to fires (EEA 2008).

International trends show that the number of forest fires is increasing and they are expected to continue to grow, especially in south-western Europe and in areas close to settlements (EEA 2016). This is mainly due, in addition to climate change, to spatial landscape structure changes. Urban growth and sub-urbanization on the one side, and rural abandonment and spontaneous reforestation processes, resulting in an increase in fuel mass, on the other side, have led to a steadily growing interface between vegetated land and settlements (the so-called "Wildland-Urban Interface", WUI). The contact zone between human infrastructures and wildland vegetation has significantly increased, establishing direct relations with the intensified risk of forest fires (Chas-Amila et al. 2013; Modugno et al. 2016). The prevention and reduction of forest fires is therefore a crucial challenge for both urban and forest areas.

With specific reference to forest areas – i.e. the main focus of this contribution – their resilience (as intended by Carpenter et al. 2005 and Davoudi 2012[1]) is significantly challenged by fires because of their vulnerability (see the above-cited factors, namely climate change and land use changes) and the pervasive environmental and

[1] Resilience "is the ability of complex socio-ecological systems to change, adapt and, crucially, transform in response to stresses and strains" (Davoudi 2012).

landscape impact of large-scale fires[2], that deeply affect forest capacity to persist, and to adapt and transform in the face of an event.

Forest fire increase is still more alarming if we consider the varied spectrum of risks deriving from fires associated with the typical multifunctional features of forest landscapes (Merlo and Croitoru 2005). These landscapes are acknowledged to carry out several functions, ranging from an environmental (e.g. watershed protection or climate regulation), socio-economic (e.g. provision of wood and non-wood forest products) and cultural (e.g. connected to tourism, but also to aesthetic and spiritual values) point of view. These functions deeply affect also urban areas, both at a local and territorial scale, depending on the proximity of urban settlements (in northern European countries forests can be easily reached by urban citizens for tourism and leisure), and they are a strategic resource in terms of enhanced life-quality. Forest degradation has therefore multiple consequences, also on urban environments, and fire risk (i.e. the main driver of forest degradation) is directly connected to other types of risks, such as hydrogeological or socio-economic ones.

Since shortcomings in fire management are also acknowledged (MEA 2005) to be an important driver of the above-mentioned forest fire increase, the urgent need for more effective forest risk prevention and management policies is evident. The "Advanced Forest Fire Fighting" (AF3) project – in which context the Interuniversity Department of Regional and Urban Studies and Planning (DIST)/Department of Architecture and Design (DAD) research group of Politecnico di Torino has developed this research[3] – addresses this need, aiming at overcoming the current operational gaps in managing forest fires through the development of new technologies and methodologies[4].

From a spatial and planning perspective, a multiscalar approach is particularly appropriate to face forest fire risks. On the one hand, regional policies that systematically address fire risk are needed (territorial and supra-local scale is the most appropriate to detect and understand environmental features and dynamics related to large forests fires). On the other hand, careful application of the subsidiarity principle is also necessary, fostering "scale matching" – namely the "alignment between decision taken, direct experience with the outcome or consequences of those decisions, and responsibility for adaptive responses" (Abrams et al. 2015) – and cross-

[2] From an environmental point of view, "in addition to the destruction of vegetation, forest fires produce other damaging effects (…). Among these are the emissions of particle and gases (e.g. CO2) into the atmosphere, outflow of mineral nutrients, the destruction of the organic layer of the soil, and the changes in the water infiltration rates in the soil, which makes burnt areas prone to erosion, soil loss, and landslides" (EEA 2008).

[3] The research activity developed by the Politecnico di Torino in the framework of the AF3 project was coordinated by Vittorio Verda, Department of Energy (DENERG).

[4] "These so-called "mega-fires" are particularly destructive and difficult to control with the technologies and systems currently available to fire fighters and emergency agencies. The AF3 project is designed to improve efficiency of current fire-fighting operations and to the protection of human lives, the environment and property by developing innovative technologies and means to ensure a high level of integration between existing and new systems" (http://af3project.eu/description/, accessed 17 March 2017).

scale linkages, in order to guarantee an effective approach to forest fire fighting at a local level.

This contribution – consistently with the Italian approach, that requires regional authorities to manage forest fire – is going to focus on the territorial scale (valuation of ecosystem services provided by forests in the Sardinia Region, see Par. 9.3), conceived as a framework for local and urban action.

## 9.2 Ecosystem Services: Meaning and Potential of a Bridging Concept

The multiple functions of forests have been the object of a growing international attention, being widely interpreted in terms of Forest Ecosystem Services (FES).

The concept of Ecosystem Service (ES) – namely the benefits people obtain from ecosystems[5] – effectively links the biophysical aspect of ecosystems and human wellbeing, bridging environmental and socio-economic spheres (for a clear review about the development of the ecosystem service concept, see Braat and de Groot 2012). The ES concept, instead of contrasting conservation and development options, allows to highlight the added value that ecosystems provide to human society and economy, thus effectively supporting sustainable development policies (Häyhä et al. 2015).

ES definition and evaluation is at the core of numerous academic studies and institutional initiatives. Among the latter, we can cite the Millennium Ecosystem Assessment (MEA 2005), The Economics of Ecosystems and Biodiversity (TEEB 2010), the Common International Classification of Ecosystem Services (CICES 2013) and the initiative Mapping and Assessment Ecosystems and their Services (MAES 2014), aimed at supporting the European Biodiversity Strategy implementation[6].

Several efforts have been made to perform a parallel assessment of the biophysical value of ES and their economic value (Krieger 2001, Schirpke et al. 2014). The opportunity of an *economic* valuation of ES, and, therefore, also of FES, is first connected to communicative objectives, namely the possibility, due to the high communicative power of economic evaluation, of raising awareness in the general public about the contribution of ES to social wellbeing (Nasi et al. 2002).[7] Economic

[5]As defined by MEA (2005) with respect to four typologies: supporting, provisioning, regulating and cultural ES.

[6]MAES is currently defining operative guidelines to support Member States to implement Action 5 of the European Biodiversity Strategy, namely: "Map and assess the state and economic value of ecosystems and their services in the entire EU territory; promote the recognition of their economic worth into accounting and reporting systems across Europe" (Target 2: "Maintain and restore ecosystems").

[7]"It is generally assumed that the incomplete valuation of the forest goods and services is one of the main reasons contributing to deforestation and forest degradation (…). If the total economic value of forests was really taken into account, then people would recognise their importance and

valuation can also support the design of policy instruments targeting the ecosystem service provision (Mavsar and Varela 2014), providing useful information to policy-makers by highlighting the economic consequences of an alternative course of action (Forest Europe 2014) and thus supporting the choice among multiple intervention options [8].

*Mapping* ES economic value is also crucial to make evaluation a genuine tool for awareness raising and for supporting planning and management policies. As highlighted by Barò et al. (Baró et al. 2016), ES mapping is gaining importance in the field of environmental science (for a review of ES mapping methods and issues, see Maes et al. 2012, Englund et al. 2017) and policy agendas (e.g. the Action 5 of the European Biodiversity Strategy and the above-mentioned MAES initiative). With the development of GIS technology, ES mapping has emerged as an important research topic in recent years, still being relatively new and not extensively studied as yet (Schägner et al. 2013). The reasons of interest in the ES spatialisation can be easily traced to the capacity of maps to act as an effective communication tool for social actors ("they have a pedagogical value in that they support efforts to explain the relevance of biodiversity and ecosystem services to the public", Hauck et al. 2013); moreover, they are assuredly essential for spatial planning.

ES valuation and mapping are therefore crucial steps in the definition of sustainable planning processes, both at regional and local scale. These processes start from the definition of ecosystem services to be evaluated, passing through their biophysical/economic valuation and spatial mapping, and eventually reaching the identification of policies or ad hoc actions to be implemented by regional and local plans.

## 9.3 A Method for Valuing and Mapping Forest Ecosystem Services

### *9.3.1 Methods*

In the context of the overall AF3 project, the DIST/DAD research group focused its activity on defining a method for valuing and mapping FES, both from a biophysical and economic point of view. The method aims at highlighting forest landscape values and defining the potential damages caused by fires, in order to support the definition of forest fire risk[9]. Moreover, the method is designed to sensitise social actors and institutions over forest landscape values and to support the design of policy

better protect and manage forest ecosystems" (Nasi et al. 2002).

[8] The practice of economic valuation has also some limitations and it has been subject to debate and criticism (see, among the others, Schägner et al. 2013, Mavsar and Varela 2014).

[9] Meaning "risk" as the intersection between Threats, Vulnerabilities and Asset Value, or potential damage (R = T*V*AV), as widely intended in literature and also by the *Piano regionale di previsione, prevenzione e lotta attiva contro gli incendi boschivi 2014–2016* by Regione Sardegna (case study of this research).

instruments that effectively address forest fire risk by targeting FES. Therefore, this method is not meant to remain a purely academic exercise, as often forest valuation studies are (Emerton 2003), but it aims at having an impact on practical planning and management.

Hence, the method is conceived to be:

- highly communicative – FES are valued not only from a biophysical point of view but also from an economic one; moreover, FES values are georeferred through Geographic Information System (GIS) procedures to define forest landscape value maps;
- easily implemented by institutional actors – FES valuation is developed without using complex digital and mathematical models but only GIS-based indicators; moreover, the sources of data used to calculate indicators are open access data. Indicators can therefore be easily calculated and monitored.

The method was applied to the Sardinia Region, which presents recurring forest fires and where more than 20 per cent of the overall territorial area is covered by forests[10].

Six indicators related to different FES types (provisioning, regulation and maintenance, cultural, according to CICES classification, 2013) have been defined to catch the complexity and multifunctionality of forest landscapes (Table 9.1).

Several methods, such as market price method (indicators 1, 2, 3, 5), replacement cost methods (indicator 4) and benefit transfer methods (indicator 6), have been used to estimate the economic value of FES

Biophysical and economic values of FES are all referred to the use of services per year. Calculation procedures and data sources of each indicator are briefly explained below.

Regarding provisioning FES, both Wood Forest Products (WFPs) and Non-Wood Forest Products (NWPFs) were considered. With specific reference to *timber and fuelwood* (WFPs), the harvest volume was valued considering various forest categories (broadleaf, coniferous and mixed forest) and forest cover (21–50%, 51–80%, >80%), on the basis of data collected from Regione Sardegna Land Use Map and from the National Institute for Statistics (ISTAT, *Utilizzazioni legnose forestali per assortimento e tipo di bosco*, 2015). The average prices provided by ISTAT, defined for forest categories (*Prezzi medi all'imposto per assortimento e tipo di bosco*, 2011), were used[11] to assign a monetary value to timber and fuelwood.

To assess *cork* production – NWPF related to the Sardinia's specific context – we referred both to cork forests with a cover of more than 25% and to cork forest mixed with grazing land (cover from 5% to 25%) as defined by the Sardinia Region Land Use Map Both typologies of cork areas play a crucial role in the overall regional

[10] "Forest" means land with tree crown cover of more than 20%. Only for cork production, areas with a lower tree crown cover were considered (see hereafter in the paragraph).

[11] Timber: 74.11 €/m$^3$ - broadleaf forest; 65.72 €/m$^3$ - coniferous forest. Fuelwood: 54.50 €/m$^3$ - broadleaf forest; 24.15 €/m$^3$ - coniferous forest.

**Table 9.1** FES biophysical and economic indicators

| Forest Ecosystem Service | | Biophysical indicator | Economic indicator (€/ha/yr) |
|---|---|---|---|
| Provisioning | 1. Timber | Volume of harvest (m3/ha/yr) | Market value of timber |
| | 2. Fuelwood | Volume of harvest (m3/ha/yr) | Market value of fuelwood |
| | 3. Cork | Amount of harvest (q/ha/yr) | Market value of cork |
| Regulation/Maintenance | 4. Hydrogeological protection | Forest areas protecting against natural hazards (ha) | Cost of bioengineering technologies |
| | 5. Carbon sequestration | Amount of carbon sequestered by tree biomass (t/ha/yr) | Carbon emission permit price |
| Cultural | 6. Tourism | Arrivals (num/yr) | Visit value |

cork production. The data source about the annual production and average price of cork (2016)[12] is the Regional Agency Agris Sardegna, whose mission is research, experimentation and innovation in the field of cork production and forestry.

The regulation and maintenance role of forests was assessed considering two key functions performed by woodland areas, namely hydrogeological protection and carbon sequestration. Forest *hydrogeological protection* was valued in relation to soil erosion, a serious threat in Sardinia due to irregular rainfall. To this end, three different average slopes were considered[13], together with different protection capacity of forest species (broadleaf, coniferous and mixed forest[14]) and different cover[15]. Costs of bioengineering technologies required to replace the protective function of forests in case they did not exist (double palisade for steeper slope areas, >70%, and hydroseeding for other areas, 40%–70%) were used to estimate the monteary value of forest hydrogeological protection (data source: Sardinia Region, *Prezziario dei Lavori Pubblici*, 2009)[16].

*Carbon sequestration* was calculated by using the indexes defined by the Forest National Inventory (Gasparini et al. 2013) to quantify the annual amount of carbon captured by the above-ground tree biomass in Sardinia, in relation to different species. Carbon monetary value was defined considering the price of emission permits regulated by the European Union Emissions Trading Scheme.[17]

[12] 70 €/q before extraction.

[13] Slope < 40%: no significant hydrogeological protection played by forests; from 40% to 70%: medium-high hydrogeological protection; >70%: very high hydrogeological protection.

[14] Mixed forests are the most protective, followed by coniferous and broadleaf forests (Arpa Sardegna, http://www.sar.sardegna.it/pubblicazioni/miscellanea/desertificazione/pag018.asp, accessed 17 March 2017).

[15] Together with other factors, such as ground cover, a high tree crown cover is associated with a high forest capacity to intercept rainfall and thus protect soil from erosion and runoff.

[16] Hydroseeding: 1.48 €/m$^2$; double palisade 160 €/m$^3$.

[17] 5.16 €/t of carbon dioxide (http://carbon-pulse.com, accessed 17 March 2017).

Finally, *tourism* was chosen as the main indicator for cultural FES and, specifically, the recreational values of forests. The impact of tourism in the Sardinian forest area could be evaluated in relation to 13 forests (corresponding to the 9% of the overall forest area) for which data on annual arrivals were available (Sardegna Foreste 2011). Through the benefit transfer method (source: Ciancio et al. 2007) a visit cost of 6.2 € (daily visit) was applied, considering visit aims, visitors' social typology and provenance, and forest area accessibility.

The method for the valuation of FES presented herein is conceived to be transferable also to other contexts, although the indicator set should be partly re-defined on the basis of local context-specific features. However, it is worth saying that some barriers to the effective application of the method could be caused by data deficiency, which is quite common regarding forest ecosystems (MEA 2005; MAES 2014).

### 9.3.2 Results

The maps reported below (Fig. 9.1) show the performance of each *economic* indicator at regional scale. The high detail of maps allows them to also be used to show FES performance on a local scale (Fig. 9.2).

Table 9.2 reports the value of each indicator. By comparing quantitative data and spatial data, we can briefly outline the economic value of the Sardinian forest landscape[18].

As the Total Economic Value clearly shows (Table 9.2), *hydrogeological protection* is by far the most valuable ES provided by forests in Sardinia. Even if only the 33% of forest areas have a protective function (Fig. 9.2), being located in regions with a slope > 40% (only 5% in regions with slope > 70%), their contribution in terms of erosion protection is remarkably high considering the significant replacement costs associated with bioengineering technologies.

The high percentage value of hydrogeological protection with respect to TEV is also explained by the other indicator performances. With regard to the second regulation/maintenance FES we considered, that is *carbon sequestration*, its value is not so significant. Though the overall regional forest surface is the fourth in Italy, its contribution to carbon sequestration is much lower (the Sardinia Region is only fourteenth in Italy, Gasparini et al. 2013) because of the presence of species with a low average carbon absorption capacity.

*Timber and fuelwood* production, in particular, is very low (corresponding respectively to 0.13% and 4% of the overall Italian production), with a more signifi-

[18] We should note that, even if the FES valuation was based on public and verified data sources, input data, so as indicators results, are, to a certain degree, affected by uncertainty. Moreover, it is important to underscore the fact that the economic value of FES is highly sensitive to the chosen valuation methods and data sources.

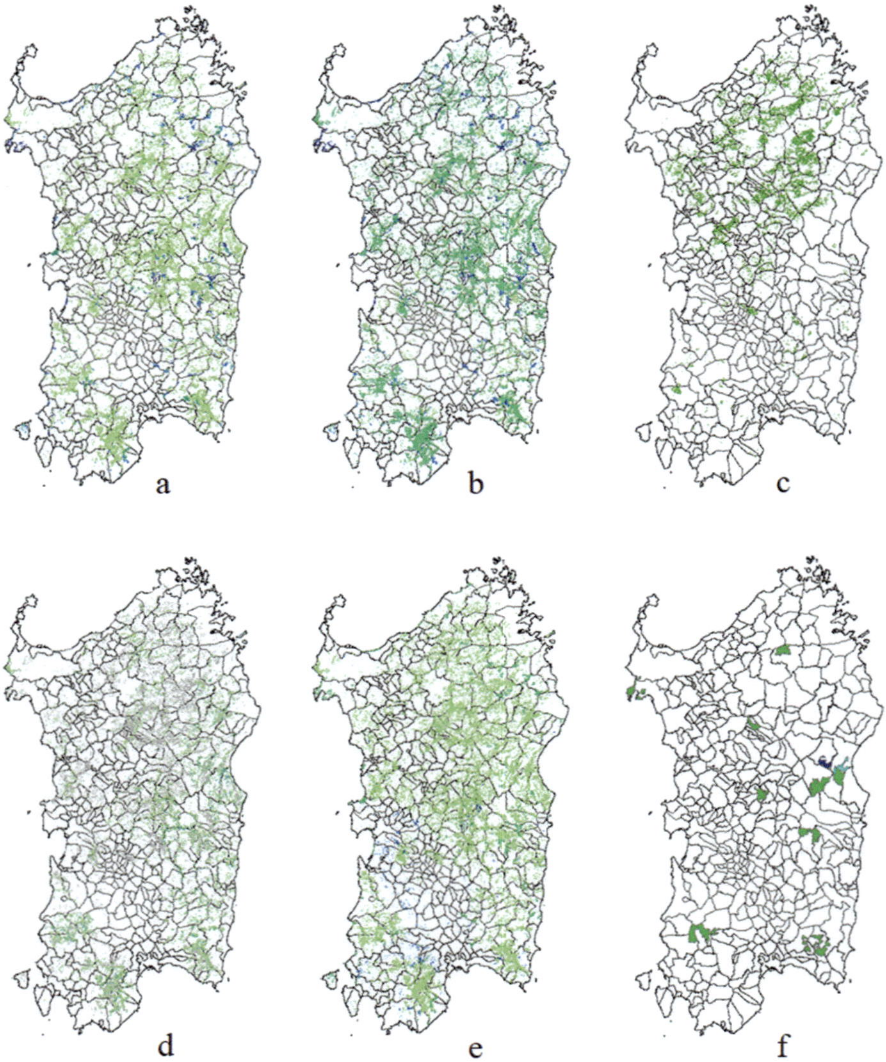


**Fig. 9.1** FES economic value
The maps show the economic value of FES, with specific reference to: (**a**) Timber, (**b**) Fuelwood, (**c**) Cork, (**d**) Hydrogeological protection, (**e**) Carbon sequestration, (**f**) Tourism. Source: author's elaboration (in collaboration with A. Cittadino, Laboratory of Territorial and Urban Research, Politecnico di Torino)

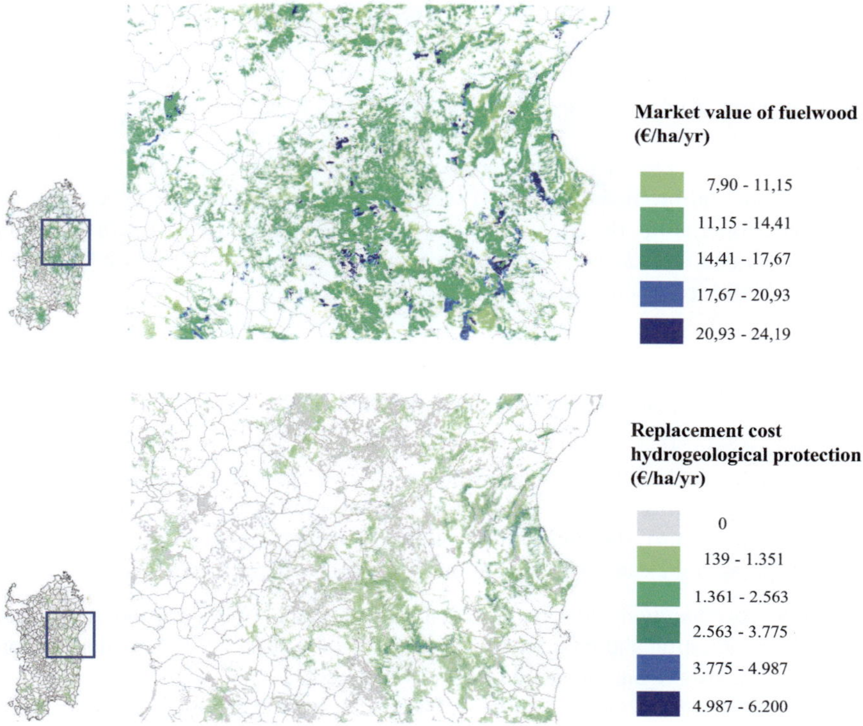


**Fig. 9.2** Fuelwood and Hydrogeological protection - economic value
The two maps show the economic value of Fuelwood (above) and Hydrogeological protection. Source: author's elaboration (in collaboration with A. Cittadino, Laboratory of Territorial and Urban Research, Politecnico di Torino)

**Table 9.2** Total Economic Value of Sardinian forests[a]

| FES | | Biophysical value | Unit | Economic value (€/ha/yr) | Economic value (€/yr) | TEV (%) |
|---|---|---|---|---|---|---|
| Provisioning | 1. Timber | 2,593 | $m^3$/yr | 0.43 | 184,711 | 0.14% |
| | 2. Fuelwood | 113921 | $m^3$/yr | 11.70 | 5,393,994 | 4.12% |
| | 3. Cork | 166,091 | q/yr | 84.00 | 11.626.376 | 8.88% |
| Regulation/ Maintenance | 4. Hydrogeological protection | 167,241 | ha | 665.06 | 111.226.897 | 84.90% |
| | 5. Carbon sequestration | 497,148 | t/yr | 4.89 | 2.565.283 | 1.96% |
| Tev | | | | | 130,997,261 | 100% |

[a]TEV does not consider Tourism since the indicator was calculated only for 13 forests (9% of the overall regional forest area).

cant role played by coniferous trees in terms of productive capacity (see blue areas in Fig. 9.2).

Conversely, *cork* production is a key economic sector (more than 11 million euros per year just considering pre-extraction prices) that plays a leading role in the Italian context.

Moreover, *Tourism*, which we could consider only with respect to 13 forests, is an important and promising sector. If we were to apply - in a purely technical yet meaningful operation - the economic value per hectare found in the 13 forests to the rest of Sardinian forests, we would obtain a value of more than 10 million € (i.e., 7% of the TEV thus re-calculated).

### 9.3.3 A Practice for Urban Resilience

In order to "operationalize" urban resilience (Crowe et al. 2016), an *adaptive management and planning approach* is strongly needed (Plummer et al. 2013; Gunderson 1999). In this regard, the method for valuing and mapping FES presented in the previous paragraphs can be a useful tool for the following reasons:

- it supports *community sensitisation*: a local or regional public administration can use FES valuation and mapping to sensitise social actors, managers and planners to better conserve and enhance forest environments;
- it promotes *institutional learning* (Walker and Salt 2006; Redman 2014; Brunetta 2016), enhancing the capabilities of those who manage fire risk thanks to indicators that can be easily implemented by institutional actors, and to data source that are easily accessible (open access data);
- as a direct consequence of the previous point, it allows a *monitoring* action of FES dynamics, fostering a *"learn-by-doing" approach* (Holling 1978) typical of adaptive management, to be applied to territorial and urban planning ("while adaptive management has been practiced successfully in natural resource management for decades, its application to urban planning and design is rare", Ahern et al. 2014);
- it strengthens *compensation actions* that, if conceived as a systemic operational framework (i.e. a compensation plan, Voghera and Negrini 2016), are an important tool for adaptive management and planning, promoting the system's capacity to favourably react to changes that significantly affect the environment (Kuiper 1997; Cowell 2003). From this perspective, a compensation plan or a compensation strategic vision oriented by ES could contribute to place environmental values at the heart of the action, overcoming local choices that, in Italy, usually favour other objectives (i.e., the design of public facilities with no ecological importance but with a strong social consensus). Moreover, specifically considering FES, their evaluation and mapping can be used to show the socio-ecological and economic effects of compensation actions that entails reforestation interventions (e.g. recovery of degraded landscapes such as quarries, caves,

industries, or creation of ecological networks in fragmented peri-urban contexts).

ES valuation and mapping can, therefore, support urban resilience, fostering "a more responsive, adaptive government that works in concert with its environment" (Redman 2014).

## 9.4 Conclusions

It is widely agreed that incorporating ES in planning and management of socio-ecological systems can effectively reduce territorial and community risk (EEA 2010) and foster resilience (McPheason et al. 2015). Providing a long-term supply of ES is crucial to ensure territorial and urban well-being (Elmqvist et al. 2014) and, more generally, a high degree of quality and diversity of territorial and urban contexts. These elements are directly linked to urban resilience (Ahern 2011).

Moreover, it is worth mentioning that, since resilience is intended as a process of co-evolution of socio-ecological systems "across temporal and spatial scales" (Meerow et al. 2016), a territorial and urban planning oriented by ES (de Groot et al. 2010) is certainly appropriate because it expresses systemic relations among different landscapes (linking forestry, rural and urban landscapes), among multiple spatial scales, and between long-term sustainability objectives and short and medium term objectives (Andersson et al. 2014).

The method presented in this paper to value and map FES goes in the direction of fostering policies (Voghera 2011) for forest fire risk management oriented by FES. Besides the general reasons mentioned above, integrating FES in forest fire risk management seems to be particularly useful in relation to the following aspects:

- FES valuation and mapping helps to underscore and define, in a more complete manner, potential damages caused by fires, thus encouraging *a deeper understanding of* forest *fire risk.* The complexity of the FES concept, which entails several dimensions (environmental, socio-economic, cultural) and functions (provisioning, regulation/maintenance, cultural), leads to an integrated approach to the many values that are threatened by fire, thus contributing to overcome sectorial interpretations of risk.
- Since ES concept effectively translates landscape "values" (Brunetta and Voghera 2008) in terms of "benefits" (also economic benefits), being moreover characterised by a strong anthropocentric perspective (i.e., the benefits provided by ecosystems to *people*), ES concept acts as an effective driver of sensitisation, among social actors and institutions, with respect to risks connected to forest fires. However, to effectively foster an *adaptive approach to fire risk* based on institutional learning and on a "learn-by-doing" approach, simple methods for valuing and monitoring ES must be defined. To this end, indicators, such as those applied in this research, rather than complex mathematical models, seem to be much more appropriate to "build institutions that learn".

Public sensitization and institutional learning processes are thus first important steps in order to reach a resilient co-evolution of socio-ecological systems, which should be based on effective community empowerment and on more responsible institutions, and that should be supported by dynamic planning processes able to interact in a proactive way with changes through nature-based actions crossing spatial and temporal scales.

## References


Abrams, J. B., Knapp, M., Paveglio, T. B., Ellison, A., Moseley, C., Nielsen-Pincus, M., & Carrol, M. C. (2015). Re-envisioning community-wildfire relations in the U.S. West as adaptive governance. *Ecology and Society, 20*(3), 34.

Ahern, J. (2011). From fail-safe to safe-to-fail: Sustainability and resilience in the new urban world. *Landscape and Urban Planning, 100*, 342–343.

Ahern, J., Clliers, S., & Niemelä, J. (2014). The concept of ecosystem services in adaptive urban planning and design: A framework for supporting innovation. *Landscape and Urban Planning, 125*, 254–259.

Andersson, E., Barthel, S., Borgström, S., Colding, J., Elmqvist, T., Folke, C., & Gren, Å. (2014). Reconnecting cities to the biosphere: stewardship of green infrastructure and urban ecosystem services. *Ambio, 43*(4), 445–453.

Baró, F., Palomo, I., Zulian, G., Pilar, V., Dagmar Haase, D., & Gómez-Baggethun, E. (2016). Mapping ecosystem service capacity, flow and demand for landscape and urban planning: A case study in the Barcelona metropolitan region. *Land Use Policy, 57*, 405–417.

Braat, L. C., & de Groot, R. (2012). The ecosystem services agenda: bridging the worlds of natural science and economics, conservation and development, and public and private policy. *Ecosystem Services, 1*(1), 4–15.

Brunetta, G. (Ed.). (2016). *Smart evaluation and integrated design in regional development. Territorial scenarios in trentino* (2nd ed.). Abingdon: Routledge.

Brunetta, G., & Voghera, A. (2008). Evaluating landscape for shared values: Tools, principles, methods. *Landscape Research, 33*(1), 71–87.

Carpenter, S. R., Westley, F., & Turner. (2005). Surrogates for resilience social-ecological systems. *Ecosystems, 8*(8), 941–944.

Chas-Amila, M. L., Touzab, J., & García-Martíneza, E. (2013). Forest fires in the wildland–urban interface: A spatial analysis of forest fragmentation and human impacts. *Applied Geography, 43*, 127–137.

Ciancio, O., Corona, P., Marinelli, M., & Pettenella, D. (2007). *Valutazione dei danni da incendi boschivi*. Firenze: Accademia Italiana di Scienze Forestali.

CICES, Common International Classification of Ecosystem Services. (2013). http://cices.eu/resources/. Accessed 17 Mar 2017.

Cowell, R. (2003). Substitution and scalar politics: negotiating environmental compensation in Cardiff Bay. *Geoforum, 34*, 343–358.

Crowe, P. R., Foley, K., & Collier, M. J. (2016). Operationalizing urban resilience through a framework for adaptive co-management and design: Five experiments in urban planning practice and policy. *Environmental Science & Policy, 62*, 112–119.

Davoudi, S. (2012). Resilience: A bridging concept or a dead end. *Planning Theory and Practice,* 13(2), 299–333.

De Groot, R. S., Alkemade, R., Braat, L., Hein, L., & Willemen, L. (2010). Challenges in integrating the concept of ecosystem services and values in landscape planning, management and decision making. *Ecological Complexity, 7*, 260–272.

EEA, European Environment Agency. (2008). *European forests - ecosystem conditions and sustainable use*. EEA Report No 3/2008. EEA EEA, Copenhagen.

EEA, European Environment Agency. (2010). *Assessing biodiversity in Europe - the 2010 report*. Copenhagen: EEA.

EEA, European Environment Agency. (2016). *Forest fire risk affecting urban areas, Data and maps*. http://www.eea.europa.eu/data-and-maps/figures/forest-fire-risk-affecting-urban-areas/forest-fire-risk-affecting-urban-areas. Accessed 17 Mar 2017.

Elmqvist, T., Barnett, G., & Wilkinson, C. (2014). Exploring urban sustainability and resilience. In P. Roberts, P. Newton, & L. Pearson (Eds.), *Resilient sustainable cities: A future* (pp. 19–28). New York: Routledge.

Emerton, L. (2003). Tropical forest valuation: Has it all been a futile exercise? In *Congress Proceedings*, Volume A, 12th Worls Forestry Congress, Quebec, Canada, 21–28 September 2003.

Englund, O., Berndes, G., & Cederberg, C. (2017). How to analyse ecosystem services in landscapes-A systematic review. *Ecological Indicators, 73*, 492–504.

Forest Europe. (2014) *Expert group and workshop on a pan-European approach to valuation of forest ecosystem services*. Final Report. Ministerial Conference on the Protection of Forests in Europe, Madrid.

Gasparini, P., Di Cosmo, L., & Pompei, E. (Eds.). (2013). *Il contenuto di carbonio delle foreste italiane. Inventario Nazionale delle Foreste e dei serbatoi forestali di Carbonio INFC2005. Metodi e risultati dell'indagine integrativa*. Trento: Ministero delle Politiche Agricole, Alimentari e Forestali, Corpo Forestale dello Stato, Consiglio per la Ricerca e la Sperimentazione in Agricoltura, Unità di ricerca per il Monitoraggio e la Pianificazione Forestale.

Gunderson, L. (1999). Resilience, flexibility and adaptive management - antidotes for spurious certitude? *Conservation Ecology, 3*(1), 7.

Hauck, J., Görg, C., Varjopuro, R., Ratamäki, O., Maes, J., Wittmer, H., & Jax, K. (2013). "Maps have an air of authority": Potential benefits and challenges of ecosystem service maps at different levels of decision making. *Ecosystem Services, 4*, 25–32.

Häyhä, T., Franzese, P., Paletto, A., & Fath, B. D. (2015). Assessing, valuing, and mapping ecosystem services in Alpine forests. *Ecosystem Services, 14*, 12–23.

Holling, C. S. (1978). *Adaptive environmental assessment and management, International Series on Applied System Analysis* (Vol. 3). Chichester: Wiley.

Ingaramo, R., Salizzoni, E., & Voghera, A. (2017). Valuing forest ecosystem services for spatial and landscape planning and design. *Valori e Valutazioni, 19*, 65–78.

Krieger, D. J. (2001). *Economic value of forest ecosystem services: A review*. Washington, DC: The Wilderness Society.

Kuiper, G. (1997). Compensation of environmental degradation by highways: A dutch case study. *European Environment, 7*(4), 118–125.

MAES, Mapping and Assessment of Ecosystems and their Services. (2014). *Indicators for ecosystem assessments under Action 5 of the EU biodiversity strategy to 2020*. European Union.

Maes, J., Egoh, B., Willemen, L., Liquete, C., Vihervaara, P., Schägner, J. P., Grizzetti, B., Drakou, E. G., La Notte, A., Zulian, G., Bouraoui, F., Paracchini, M. L., Braat, L., & Bidoglio, G. (2012). Mapping ecosystem services for policy support and decision making in the European Union. *Ecosystem Services, 1*, 31–39.

Mavsar, R., & Varela, E. (2014). Why should we estimate the value of ecosystem services? In B. G. Thorsen, R. Mavsare, P. I. Tyrväinen, & A. Stenger (Eds.), *The provision of forest ecosystem services. Volume I: Quantifying and valuing non-marketed ecosystem services*. Joensuu: European Forest Institute.

McPheason, T., Andersson, E., Elmqvist, T., & Frantzeskaki, N. (2015). Resilience of and through urban ecosystem services. *Ecosystem Services, 12*, 152–156.

MEA, Millennium Ecosystem Assessment. (2005). In R. Hassan, R. Scholes, & N. Ash (Eds.), *Ecosystems and human assessment. Current state and trends. Forest and Woodland systems*. Washington, DC/Covelo/London: Islandpress.

Meerow, S., Newell, J. P., & Stults, M. (2016). Defining urban resilience: A review. *Landscape and Urban Planning, 147*, 38–49.

Merlo, M., & Croitoru, L. (Eds.). (2005). *Valuing Mediterranean forests. Towards total economic value*. Wallingford: CABI Publishing.

Modugno, S., Balzter, H., Cole, B., & Borrelli, P. (2016). Mapping regional patterns of large forest fires in Wildlande Urban Interface areas in Europe. *Journal of Environmental Management, 172*, 112–126.

Nasi, R., Wunder, S., & Campos, J. J. (2002). *Forest ecosystem services: Can they pay our way out of deforestation?* Bogor: CIFOR for the Global Environmental Facility (GEF).

Plummer, R., Armitage, D. R., & de Loë, R. C. (2013). Adaptive comanagement and its relationship to environmental governance. *Ecology and Society, 18*(1), 21.

Redman, C. (2014). Should sustainability and resilience be combined or remain distinct pursuits? *Ecology and Society, 19*(2), 37.

Sardegna Foreste. (2011) *Monitoraggio del flusso turistico nei ComplessiForestali gestiti dall'Ente Foreste della Sardegna*. Regione Autonoma della Sardegna.

Schägner, J. P., Brander, L., Maes, J., & Hartje, V. (2013). Mapping ecosystem services' values: Current practice and future prospects. *Ecosystem Services, 4*, 33–46.

Schirpke, U., Scolozzi, R., & De Marco, C. (2014). *Modello dimostrativo di valutazione qualitativa e quantitativa dei servizi ecosistemici nei siti pilota. Parte1: Metodi di valutazione*. Report del progetto Making Good Natura (LIFE+11 ENV/IT/000168), EURAC research, Bolzano.

TEEB. (2010). The Economics of Ecosystems and Biodiverstity. In K. Pushpam (Ed.), *Ecological and economic foundations*. London/Washington, DC: Earthscan.

Voghera, A. (2011). Land use indicators for landscape assessment. In C. Cassatella & A. Peano (Eds.), *Landscape indicators. Assessing and monitoring landscape quality* (pp. 141–165). Dordrecht: Springer.

Voghera, A., & Negrini, G. (2016). Parks and landscape: Land use plan experimentation for biodiversity. In T. Hammer, I. Mose, D. Siegrist, & N. Weixlbaumer (Eds.), *Parks of the future. Protected areas in Europe challenging regional and global change* (pp. 71–84). Munchen: Oekom.

Walker, B., & Salt, D. (2006). *Resilience thinking: Sustaining ecosystems and people in a changing world*. Washington, DC: Island Press.

# Chapter 10
# Territorial Resilience and Flood Vulnerability. Case Studies at Urban Scale in Torino (Italy) and Porto/Vila Nova de Gaia (Portugal)

**Stefano Ferrari, Sandra Oliveira, Giulia Pautasso, and José Luís Zêzere**

**Abstract** This paper presents the implementation in QGIS of spatial informations for two case studies in Italy and Portugal by the conceptualization proposed in MOVE (Methods for the Improvement of Vulnerability assessment in Europe, www.move-fp7.eu). The chosen areas are in Turin (Italy) and Porto – Vila Nova de Gaia (Portugal). The conceptual model MOVE describes territorial vulnerability along three dimensions: exposure, susceptibility and lack of resilience. Each dimension can be described through a variety of indicators, such as physical, social, economic, cultural and others. A set of 59 variables to include in the analysis was drawn from the spatial informations available in publicly accessible databases. In addition to the indicators already presented in similar studies, the assets of cultural and historical heritage are introduced. Because of a rather complex measurement and research work, all indicators were standardized and mapped in the European grid (1 km x 1 km). Finally, maps of exposure, susceptibility and lack of resilience were obtained, from which it was possible to draw the maps of the local variation of flood vulnerability. The enphasis is on local variation because maps result from normalization of variables into a local range. Therefore, maps show local variability of vulnerability and can not be referred to either a national or European standard scale. The latter will be possible if further research identifies appropriate ranges for each indicator, representative for larger areas. The procedure, however, shows clearly how urban inequalities translate in inequalities in flood vulnerability and resilience. In this sense, if adopted to simulate the effects of planning, this procedure could help in designing more resilient territories.

S. Ferrari (✉)
Politecnico e Università di Torino, Dipartimento Interateneo di Scienze, Progetto e Politiche del Territorio, Torino, Italy
e-mail: stefano.ferrari@unito.it

S. Oliveira · J. L. Zêzere
Instituto de Geografia e Ordenamento do Território (IGOT), Lisboa, Portugal
e-mail: sandra.oliveira1@campus.ul.pt; zezere@campus.ul.pt

G. Pautasso
Politecnico di Torino, Torino, Italy

 
G. Brunetta et al. (eds.), *Urban Resilience for Risk and Adaptation Governance*, Resilient Cities, https://doi.org/10.1007/978-3-319-76944-8_10

**Keywords** Flood · Resilience · Vulnerability · Risk · MOVE project · Planning · Spatial indicators · Exposure · Susceptibility

## Case Study of Torino and Porto - Vila Nova de Gaia

| | |
|---|---|
| Image or graphic | 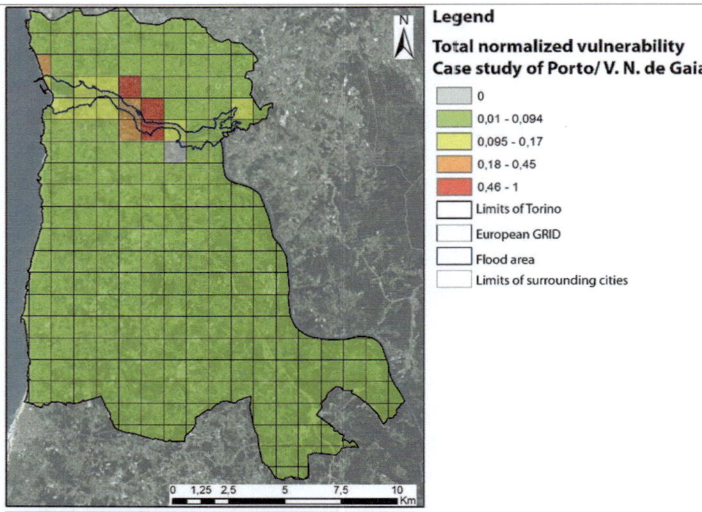<br>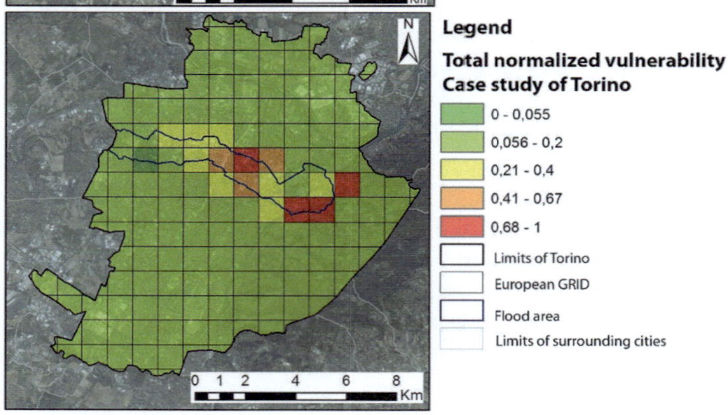 |
| Locations | Europe, Italy, Region Piemonte, Torino |
| | Europe, Portugal, Distrito de Porto, Porto/Vila Nova de Gaia |
| Duration | 11 months |
| Time Frame | Start date: 4/2016 - End date: 4/2017 |
| Consortium | DIST-Interuniversity Department of Regional and Urban Studies and Planning-Torino, IGOT-Institute of Geography and Spatial Planning-Lisbon |
| Cost | No direct cost, the study was part of a research thesis exchange. |
| Resources | 11 person/month, free and institutional software |
| People affected | 500.000 estim. |
| Donors | No donors |

| | |
|---|---|
| Urban Sectors | Water, building, transport, naturalistic ecology, population, economy and heritage |
| Resilience dimension | Coping capacity and recovery |
| Strengths | Holistic and multidisciplinary approach and a good tool to link schools of thought, in addition to develop a tool to validate policies regarding spatial planning. |
| Weaknesses | Absence of open data and difficulties in defining vulnerability and resilience concepts and modelling approaches |
| Impact | At local scale, impact in spatial planning and at European scale concerning a common standardization |
| Stakeholders List | City and Regional administrations, decision makers, Civil Protection Agencies. |
| Challenge | This study analyses vulnerability and resilience at urban scale in Italy and Portugal through the implementation of the conceptual approach from the MOVE framework, considering the assets of cultural and historical heritage. |
| Project description | After identifying the MOVE framework and appropriate indicators, we have chosen a set of 59 suitable variables in Italy and Portugal. As a result of a rather complex measurement and research work, the variables were standardized and mapped in European grid (1 km x 1 km). |
| | Finally, we obtained maps of exposure, susceptibility and lack of resilience, from which it was possible to draw a map of the local vulnerability levels. |
| Urban resilience elements | Transport stations, health facilities, hospitals, pharmacies, hospital beds, public security services for coping capacity and industrial buildings and agricultural areas for recovery capacity. |
| List of innovation elements. | Implementation of cultural and historical assets into a holistic and multidisciplinary approach that considers resilience in the assessment of vulnerability. |
| Lessons learned | There is still a need to define a common language, also to make spatial information easier to manage in the European context. There is an important commitment to researchers: to extend the set of variables for each dimention of vulnerability and fix a range of variability of the same variables, to have comparable normalization in different territories. |
| Upscaling and replication | To upscale the implementation, it is necessary to define a set of variables (also broader than one used) for each dimention of the vulnerability and a range of standard variation for each single indicator, suitable for standardization in larger territory. |
| Web page | |
| Reference | |

## 10.1 Flood Hazard Vulnerability and Territorial Resilience in Torino and Porto/Vila Nova de Gaia

The purpose of this chapter is to analyse flood vulnerability and territorial resilience at local level based on the European MOVE framework in two specific flooded areas in Torino (Italy) and Porto - Vila Nova de Gaia (Portugal).

It is recognized that, under conditions of global environmental change in Europe, the intensity and frequency of extreme natural events is expected to increase (Beniston et al. 2007). Moreover, other factors that make it necessary to improve knowledge about vulnerability are: growing urbanization of hazardous areas, high levels of dependency on critical infrastructure and the increasing vulnerability of some social groups.

Regarding the reduction and measurement of vulnerability, there are now relevant studies. However, but the vulnerability concept remained, until a few years ago, and remains sometimes, even now, unclear and under-researched.

There are two important shortcomings:

- Absence of common ground between different disciplines and schools of thought about the conceptual approaches of vulnerability;
- Lack of common framework for measuring the vulnerability, especially on its physical, social, economic, environmental, cultural and institutional dimensions.

In the legislative and institutional context in Italy as in Portugal, arising both from the EU 60/2007 Directive, flood risk "means the combination of the probability of a flood event and of the potential adverse consequences for human health, the environment, cultural heritage and economic activity associated with a flood event" (EU 60/2007 Directive).

The words "vulnerability" and "resilience" do not occur in the EU Directive text. However, according to the Directive, the Member States should provide for hazard and risk maps.

Flood hazard means the probability of occurrence within a specified period of time and within a given area of a flood of assigned intensity or frequency. The Directive requires that the maps represent the geographical areas flooded with high, medium and low probability.

The concept of vulnerability could emerge in the flood risk maps, but in practice they are obtained by overlaying the map of flood hazard to the land use map. This approach translates in directing the urban planning focus on hazard reduction, overlooking copying capacity and recovery.

### *10.1.1 Short Remainder on Terminology*

In the literature, the existing conceptualizations and frameworks on vulnerability differ according to schools of thought and scientific disciplines, such as social, ecological and natural, engineering, corporate sciences or related to disaster man-agement. Some of these are based on holistic conceptual approaches about vulnerability and resilience, including multidisciplinary spheres of analysis, while others focus the attention on specific relevant factors or dimensions of vulnerability. Alexander (2013) and Birkmann et al. (2013) give an interesting review. Despite these differences, in the various frameworks about vulnerability (and resilience), there are some conditions to follow:

- Vulnerability must be determined "for" the territory, not "on" the territory (Cutter 1996, Research and Assessment System for Sustainability Program 2001, Steinfuhrer and Kuhlicke 2007, cited by Fekete 2009);
- Vulnerability has to be determined for specific temporal and spatial scales;
- Vulnerability is caused by a specific natural hazard and affects specific exposed elements.

In addition to this, vulnerability can be potential or revealed. Indeed, one of the critical issues about the concept of vulnerability is related to time, wheter to observe it before, during or after a disaster. The potential vulnerability is defined for the predisaster conditions, considering the components of exposure, susceptibility and coping capacity and helps to address the prediction of potential disasters that can eventually occur. The revealed vulnerability, instead, is focused on post-disaster situations and is a measure of inequality regarding the distribution of losses and hardships amongst social groups, for instance (Fekete 2009).

For decades, the study of natural risks was focused on geophysical processes, overlooking aspects like territorial vulnerability and resilience. Historically, indeed, the concept of vulnerability was a direct criticism of dominant paradigms about disaster's analysis, where Hewitt (1983) offered the more complete criticism. The similar research priority was followed by Bolin (2006), who examined economic and spatial processes which develop a situation of exclusion for some social groups in high hazard areas.

Regarding the factors driving vulnerability, Wisner et al. (2004), defined vulnerability as a process consisting of components that involve deep causes (historical, political, economic, environmental, etc...), dynamics (specific social processes, for instance the growth and speed of urbanization) and uncertain influences (irregular exposure to risk). In this conceptualization, vulnerability was related to the degree of exposure to risk and natural and human extreme events, based on the resilience of a community and affected individuals.

A much broader research perspective, is the vulnerability concept of Maskrey (1989) and Smith-Oliver (2004), cited by Mendes et al. 2011), which added the biophysical and socio-cultural factors in the analysis. Their vulnerability doctrine can be defined as a concept about political ecology that includes the multi-dimensionality of risks and disasters. This view is also expressed by Perrow (2006) who discussed the structural factors of social vulnerability and its mainstreaming in spatial planning.

Overall, the concepts of vulnerability and territorial resilience are often considered as eparated, and only in certain studies the latter is involved in the former. According to Kuhlicke et al. (2011), vulnerability is the product of a certain spa-tial, demographic, socio-economic, cultural and institutional background, so the results of this analysis have to be integrated in a facet not only of vulnerability, but also of resistance of the potentially affected population. This notion about vulnerability is present also in the Hufschmidt's studies (2011), as the capacity of adaptation of individuals and communities. Moreover, other authors such as Wisner et al. (2004) claimed that vulnerability is the degree of re-silience and resistance of people and communities, exposed to hazardous events.

Bogardi (2006, cited by Fekete 2009) added further observations concerning how hazard has considerable effects on vulnerability and how territorial systems react after event. This concept shows a specific difference between the resistance (and here also resilience) and the performance of territorial systems after an event due to a stress factor in a certain time.

In the literature, vulnerability and resilience definitions are often associated with social vulnerability. In fact, for UNISDR (2011), this term is related to the pre-disposition of the population exposed to a hazard or of the economic resources to suffer losses, considering the territorial link among physical, social, political, economic, cultural and institutional dimensions.

An important social vulnerability application is also seen in the structuring of so-cial vulnerability related to Cutter (1996) and Tapsell et al. (2010), who includes the personal capital (individuals) and social capital (groups, communities and organizations) in a framework where there are several classes of factors: quality and security (homes, public facilities, schools, hospitals, etc.…), economic factors affecting recovery capacity (wealth and income inequality) and social factors (age composition, gender, family structure, occupation, etc.…). Following a spread definition from UNISDR Terminology, vulnerability includes all the characteristics and circumstances of a community, system or asset that make it susceptible to the damaging effects of a hazard, (UNISDR 2009). This is the meaning adopted in this study. From the same source came an effective definition of resilience as "the ability of a system, community or society exposed to hazards to resist, absorb, accommodate to and recover from the effects of a hazard in a timely and efficient manner, including through the preservation and restoration of its essential basic structures and functions." (UNISDR 2009). It remains to be clarified, however, how to express the rapidity with which this ability is acquitted.

The factors required to define vulnerability and resilience can be different, but there is an overall agreement regarding the concept of vulnerability as the result of social inequalities (for instance caused by hardship, age, gender, social class, etc.…), and not only of exposure to hazard. Additionally, in all of these frameworks vulnerability is considered essential to understand how to reduce risk and to promote a culture about resilience (Birkmann et al. 2013).

Another line of thought is found in Cutter et al. (2008), who stated that the components of vulnerability can change on the basis of the characteristics of communities and that the territorial elements included in the analysis are not the ones exposed (or not) to hazards, but those in a condition of marginalization, addressed towards a permanent emergency.

## 10.2 The MOVE Framework and Its Implementation for Flood Hazard

The European MOVE Project was cofinanced by the European Commission and it is part of *"7th Framework Programme for Research and Technological Development*". The MOVE model analyses the vulnerability through a multidisciplinary, holistic and heuristic framework at different territorial scales: international, national, sub-national and local. For this application, the local urban scale will be used.

The possible types of landforms and hazard phenomena to which the MOVE framework could applie are: avalanches, coastal erosions, droughts, earthquakes, flash floods, floods, forest fires, heat waves, landslides and debris flow. In the following study cases, the specific hazard will be the floods.

The MOVE model harmonises existing concepts and frameworks of vulnerability assessment within a generic structure that supplies general guidance for a vulnerability assessment in Europe. In this study, the framework was applied to floods vulnerability assessment (Fig. 10.1).

The framework is built in an environment where there are coupling interactions between the hazard context and the social system at different spatial scales, defining vulnerability as the product of the exposure, the susceptibility (or fragility)

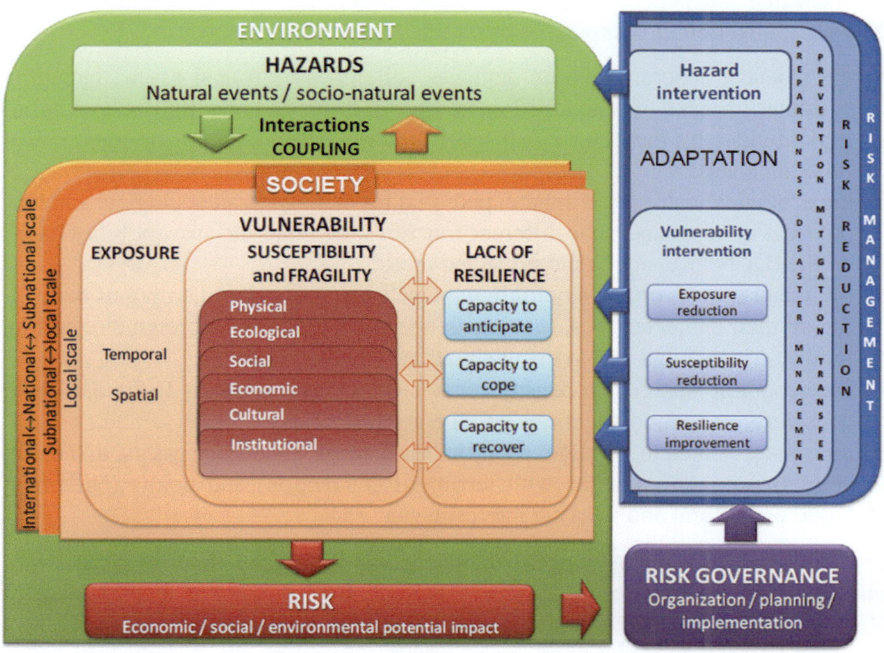


**Fig. 10.1** MOVE framework (from Birkmann et al. 2013)

and the lack of resilience. Successively, calculated vulnerability is part of risk and risk governance involves risk management, risk reduction and disaster management (through prevention, mitigation, transfer and preparedness) in the adaptation's dimension present in the framework.

Resilience is managed as "lack of resilience" in vulnerability and as "resilience improvement" in adaptation. Especially, if the resilience is related to the processes about adaptation and adaptive capacities, it may include potential practices for future changes concerning hazard and vulnerability. Differently, the resilience in the strict sense (as lack of resilience in MOVE vulnerability) examines interventions on the existing territorial assets.

Below, the concepts concerning exposure, susceptibility and lack of resilience are explained, since the framework is applied based on these components.

**Exposure** describes the density and coverage of elements present in an area that can be affected by the flood hazard. It includes the territorial elements of human systems. Its classes in the study are: physical (buildings, roads, critical elements), ecological naturalistic, social and heritage.

These elements are examined within the areas affected by medium frequency floods, as determined by national Authorities as a result of application of EU 60/2007 Directive.

The **susceptibility** (or fragility) defines the predisposition of the exposed elements to be harmed, understanding their weakness and lack of reactionum. It could be measured independently from the exposure and the lack of resilience. In the MOVE application, it is expressed through:

- the physical sphere: potential damages suffered by physical assets, including buildings (divided by age, function, number of floors and material), school buildings (divides by the educational level) and roads (divided by the hierarchical function level);
- the ecological naturalistic sphere: potential for damage to all ecological and biophysical systems and their different functions, defining the susceptibility based on the IUCN level applied to protected areas;
- the social sphere: propensity for human well-being to suffer disruptions depending on their characteristics, (namely gender, age, education, foreign citizenship, family structure and employment situation);
- the economic sphere: propensity for socio-economic systems to be damaged, considering unemployment rate and the housing costs;
- the sphere related to heritage: potential for damage to heritage assets, divided in areas (roads and elements with territorial connection) and points (buildings) based on cultural and touristic interest level.

Finally, the **lack of resilience** is defined as the limitations in terms of access to and mobilization of the resources of a community in responding to a precise hazard (floods in this study) and includes in-time coping (transport, health, public security services, hospital beds and pharmacies) and post-event response (industrial buildings and agricultural areas).

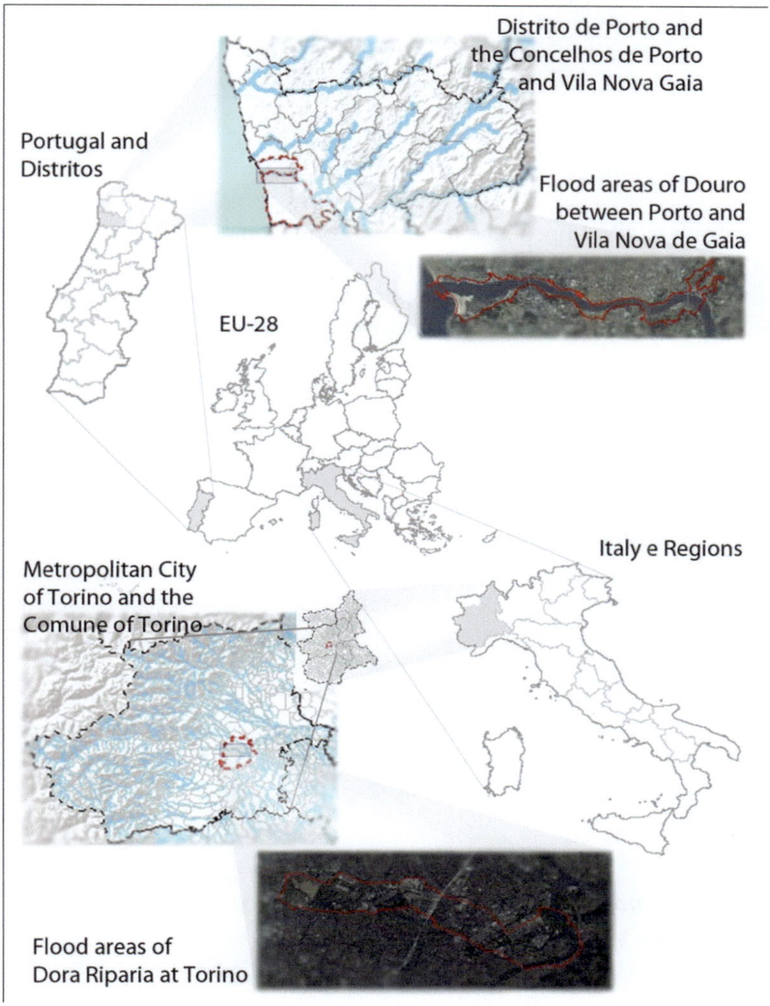


**Fig. 10.2** Locations of studies

### *10.2.1 Study Sites*

The study was implemented into two territory, see Fig. 10.2.

In Italy: areas flooded by Dora Riparia river in the municipality of Torino in the Italian case study. Torino is in Piemonte region, at a mean altitude of 239 m a.s.l., covering about 130,08 km2.

In Portugal: areas flooded by Rio Duro in Porto – Vila Nova de Gaia in Portugal. The municipalities of Porto and Vila Nova de Gaia are in Distrito of Porto in the southwest that is in the Sub-região di Grande Porto and Porto stand for a strategic site in Portugal. Porto is located at an altitude of about 104 m a.s.l. on a surface of 41,66 km2, while Vila Nova de Gaia (the municipality south of Porto) covers 168,7 km2.

#### 10.2.1.1 Methodology

The low availability of open-source data in Portugal and Italy required additional efforts to spatially represent the elements integrated in the analysis. Whenever possible, the variables were mapped corresponding to about twenty-five/thirty different variables. These variables were obtained from different data sources, as shown in Table 10.1.

#### 10.2.1.2 Implementing Data in MOVE Model

Variables were chosen for each dimension of vulnerability (exposure, susceptibility and lack of resilience) and for every class of their dimensions, based on the real possibility of comparing different data in Italy and Portugal through the punctual exam of data meanings. Variables not available in existing database were retrived from original sources or by digitalization of existing maps. This affected 25 variables on the total 59 adopted.

To calculate the indicators, it was required to define the spatial unit to aggregate data in GIS. In this study, the spatial unit was the European GRID (1 km x 1 km) limited to the flood area for exposure and extended to the entire territory in the susceptibility and lack of resilience analysis (Fig. 10.3).

### *10.2.2 The Indicators in Three Dimensions of the Vulnerability*

Following the framework, Italian and Portuguese case studies undertook a flood vulnerability assessment, looking at the causal factors of exposure, susceptibility (and fragility) and lack of resilience.

All the indicators (in total 59) are quantitative and the sphere of interest is multidisciplinary. The thematic areas assessed in the exposure and the susceptibility were: physical, ecologic naturalistic, social, economic (just in the susceptibility) and related to heritage. Instead, in the lack of resilience, the elements are especial-ly related to public facilities and services.

The indicators are presented in Table 10.2, that shows for each dimension the name assigned to the variable and a short description.

**Table 10.1** Data sources for variables

| Description of useful elements in the study | Case study of Torino | Case study of Porto / Vila Nova de Gaia |
|---|---|---|
| Vectors about flood areas, river basins and water bodies | Database of Autorità del Bacino Distrettuale del Fiume Po | Interactive map of Agência Portuguesa do Ambiente |
| Vectors about spatial units at municipal level | Geoportale Piemonte | Dg Territorio (Direção-Geral do Territorio) |
| Vectors about population and housing census | ISTAT_ XV Censimento della popolazione e delle abitazioni 2011 | INE_XV Recenseamento general da populacão. V Recenseamento general de habitacão 2011 |
| Vectors about roads | Region Piemonte_ Graphic ““Nuovo Grafo Viabilità – Grafo di sintesi GDF2 2012″, updated 2016 with road graphic of BDTRE (Region Piemonte), and classified on the basis of road function through the “PUMS - PIANO URBANO DELLA MOBILITA SOSTENIBILE, 2008, Tavola “Infrastrutture stradali _ Gerarchia viaria” | Instituto de Geografia e Ordenamento do Território (IGOT)_2010 |
| Vectors or information about railways, transport stations, health facilities and buildings | BDTRE - Database GeoTopografico – 2016 | Self-handling (mapping), through orthophoto maps (2005), related to IGOT's cartographic service |
| Vectors or information about critical elements, health facilities, public security services, schools, industrial buildings and agricultural areas | Geoportale of Torino and aperTO<br>Corine Land Cover 2012 (from Copernicus programme) | Self-handling (mapping), through “MIPWEB - Portal de Informação Geográfica do Porto” and “Geoportal - Portal Digital de Informação Geográfica e Geostatística de Gaia”, in addition to information about water treatment plants ETAR in web sites of Porto and Vila Nova de Gaia and PDM maps of Porto (“PDM do Gaia - Carta da Estrutura Ecológica Municipal” e “Geoportal - Portal Digital de Informação Geográfica e Geostatística de Gaia”) and V. num. De Gaia (“PDM do Porto - Carta de Qualificação do solo” e “MIPWEB - Portal de Informação Geográfica do Porto”) |

(continued)

**Table 10.1** (continued)

| Description of useful elements in the study | Case study of Torino | Case study of Porto / Vila Nova de Gaia |
|---|---|---|
| Vectors or information about protected areas and urban green areas | The World Database on Protected Areas (WDPA) | The World Database on Protected Areas (WDPA) |
| | Geoportal of Torino | Self-handling (mapping) of urban green areas and REN areas through municipal geoportals and PDM map of Porto ("PDM do Gaia - Carta da Estrutura Ecológica Municipal" e "Geoportal - Portal Digital de Informação Geográfica e Geostatística de Gaia") and V.Num. de Gaia ("PDM do Porto - Carta de Qualificação do solo" e "MIPWEB - Portal de Informação Geográfica do Porto") |
| Vectors or information about schools | Geoportale di Torino | Self-handling (mapping) through the "DGEstE_Direçao-General dos Estabelecimentos Escoles", "Universia", "MIPWEB - Portal de Informação Geográfica do Porto", "Geoportal - Portal Digital de Informação Geográfica e Geostatística de Gaia" |
| | BDTRE - Database GeoTopografico – 2016 (from LARTU) | |
| Vectors or information about areal and point heritage | PTC2 - Piano vigente (Tavola 3.2 - Sistema dei beni culturali), and exam through the PPR- Piano Paesaggistico Regionale vigente (Tavola P5 Rete di connessione paesaggistica) | Self-handling (mapping) through the "Atlas do Património (Cultural) Classificado e em Vias de Classificação, Pesquisa Georreferenciada" |
| | Visualizzatore SIT (voice about areal heritage) in Geoportal of Torino | |
| Vectors about housing cost | OICT_ Osservatorio immobiliare Città di Torino (data in pdf format, then converted in excel format e joined to specific units) | INE_ XV Recenseamento geral da populacão. V Recenseamento geral de habitacão 2011 |

#### 10.2.2.1 Comments on Physical Dimension Indicators

The elements considered for the physical dimension were: buildings, school buildings (just in the susceptibility), roads and critical elements (just in the exposure). The variables were presented as densities or percentages.

In **exposure**, the building density is closely linked to population density, because their high values define difficulties in coping with a flood event. Also, the roads density is important to understand the degree of damage related to communication and rescue axes. Finally, specific elements in flood areas are considered critical for their exposure or susceptibility levels. Therefore, it is necessary to identify them,

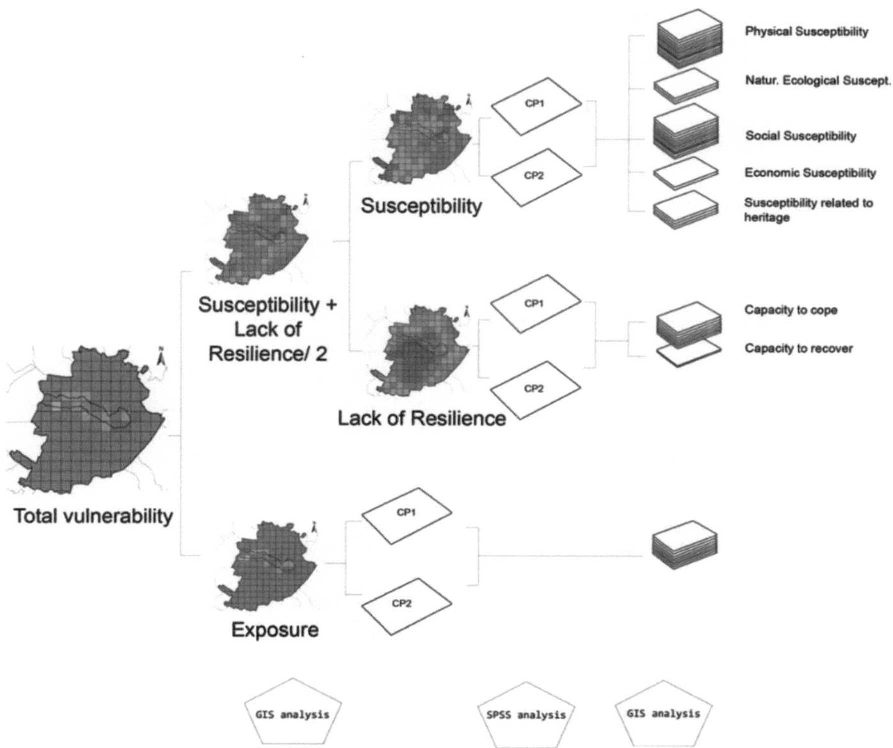


**Fig. 10.3** Layout of implementation of data in MOVE model

considering the greater exposure if the structures host sick people, students, services fundamental for the rescue phase, and if the buildings or areas that can spread solid and liquid pollutants. Therefore, in this study the critical elements considered were: hospitals, health facilities, schools, public security services, dumps and water treatment plants.

In **susceptibility**, the buildings were analyzed on the basis of age, function, number of floors and construction materials. According to Tapsell et al. (2010), the building age allows to determine building codes and their applications. In order to have a similar indicator in Italy and Portugal, two classes were defined: percentage of building constructed prior to 1980 and after 1980, because safety regulations were implemented for buildings in Portugal in 1980, while the first regional urban development law in the Torino case study dates back in 1977. Moreover, the residential buildings are more susceptible than non-residential ones, as the upper floors of buildings can be used to protect people during a flood event and there are some construction materials which are less susceptible to damage, such as the reinforced concrete.

The roads are defined according to their hierarchical function, with the following classes: highways, urban express roads, urban connecting district roads and urban

Table 10.2 Indicators adopted in this study

| **Exposure** | | |
|---|---|---|
| V1 | Density of building | num. buildings/total municipal area in $km^2$ |
| V2 | Density of road | km total roads/total municipal area in $km^2$ |
| V3 | Density of railway | km total railway lines/total municipal area in $km^2$ |
| V4 | Density of subway | km total subway lines/total municipal area in $km^2$ |
| V5 | Density of critical elements | num. critical elements/total municipal area in $km^2$ |
| V6 | Percentage of protected areas | $km^2$ total protected areas * 100/total municipal area in $km^2$ |
| V7 | Density of population | num. residents/total municipal area in $km^2$ |
| V8 | Percentage of areal heritage | $km^2$ total heritage areas * 100/total municipal area in $km^2$ |
| V9 | Density of point heritage | num. heritage points/total municipal area in $km^2$ |
| Susceptibility | | |
| V10 | Percentage of buildings previous 1980 | num. buildings previous 1980 × 100/num. total buildings |
| V11 | Percentage of buildings after 1980 | num. buildings after 1980 × 100/num. total buildings |
| V12 | Percentage of residential building | num. residential buildings × 100/num. total buildings |
| V13 | Percentage of not residential building | num. not residential buildings × 100/num. total buildings |
| V14 | Percentage of buildings with floors <3 | num. buildings with floors < 3 × 100/num. total buildings |
| V15 | Percentage of buildings with floors ≥3 | num. buildings with floors ≥ 3 × 100/num. total buildings |
| V16 | Percentage of buildings in reinforced concrete | num. buildings in reinforced concrete × 100/num. total buildings |
| V17 | Percentage of buildings in bricks | num. buildings in bricks × 100/num. total buildings |
| V18 | Percentage of buildings in other materials | num. buildings in other materials × 100/num. total buildings |
| V19 | Densidty of highways | km total highways/total municipal area in $km^2$ |
| V20 | Density of main urban roads | km total main urban roads/total municipal area in $km^2$ |
| V21 | Density of urban district roads | km total urban district roads/total municipal area in $km^2$ |
| V22 | Density of urban local roads | km total urban local roads/total municipal area in $km^2$ |
| V23 | Density of nursery schools | num. nursery schools/total municipal area in $km^2$ |
| V24 | Density of primary schools | num. primary schools/total municipal area in $km^2$ |

(continued)

**Table 10.2** (continued)

| **Exposure** | | |
|---|---|---|
| V25 | Density of other high-level schools | num. other high-level schools/total municipal area in $km^2$ |
| V26 | Percentage of I IUCN protected areas | $km^2$ I IUCN protected areas ×100/total municipal area in $km^2$ |
| V27 | Percentage of III IUCN protected areas | $km^2$ III IUCN protected areas ×100/total municipal area in $km^2$ |
| V28 | Percentage of V IUCN protected areas (urban parks and gardens) | $km^2$ V IUCN areas ×100/total municipal area in km2 |
| V29 | Percentage of female population | num. female residents × 100/num. total residents |
| V30 | Percentage of male population | num. male residents × 100/num. total residents |
| V31 | Percentage of population under 15 years old | num. residents under 15 years old × 100/num. total residents |
| V32 | Percentage of population over 64 years old | num. residents over 64 years old × 100/num. total residents |
| V33 | Percentage of illiterate population | num. residents under 15 years old × 100/num. total residents |
| V34 | Percentage of population with elementary school certificate | num. illiterate residents × 100/num. total residents |
| V35 | Percentage of population with secondary school certificate and high school diploma | num. residents with secondary school certificate and high school diploma × 100/num. total residents |
| V36 | Percentage of population with university degree | num. residents with university degree × 100/ num. total residents |
| V37 | Percentage of foreign population | num. foreign residents × 100/num. total residents |
| V38 | Percentage of population in one member's families | num. residents in one member's families × 100/ num. total residents |
| V39 | Percentage of population in families with members ≥2 | num. residents in families with 2 or more than 2 members × 100/num. Total residents |
| V40 | Percentage of working population | num. working residents × 100/num. total residents |
| V41 | Percentage of unemployed over total population | num. unemployed residents × 100/num. total residents |
| V42 | Percentage of population out of labour forces | num. residents out of labour forces × 100/num. total residents |
| V43 | Housing costs | housing cost €/$m^2$ housing |
| V44 | Unemployed over total employed rate | num. unemployed residents × 100/(num. working residents + num. unemployed residents) |
| V45 | Percentage of areal heritage with international interest | $km^2$ international heritage areas ×100/total municipal area in $km^2$ |
| V46 | Percentage of areal heritage with national interest | $km^2$ national heritage areas ×100/total municipal area in $km^2$ |

(continued)

**Table 10.2** (continued)

| **Exposure** | | |
|---|---|---|
| V47 | Density of point heritage with international interest | num. international heritage points/total municipal area in km$^2$ |
| V48 | Density of point heritage with national interest | num. national heritage points/total municipal area in km$^2$ |
| V49 | Density of point heritage with local interest | num. local heritage points/total municipal area in km$^2$ |
| Lack of Resilience | | |
| V50 | Distance from railway stations | distance in m from railway stations out of flood areas |
| V51 | Distance from subway stations | distance in m from subway stations out of flood areas |
| V52 | Distance from hospitals | distance in m from hospitals out of flood areas |
| V53 | Distance from health facilities | distance in m from health facilities out of flood areas |
| V54 | Density of hospitals beds on population | num. hospitals beds/num. total residents |
| V55 | Density of pharmacies on area | num. pharmacies/total municipal area in km2 |
| V56 | Density of pharmacies on population | num. pharmacies/num. total residents |
| V57 | Distance from public security services | distance in m from public security services out of flood areas |
| V58 | Density of industrial buildings | num. industrial buildings/total municipal area in km$^2$ |
| V59 | Percentage of agricultural areas (SAU) | km$^2$ agricultural areas (SAU) × 100/total municipal area in km$^2$ |

district roads. Indeed, according to "Traffic Engineering Company of Rio de Janeiro" (CET-Rio), cited by Zonensein et al. (2008), it is possible to classify the urban road system through the connections created and the physical and operative characteristics of each road type.

Finally, the school buildings are considered for their education level, defining the classes related to: kindergardens, primary schools and schools of higher level. The first two levels are most susceptible because of the age of the children and their lower level of self-protection mechanisms.

#### 10.2.2.2 Ecological-Naturalistic Dimension

The ecologic-naturalistic dimension was determined through the protected areas, defined by the classes of "The International Union for Conservation of Nature" (Dudley 2008). In the exposure, the indicator is expressed as the percentage of total ecological and naturalistic areas. In the susceptibility, three classes are made for the first IUCN category, the third IUCN category and the fifth IUCN category. The most susceptible areas are related to the first IUCN level.

#### 10.2.2.3 Social Dimension

In the exposure, the density of population is one of most important indicators.

In the susceptibility, instead, the population is examined through the characteristics that can affect their level of losses and their coping capacity, namely gender, age, education level, immigration, family structure and employment conditions.

On the basis of literature (e.g. European Commission and Cooperation Theme 6 2011, Fekete 2009, Fernandez et al. 2015, Müller et al. 2011, Rufat et al. 2015), the most susceptible people in a flood event are women, people under 14 years old and over 64-year-old, illiterate people or people with only elementary school certificate, foreign individuals, people living alone in a family with just one individual, and unemployed workers.

#### 10.2.2.4 Economic Dimension

The economic dimension was used only for the susceptibility analysis and it was defined through the unemployment rate (European Commission and Cooperation Theme 6 2011 and Fernandez et al. 2015) and the housing costs.

#### 10.2.2.5 Heritage Dimension

In the literature, there are few examples of models that analyze the heritage dimension. As a result, this part is an experimental novelty. These assets were identified through the heritage buildings, the heritage areas (ancient city and roads, represented as polygons), and by calculating the density of heritage buildings and the percentage of heritage areas in each grid cell for exposure analysis.

For susceptibility, instead, these elements (points and areas) were classified according to the interest and recognition level (international, national, regional and local). The more susceptible elements are those with the international level.

#### 10.2.2.6 Dimension Related to Capacities to Cope and to Recover

This last dimension described the lack of resilience through the capacity to cope and to recover in case of a flood event. For the capacity to cope, the variables in-cluded were the distance from the transport stations, hospitals, health facilities and public security services. In addition, the density of hospital by the resident population (Mendes et al. 2011), the density of pharmacies (in relation to surface area) and the density of pharmacies (with regards to resident population) were also integrated. Instead, for the capacity to recover, the density of industrial buildings and the percentage of agricultural area can be proxy variables to understand the economic welfare of a community, the potential losses and the difficulties that the communities have to face in the rescue phase. Moreover, the percentage of agricultural areas is also an indicator of the territorial resilience.

### 10.2.3 Statistical Treatment of Indicator Variables

Once the indicators for each case study were measured and collected, the value that they assumed in each spatial unit of the grid was normalized in scale 0–1 using the standard formula:

$$x_{ST} = \frac{x - x_{\min}}{x_{\max} - x_{\min}} \tag{10.1}$$

where:
*xst* = value of standardised variable
*x* = value of variable in a single cell
*xmin* = minimun value of variable into total units of single case study
*xmax* = maximum value of variable into total units of single case study

For subsequent considerations, it is important to note that we have adopted the local range of each variable.

The next step regarded the choice of the statistical method and the related geo-statistical software. Principal Component Analysis (PCA) was applied to reduce the large set of initial variables, using the software IBM SPSS through the func-tion "Reduce dimension" and the correlation matrix as the mathematical proce-dure.

The PCA was used separately for every component of vulnerability (exposure, susceptibility and lack of resilience) on the standardized variables aggregated at grid scale. The main result was the definition of weights for each indicator based on their relevance for the overall model.

The scores of the first two principal components were applied individually to eve-ry variable in the three dimensions of vulnerability (exposure, susceptibility and lack of resilience), to get their spatial representation in maps, using the formulations shown in Table 10.3.

Then the first two principal components for each dimension were added to have a measure of exposure, susceptibility and lack of resilience (Table 10.4).

The expression suggested in European Commission and Cooperation Theme 6 - Environment (2011), "D.4.2. Handbook of Vulnerability_Assessment in Europe", was used to join the three dimensions of vulnerability into a single map:

$$Vulnerability = ESP \times \left( \frac{SUSC + LoR}{2} \right) \tag{10.2}$$

The last step concerned the normalization of the vulnerability map in scale 0–1 through expression [10.1].

**Table 10.3** Layout of computation for 1st and 2nd component

| Dimension of the vulnerability | Formulations for first and second CP | Notes |
|---|---|---|
| Exposure | $EXP1 = \sum_{i=1}^{9}(\alpha_{1,i}{}^{*}v_i)$ | $EXP1$: 1st component of exposure |
| | | $EXP2$: 2nd component of exposure |
| | | $\alpha_{1,i}$ : 1st PCA weight in the exposure |
| | $EXP2 = \sum_{i=1}^{9}(\alpha_{2,i}{}^{*}x_i)$ | $\alpha_{2,i}$ : 2nd PCA weight in the exposure |
| Susceptibility | $SUSC1 = \sum_{i=10}^{49}(\alpha_{1,i}{}^{*}v_i)$ | $SUSC1$ : 1st component of susceptibility |
| | | $SUSC2$ : 2nd component of susceptibility |
| | $SUSC2 = \sum_{i=10}^{49}(\alpha_{2,i}{}^{*}v_i)$ | $\alpha_{1,i}$: 1st PCA weight in the susceptibility |
| | | $\alpha_{2,i}$ : 2nd PCA weight in the susceptibility |
| Lack of resilience | $LoR1 = \sum_{i=50}^{59}(\alpha_{1,i}{}^{*}v_i)$ | $LoR1$ : 1st component of lack of resilience |
| | | $LoR2$ : 2nd component of lack of resilience |
| | $LoR2 = \sum_{i=50}^{59}(\alpha_{2,i}{}^{*}v_i)$ | $\alpha_{1,i}$ : 1st PCA weight in the lack of resilience |
| | | $\alpha_{2,i}$ :2nd PCA weight in the lack of resilience |

**Table 10.4** final expression for computing dimentions of vulnerability

| Dimension of the vulnerability | Final formulations |
|---|---|
| Total exposure | $ESP = ESP1 + ESP2$ |
| Total susceptibility | $SUSC = SUSC1 + SUSC2$ |
| Total lack of resilience | $LoR = LoR1 + LoR2$ |

## 10.3 Results

### 10.3.1 Exposure Maps

In the Torino case, for the 1st principal component of exposure, the most relevant indicators were the density of building (V1), of the population (V7), of the critical elements (V5) and of the infrastructure (road V2 and railway V3 density), all included in the first principal component of exposure.

For the 2nd principal component of exposure, instead, the most relevant variables were the density of roads (V2) and percentage of area occupied by heritage (V8).

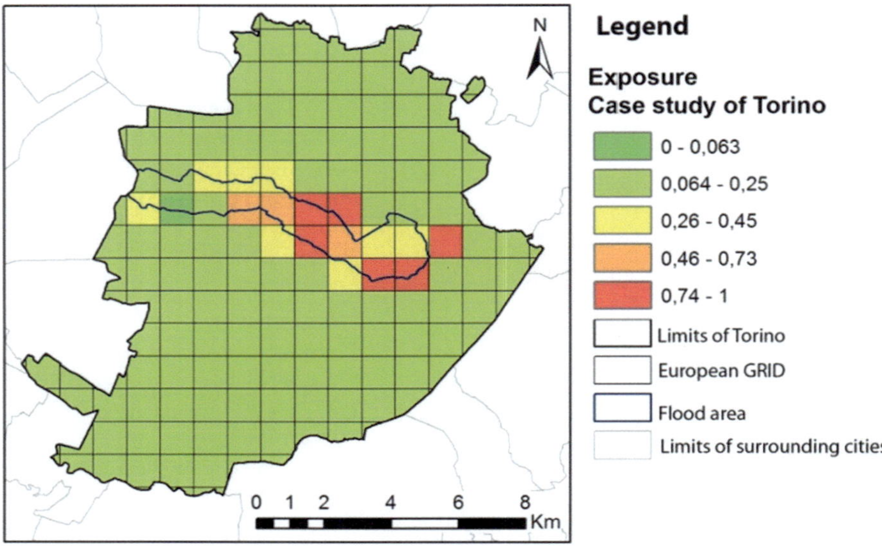


**Fig. 10.4** Exposure in Torino

In the final exposure map, a greater importance of variables belonging to the 1st principal component can be seen in Fig. 10.4.

For the Italian case study, the higher exposure levels were concentrated in the central and east part of the flood area.

In the Porto study the most important indicators were density of building (V1), of population (V7) and of infrastructure (road V2 and railway V3 density).

The 2nd principal component has density of subway (V4) and of punctual heritage (V9) as relevant variables.

The higher exposure levels are more concentrated in few grid cells for the Portuguese case study, see Fig. 10.5, showing higher levels in the central part of the flood area.

## *10.3.2 Susceptibility Maps*

Regarding susceptibility, the most relevant variables in the model for both case studies were related to the physical and social spheres in the first two principal components.

Indeed, for the 1st principal components, in the Italian case study, the indicators that affected mainly were percentage of ante 1980 building (V10), of residential (V12), of brick buildings (V17), of women (V29) and of residents with university degree (V36).

For the 2nd principal component, the greater importance was given to the percentage of buildings constructed prior to 1980 (V10), of residential (V12), of

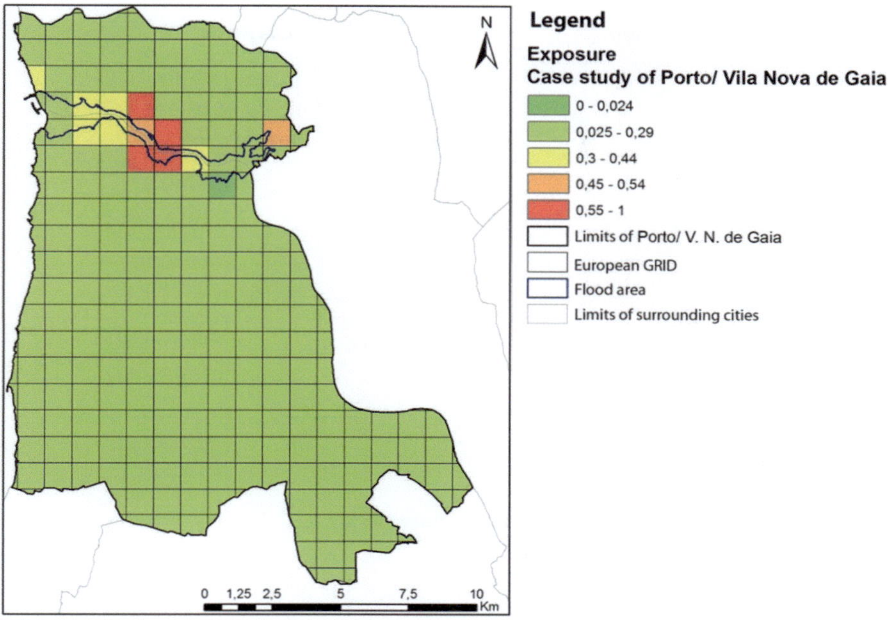


**Fig. 10.5** Exposure in Porto - Vila Nova de Gaia

one-two floors buildings (V14) and above all the housing cost (V43), as it was possible to notice in the Torino hill (S-E) sector.

In susceptibility map, see Fig. 10.6, the Torino area showed the greater relevance of variables present in the second principal component.

In the Porto case study, for the 1st principal component the most relevant indicators were the percentage of not residential (V13) and of over 3 floors buildings (V15), the percentage of nursery school buildings (V23) and of the residents with university degree (V36).

Instead, for the 2nd principal component, most relevant were the percentage of ante 1980 buildings (V10, like in Torino), of over 64 years old (V32), of illiterate people, of people with elementary school certificate, of residents (V33 and V34), the unemployment rate (V44) and the percentage of national areal heritage (V46).

In Porto, as for Torino case study, the final susceptibility map, in Fig. 10.7, is dependent upon the most relevant variables of the 2nd principal component.

### 10.3.3 *Lack of Resilience Maps*

Finally, the lack of resilience was characterized by the variables related to distance from services (transport, health and public security ones).

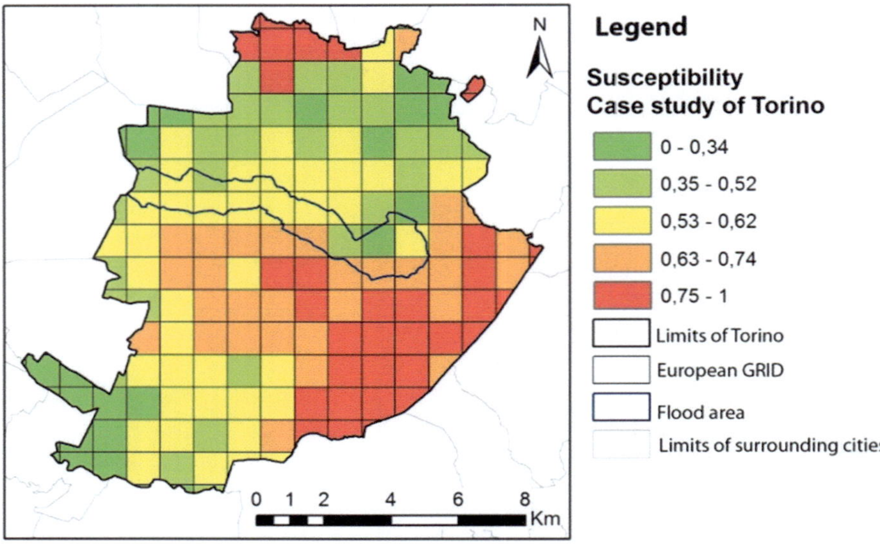


**Fig. 10.6** Susceptibility in Torino

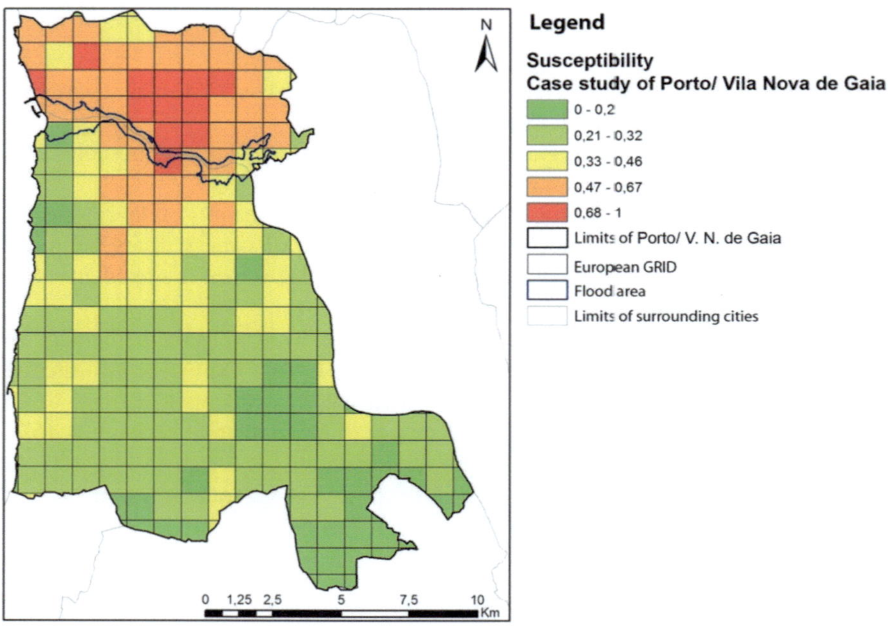


**Fig. 10.7** Susceptibility in Porto - Vila Nova de Gaia

In Torino, for the 1st principal component, the variables related to the distance from the subway stations (V51), from hospitals (V52), from health facilities (V53) and from public security services (V57) were the most relevant.

However, the 2nd principal component was defined also by density variables, like density of industrial buildings (Figs. 10.8 and 10.9).

In Porto for the 1st principal component, most relevant variables were the distance from hospitals (V52), from subway stations (V51) and from railway stations (V50). The 2nd component shows as rrelevant the density of pharmacies (V56).

In the final maps on the lack of resilience, both contexts showed a greater relevance of variables related to the 1st principal component.

### 10.3.4 Vulnerability Maps

The three dimensions are aggregated according to expression [10.2] to obtain the vulnerability maps. There were two fundamental factors making the total vulnerability in [10.2], which are exposure (see Figs. 10.4 and 10.5) and a composite index of susceptibility and lack of resilience. The second addendum is mapped in Fig. 10.10 for Torino and Fig. 10.11 for Porto.

In both case studies, a stronger focus was placed on variables belonging to the exposure, while the second factor has a limited effect in reducing the maximum values in the total vulnerability.

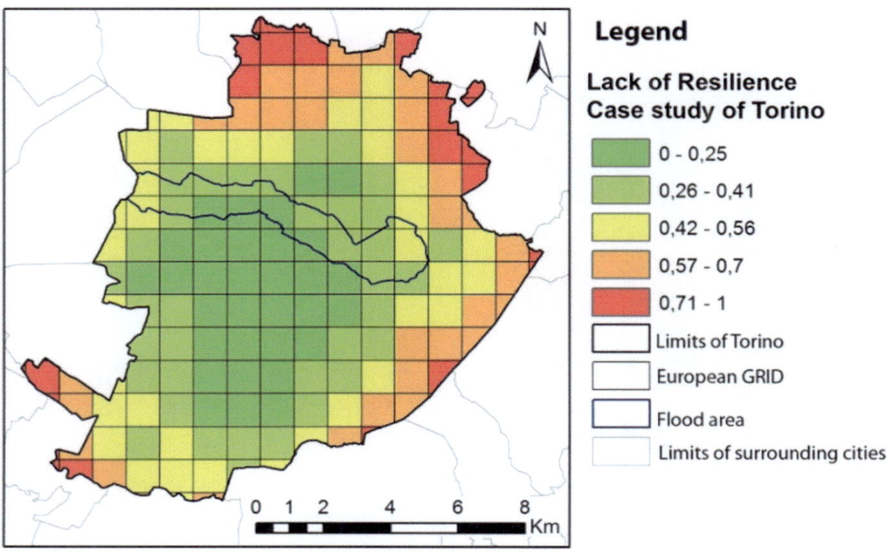


**Fig. 10.8** Lack of resilience in Torino

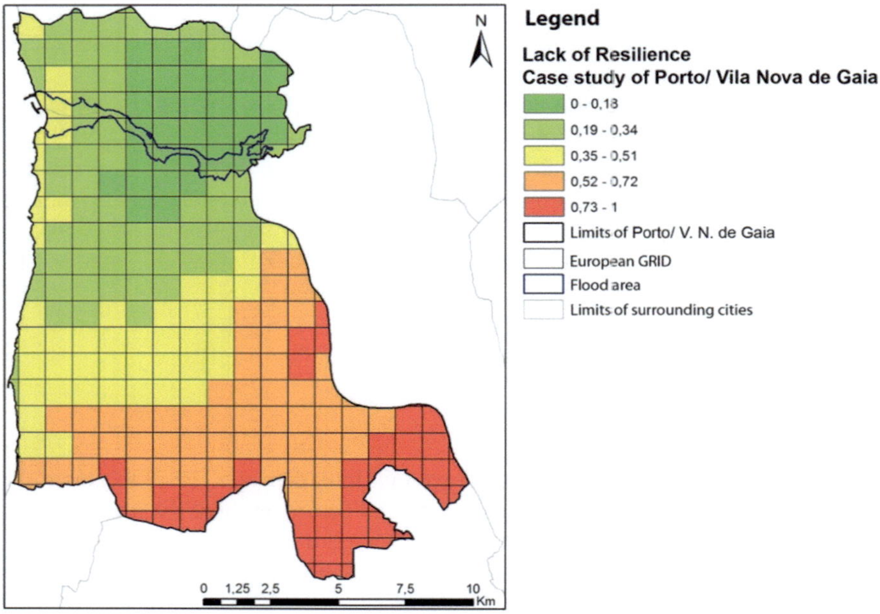


**Fig. 10.9** Lack of resilience in Porto - Vila Nova de Gaia

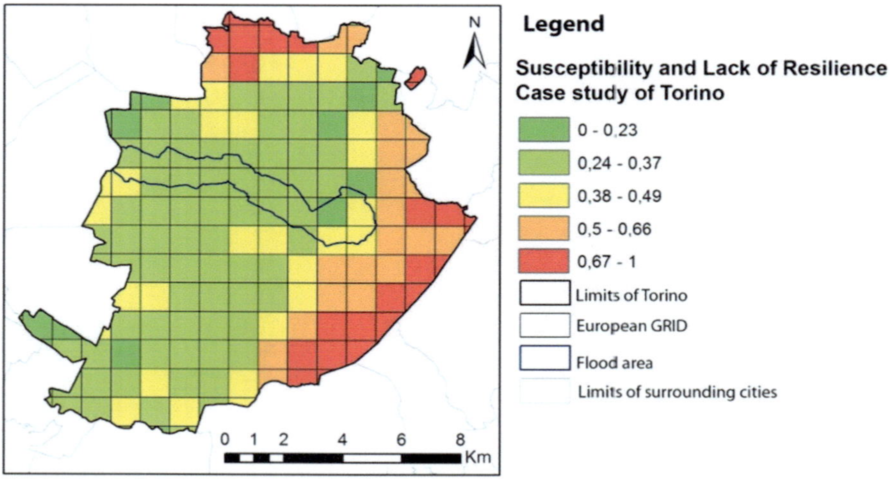


**Fig. 10.10** Mean of susceptibility and Lack of resilience in Torino, recurring in [10.2]

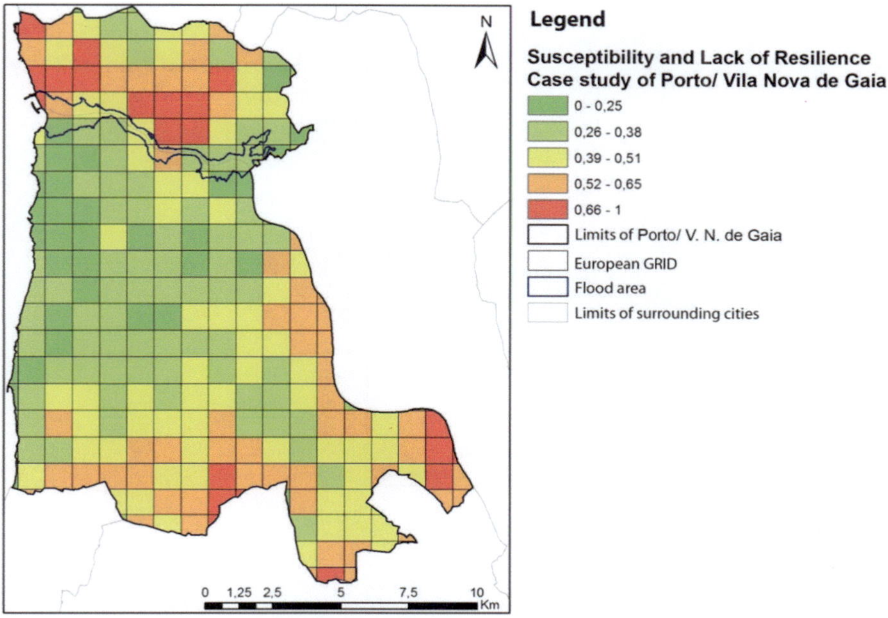


**Fig. 10.11** Mean of susceptibility and lack of resilience in Porto, recurring in [10.2]

In addition to this, the second factor (mean of susceptibility and lack of resilience) was influenced by several variables in the case study. Indeed, in the Porto study, there was a greater importance of variables related to the susceptibility, while in the Torino one to the lack of resilience.

Finally, in Figs. 10.12 and 10.13, normalized maps of vulnerability are shown.

## 10.4 Conclusions

This study has shown two different realities about data availability in Italy and Portugal, influenced by available open data and personal knowledge of territories.

Data types mapped were related to: school buildings, industrial buildings, hospital beds, hospital and health facilities, urban public green areas, etc...

Another difficulty arose, connected to the lack of available data, which was the lack of homogenization about databases in Italy and Portugal. Indeed, all the indicators in both case studies were obtained through a conversion of the initial data to the European grid.

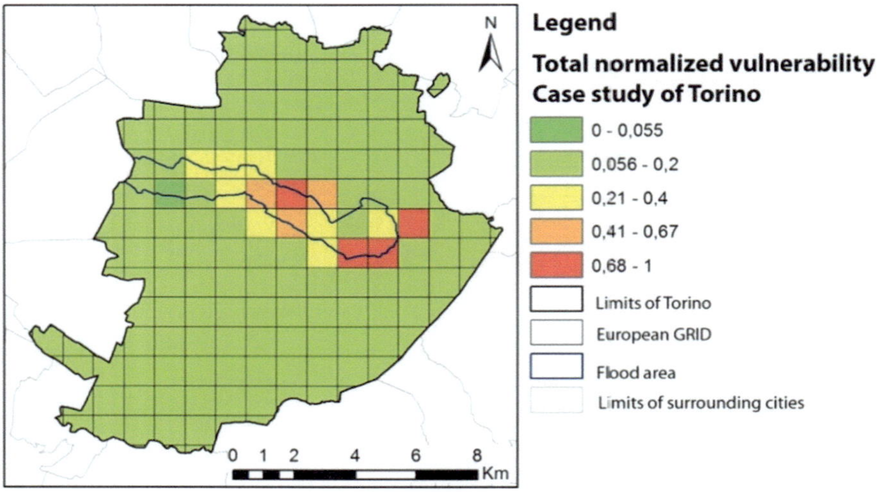


**Fig. 10.12** Normalized vulnerability in Torino

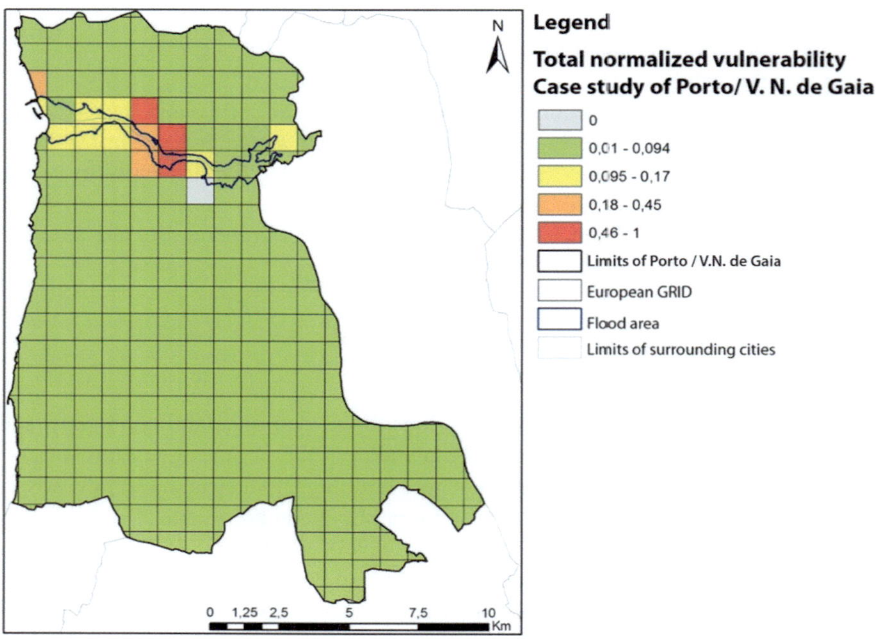


**Fig. 10.13** Normalized vulnerability in Porto

The resulting maps show, in 0–1 scale, the local variation of exposure, susceptibility, lack of resilience and vulnerability. This is because they result from normalization of variables into the local range. The choice to report the normalization at the local scale is made as a first step in the implementation of the model, motivated by the fact that the study was referring to very different territories. Therefore, maps show local variability of, e.g. vulnerability, and can not be referred to other national or European standard scale.

This next step will be possible if further research identifies appropriate ranges for each indicator, representative of larger areas.

To overcome the local level, it would be necessary to work on two different fields of research: first, by defining a variable set (also broader than the one used) for each dimension of the vulnerability, shared with experts of the sectors; secondly, by identifying a range of variation for each variable, suitable for standardization in the area which is object of governance and planning.

From the methodological point of view, it should be noted that expression [10.2], which was adopted to estimate the vulnerability, in fact does not rest upon any theoretical or empirical basis, resulting only in an algebraic way to aggregate information. Therefore, a very interesting, but not easy, research area would be to estimate relationships between the three dimensions and empirical measures of territorial vulnerability.

Furthermore, implementing the spatial indicators, aggregated by the framework MOVE, allows to compare the security of society in different parts of the territory. This aspect can also be very uselful to the spatial planning. It shows clearly how urban inequalities translate into inqualities in flood vulnerability and resilience.

In this sense, if adopted to simulate the effects of planning, this procedure could help in designing more resilient territories.

This approach lends itself well to simulate future scenarios and prospectives, following policies about increasing the territorial resilience with a view to assess the real impact for the better. As a result, this model can stand for the analytical base to validate territorial policies.

## References

Alexander, D. E. (2013). Resilience and disaster risk reduction: An etymological journey. *Natural Hazards and Earth System Sciences, 13*, 2707–2716.

Beniston, M., Stephenson, D. B., Christensen, O. B., et al. (2007). Future extreme events in European climate: An exploration of regional climate model projections. *Climatic Change, 81*(Suppl 1), 71. https://doi.org/10.1007/s10584-006-9226-z.

Birkmann, J., Cardona, O. D., Carreno, M. L., Barbat, A. H., Pelling, M., Schneiderbauer, S., Kienberger, S., Keiler, M., Alexander, D., Zeil, P., & Welle, T. (2013). Framing vulnerability, risk and societal responses: The MOVE framework. *Natural Hazards, 67*, 193–211.

Bogardi, J. J. (2006) in Fekete A. (2009). *Assessment of Social Vulnerability for River –Floods in Germany*. Master thesis, University of Bonn, Bonnum.

Bolin, B. (2006). In Rodriguez H., Quarantelli E., & Dynes R. (2007). *Handbook of disaster research*. New York: Springer.

Cutter. (1996). Research and Assessment System for Sustainability Program 2001, Steinfuhrer and Kuhlicke 2007. In A. Fekete (2009). *Assessment of Social Vulnerability for River –Floods in Germany*. Master thesis, University of Bonn, Bonnum.

Cutter, S. L., Barnes, L., Berry, M., Burton, C., Evans, E., Tate, E., & Webb, J. (2008). A place-based model for understanding community resilience to natural disasters. *Global Environmental Change, 18*, 598–606.

Dudley, N. (2008). *Guidelines for applying protected area management catego-ries*. IUCN.

European Commission and Cooperation Theme 6. (2011). *Assessing vulnerabilità to natural hazards in Europe: From principles to practice_ A manual on concept, methodology and tools*. MOVE Project.

Fekete, A. (2009). *Assessment of social vulnerability for river –floods in Germany*. Master thesis. Bonnum: University of Bonn.

Fernandez,, P., Mourato, S., & Moreira, M. (2015). Social vulnerability assessment of flood risk using GIS-based multicriteria decision analysis. A case study of Vila Nova de Gaia (Portugal), *Geomatics, Natural Hazard and Risk, 2016,* 7(4), 1367–1389.

Hewitt, K. (1983). *Interpretations of calamity from the viewpoint of human ecology*. Londonum: Allen and Unwin.

Müller, A., et al. (2011) (Modified after Dwyer et al., 2004) in Ilesanmi Oluwatosi Taiyet (2014). *Assessment of urban vulnerability towards flood: A case study of Kosofe Lga*. Facolty of Geography, University of Lagos, Lagos.

Kuhlicke, C., Scolobig, A., Tapsell, S., Steinfuhrer, A., & De Marchi, B. (2011). Contextualizing social vulnerability: Findings from case studies across Europe. *Natural Hazards, 58*, 789–810.

Maskrey, A. (1989). *Disaster Mitigationum. A Community Based Approach*. Oxford: Oxfam.

Mendes, J. M., Oliveira, T. A., Cunha, L., & Freiria, S. (2011). A vulnerabilidade social aos perigos naturais e tecnológicos em Portugal. *Revista Crítica de Ciências Sociais, 93*, 95–128.

Perrow, C. (2006). In H. Rodriguez, E. Quarantelli, & R. Dynes (Eds.), *Handbook of disaster research*. New York: Springer.

Rufat, S., Tate, E., Burton, C. G., & Maroof, A. S. (2015). Social vulnerability to floods: Review of case studies and implications for measurement. ELSEVIER _ *International Journal of Disaster Risk Reduction, num.14*, 470–486.

Smith-Oliver. (2004). In Mendes J. M., Oliveira Tavares A., Cunha L., Freiria S. (2011). A vulnerabilidade social aos perigos naturais e tecnológicos em Portugal, *Revista Crítica de Ciências Sociais, 93*, 95–128.

UNISDR. (2009, May). *UNISDR terminology*. Geneva: United Nations Geneva: International Strategy for Disaster Reduction.

UNISDR. (2011). *2011 Global assessment report on disaster risk reduction: Revealing risk, redefining development*. Geneva: International Strategy for Disaster Reduction Geneva, Switzerland.

Tapsell, S., Mccarthy, S. S., Faulkner, H., Alexander, M. et al. (2010). *Social vulnerability to natural hazard*. CapHaz-Net_ Social Capacity Building for Natural Hazard Toward More Resilient Societies.

Wisner, B., Blaikie, P., Cannon, T., & Davis, I. (2004). *At risk – Natural hazards, people's vulnerability and disasters*. Londres: Routledge.

Zonensein, J., Miguez, M. G., De Magalhães, L. P. C., Valentin, M. G., & Mascarenhas F. C. B. (2008). Flood risk index as an urban management tool. In *11th international conference on urban drainage*, Edinburgh.

# Chapter 11
# Climatic Change, Management of Water Rain and Flood Risk in the Metropolitan Area of San Luis Potosí, México

**Adrián Moreno-Mata, Ricardo Villasís-Keever, and Jordi Morató**

**Abstract** This paper presents two approaches to interpreting urban system dynamics: (i) the complex systems approach and (ii) the resilient city's approach. It begins with the consideration of cities as a complex and dynamic social system - far from equilibrium and open to their environment - that is composed of elements and actors, that is, endogenous resources, both natural as artificial - elaborated by human societies - that put in value the territorial resources. These are related and are part of other territorial systems of higher order (municipal, state, regional), which, in turn, are within a national system connected to the global scale.

The incorporation of this analytical perspective to the study of the Metropolitan Area of San Luis Potosí (Mexico), and specifically the relationships between climate change, rainwater management and flood risk, would help to create more viable structures and networks, plastic and flexible, with a greater capacity to adapt and evolve to the turbulences and changes in their environment, and to have more viable, effective and efficient tools, within the framework of urban sustainability. The objective of the document is to generate knowledge to identify problems, elucidate dynamics and present tools that allow a better understanding of the urban systemic environment into the case study.

The first part describes the fundamental elements of the approaches of complex systems and resilient cities and culminates with the methodological approach of the investigation and the proposal of an analytical model. The second part shows the application of the theoretical - methodological framework to the case study. A diagnosis and a systemic analysis of the case studied are made. The third section presents a proposal for intervention based on strategic and participatory planning approaches, supported by the scenario building tool, with measures that aim to solve the development and sustainability problems that this metropolitan area presents.

---

A. Moreno-Mata (✉) · R. Villasís-Keever
Autonomous University of San Luis Potosí, San Luis Potosí, Mexico
e-mail: adrian.moreno@uaslp.mx

J. Morató
UNESCO Chair of Sustainability, U.P.C, Barcelona, Spain
e-mail: jordi.morato@upc.edu

 
G. Brunetta et al. (eds.), *Urban Resilience for Risk and Adaptation Governance*, Resilient Cities, https://doi.org/10.1007/978-3-319-76944-8_11

**Keywords** The city as a complex system · Resilience city · Citi's resistance to flooding

## 11.1 Introduction

Climate change is considered one of the main environmental issues challenging contemporary cities. Meanwhile, urban development patterns and the growth of urban population represent the main contributors to climate change, affecting the total energy consumptions and the related greenhouse gas emissions. Therefore, a breakthrough in current urban development patterns is required to counterbalance the climate-related issues.

The urban flood represents a danger which has been gaining greater importance at the present time, mainly because the cities are in growth and because the processes of urbanization alter the components of the hydrological cycle (Tucci and Bertoni 2006), causing episodes of rain that previously were not significant, now have dangerous consequences or disastrous, causing widespread devastation, economic damage and even loss of human lives. If this is associated with the impact of climate change we are faced with a phenomenon whose consequences can reach disaster dimensions. In fact, natural climate variability coupled to the process of climate change is generally recognized as the main reason for the increase in the frequency and intensity of extreme events, which are one of the main factors of floods (SADSN 2014).

In practice, there is a close relationship between the hydrological cycle, urban ecosystems and society. This relationship is given by the interaction between the structure and functionality of the different ecosystems present in a river basin and is accentuated by the presence of social, political and economic processes own of the urban territories. There is a large number of factors in those interrelationships, configuring particular problems for each city: the degradation of water resources, the depletion of water sources and waste of large amounts of waste water, derived from the domestic, industrial uses and services, as well as of the runoff, which elapse periodically or seasonally on the surface urbanized without any possibility of retaining and much less of utilization.

So prepared are cities to cope with the changes imposed by rapid urbanization, population growth and climate change? How these processes affect the urban ability to deal with flooding? During the recent decades, both the context and causes of urban flooding have been modified, and in many cases, their impact is more severe and it is accelerating (Shelfaut et al. 2011).

The understanding of these problems has opened a wide field of discussions at local, regional and global levels, generating diverse paradigms, strategies, policies and approaches whose challenge is to open up paths toward the transformation of this reality. A consensus has been reached on the need to question the classic paradigm of the handling or management of these disasters to make way for the adoption of new paradigms focused on systemic, integrated, and dynamic approaches.

## 11.2 Theoretical and Methodological Approaches

The first issue to be addressed is what is meant by a complex system, to further deepen the study of the city as a complex and open system, and specifically, how the systemic approach can connect with the concept of resilient cities and with the perspective of Sustainability. A basic point of the theoretical-conceptual approach refers to the identification of the limits or territorial demarcation of the system and to confirm the pertinence of studying it as a complex system. Finally, the methodological strategy of the research is presented.

*What Is a Complex System?*
Complex systems are structured systems that are composed of a series of elements or variables that are closely interrelated and interact with each other to contribute to overall system behavior (García 2006). This interaction of its different elements produces an emergent behavior in the system, which makes it almost impossible to have a generalized perspective to predict its behavior in view of the large number of elements that can make them conform, since this requires an individual analysis of each one from them. Therefore, it is not possible to limit its description to a specific number of parameters or variables that characterize it without losing its functional properties, and according to the definition of systems is composed of several interconnected parts whose links contain additional information and communication between them that may be hidden from the observer (Prigogine and Stengers 1994).

Complex systems present a series of characteristics: feedback, emergency, adaptation, recursion, irreducibility, self - organization and circular causality (Vergara and Zureck Varela 2013: 91–93). Complex systems, in turn, are: (i) emerging systems, because they develop emerging behaviors that can evolve from simple rules to more complex ones, depending on the needs that the environment requires; (ii) self-organizing systems because they tend to modify and organize their structures through internal processes, which ensure their survival according to the demands imposed by the environment (Vergara and Zureck Varela 2013: 104) and part of the viability of change comes from the internal capacity of these structures to adapt to the moment, and generate emergent actions, and (iii) complex adaptive systems because they are composed of the interaction between the parts, in which it gives rise to patterns of emergent behavior, that is to say, behavioral changes to adapt to the changes in their surroundings (Sanders 2003).

Complex systems are characterized by their robustness, stability and resistance and are located in a territorial space. Therefore, it can be said that local spaces, including cities, contain other complex systems that are interrelated. Consequently, there are four basic aspects of complexity: (i) hierarchy and structure, pointing out that complex systems are composed of subsystems and these, in turn, have their own subsystems; (ii) the evolution of hierarchical systems; (iii) the dynamic properties of the organized structures of the systems and their decomposition into subsystems, and iv) the relationship between complex systems and their description.

### 11.2.1 The City as a Complex System

For many years, various studies have been made concerning the dynamics of the different urban systems in which societies develop (Alberti et al. 2003; Conte and Monno 2016; Papa et al. 2015; Rueda 2005, among others). The systemic approach, which understands the city as a complex system in development, has been proposed as an interpretive and an instrumental framework of the city from geography (Harvey 1977), sociology (Luhmann 1998) and urbanism (Batty 2008; Salingaros 2005; Soja 2008). From the perspective of complex systems, the essence of cities lies in their internal processes (Jacobs 1971: 460).

As an open system, the city interacts with its socio-territorial context, which, however, has internal intrinsic logic. However, cities are local spaces that have definite and / or definable limits that allow identifying its components and their relations from three approaches, structural, relational and functional. Consequently, a local space can be a system if it has composition, structure and environment. In addition, they have a society that is identified with the territory of which they are part, and are composed of autopoietic systems of first order (cells), the second order (organisms) and third order (ecosystem, society). For this reason, the introduction of this knowledge at the local level contributes to its evolution in a sustainable way.

In sum, the systemic approach conceives of the city as "a pattern in time" (Johnson 2003: 94), defined by its internal structure, which can only be explained as a non-linear process. The nature of cities is unstable, continually giving rise to continuous processes of change, crisis and adaptation. However, we can recognize the same city in all its stages; even when it is completely moved to another place, due to natural disasters or infrastructure projects, epidemics or wars. The hypothesis that guides the systemic analysis implies a change in the understanding of the way in which a city works in the long term, the key of this is that we intuitively already understand to the cities like a complex structure that stays stable in spite of being away of a situation of equilibrium, because it preserves particular patterns of development.

Conceptualizing the city as a complex system implies an alternative vision to the analytical approach that characterizes the discipline of urbanism, forged in a partial, deterministic, reductionist view. In contrast to this traditional approach, systemic analysis constructs a perspective to define the urban, understanding the simultaneity and complex interrelation of the social, economic, political and spatial dimensions, as well as the diversity among actors, resources and exchange flows. The mechanistic view that saw the city as a great machine and the deterministic approach that presented the city as a mechanism that worked according to criteria set a priori and external, through urban design and zoning in order to organize, modeling and controlling certain social and productive processes, systemic analysis studies the urban phenomenon in relation to a larger context, with emphasis on synergistic processes that generate and reproduce dynamically over time, in an interaction between the components themselves of the city - both internal and external - and its environment (Boccolini 2016).

### 11.2.2 The Resilience of the Urban System

In the logic of a complex system cities, according to their circumstances, simultaneously generate processes of homeostasis, which determine the resilience of the urban system (compared to a fluctuation, the system tends to maintain its initial condition of equilibrium through negative feedback mechanisms that reduce the deviation of the system) and processes of learning, adaptation and evolution (where fluctuations are so strong that they push the system through a situation of instability towards new structural properties, relatively more stable in that situation). This explains the hyper stability developed by complex systems, such as cities and social structures, which can withstand very intense changes - quantitatively and quantitatively - in context or internally, and yet preserve the organizational properties that characterize them (Boccolini 2016).

Although the concept of Resilient City has recently come to the fore, the studies on resilience have been developed since the Fifties through different disciplinary fields, from physics to psychology, from ecology to management science. Holling (1996) provided a central and a clear distinction between two main approaches: engineering and ecological. According to Holling (1996) the engineering approach that refers to stability, efficiency, constancy, predictability, return time to a previous state and, above all, to the idea of a single, stable equilibrium, using "resistance to disturbance and speed of return to equilibrium (...) to measure the property; On the opposite, ecological approach, emphasizes "conditions far from any equilibrium steady state", recognizes the existence of multiple equilibrium states and can be absorbed before a system changes its structure". Thus, ecological resilience focuses on the twofold possibility for a system to absorb changes, maintaining its main features, below a given threshold of disturbance, or change its state, moving towards a different one, not necessarily better than the previous one, above such a threshold (Papa et al. 2015: 28).

There are numerous and heterogeneous definitions of the concept of resilient cities; some of them have been provided by scholars (Fusco Girard et al. 2012; Newman et al. 2009) others by institutions (UNISDR 2012; World Bank Group 2011) or private foundations (Rockefeller Foundation 2015). According to these definitions, the resilient city is one that supports the development of greater resistance in its institutions, infrastructure and social and economic life. Resilient cities reduce vulnerability to extreme events and respond creatively to economic, social and environmental changes in order to increase their long-term sustainability.

A resilient city is characterized by: (i) its ability to withstand or absorb the impact of a hazard through resistance or adaptation, which allows certain basic functions and structures to be maintained during a crisis and recovered or recovered from an event; (ii) having integrated systems that can adapt to change, diversity of transport and land use systems and multiple renewable energy sources that allow a city to survive fuel supply shortages and (iii) a vision of the city where efforts are being made to increase the city's capacity to respond to heterogeneous pressure factors (climate, environment, energy and economy) with the aim of ensuring a better quality of life and urban development sustainable (Boccolini 2016).

Recent works have further extended the concept of resilience, defining the latter as a dynamic interplay of persistence, adaptability and transformability across multiple scales (Batty 2008; Papa et al. 2015). Moreover, some scholars have pointed out the importance of "continual learning", providing an idea of resilience as 'bouncing forward', which includes the idea of 'improvement' essential structures and functions (Folke et al. 2010; IPCC 2012; Cutter et al. 2008; cited by Papa et al. 2015: 28). Therefore, learning capacity, persistence, adaptability and transformability have been classified as the key properties of a resilient city or, better, as the main goals to which strategies and measures have to be addressed for improving cities' response in the face of climate change (Papa et al. 2015).

### *11.2.3 The Capacity of the Cities as Complex Systems for Resistance to Flooding*

The urban resilience approach to assume that the human being not completely dominates nature, allows adopting new ideas in management or management of the risk of flooding and, therefore, focusing in a different way the vulnerability of cities to these phenomena. Eventually, the concept of resilience applied urban planning and design of urban systems, facilitates the adoption of an integrating vision of the different components that must be considered in the management of flood risks and in the generation of strategies and projects that reduce the vulnerability of cities to its impact. In sum, the urban resilience arises as one of the media with the greatest potential for urban systems can cope with unexpected shocks and achieve sustainability in time (Godschalk 2003).

However, to reorient the planning, administration and management of the cities towards urban resilience presents some operational and methodological limitations recognized in the literature on the subject (Toubin et al. 2014). Despite the difficulties that interfere with its implementation and execution the concept of resilience is becoming more and more important. Both theoretically and in practice, it is important to recognize that this approach offers great opportunities to promote alternative forms of planning and design of cities, new dynamics and possibilities of urban development and transformation of the urban fabric, and a greater capacity for resistance to flooding.

Basically, Serre et al. (2012) identify three capacities of flood resistance: resistance, absorption and recovery. The capability of resistance is defined as "...the ability of a city to absorb disruptions and recover its functions after the riots" (Serre et al. 2013). The absorption capacity is considered the ability of a system to operate despite the occurrence of negative events. Ability to recovery provides the ability of a system to put its components damaged in service as a line space - temporal (Vale

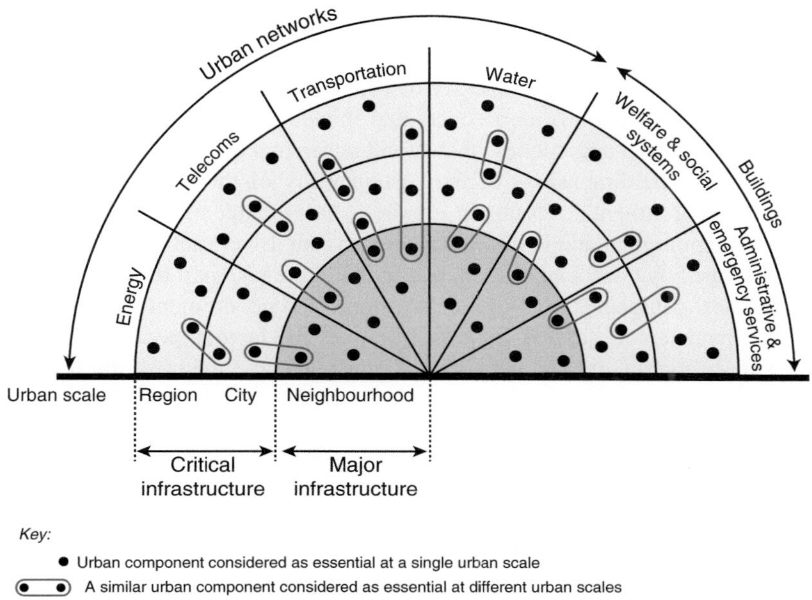


**Fig. 11.1** Resilient City. Urban components and scales (Source: Balsells et al. 2013; taken from Serre et al. 2016)

and Campanella 2005; cited by Serre et al. 2012). This ability is the most representative of the concept of urban resilience (Serre et al. 2012).

In sum, urban resilience is a concept that allows reorient possible answers to the questions raised above. First, because in the case of cities, allows to observe them as open and complex systems that are characterized by processes of change within your environment (Lhomme 2012). These processes of change involves multiple dimensions, components and scales. The main dimensions of the urban system are physical; socio economical and technical. Urban components of the system include buildings, transport networks and energy networks (Balsells et al. 2013; Serre et al. 2016), see Fig. 11.1.

In consequence, as an open and complex system, the city is composed of subsystems of different types: territorial (including intra urban system, urban, regional and more), natural, economic, social, residential, physical, institutional and cultural, which are composed of other subsystems (e.g. urban territory includes subsystems such as functional, productive, financial, etc.). In addition, its components (elements and actors), included in each subsystem, are related to each other and to those of other territorial systems of higher order, of the same order or of lower order. Therefore, the study of cities, being composed of systems and subsystems), is more complex due to the relationships existing within the subsystems and between the subsystems that form it, although the existence of subsystems facilitates the study of the global system and helps to understand their behavior.

### 11.2.4 The Conceptual Model – Framework to a Resilient City

Since we saw, the resultant complexity of considering the diverse dimensions and notable processes departs from a conceptual model of relations in the one that is considered to be two approaches: that of character macro, in whom the principal conductive axes or dimensions of the sustainability of the cities are the natural resources - environmental, economic trends, social – residential pattern, city management, technological and institutional. In turn, the drivers act in three levels: the top level shaped by the social system, and in that there take part the principal social actors (homes, companies, real-estate developers and government), the urban markets (employment, soil and housing), the resources and factors (population, employments, soil, infrastructure and networks of transport), and the institutions (NGO's; official organisms, etc.).

The conceptual model - provides a solid theoretical basis for building resilient cities in the face of climate change and its impacts. It also allows locating the strategic characteristics necessary to achieve it. Within the model, all the characteristics selected, according to their meaning and relevance, have been hierarchized and related to one or more of the key properties identified in the previously, namely learning ability, persistence, adaptability and transformability (see Fig. 11.2).

These key properties can be enhanced by other subordinate features that may be related to more than one key property, such as the efficiency that is common to persistence and adaptability. In detail, learning capacity can be improved through strategies and actions aimed at improving: (i) the capacity of networks, which allows to connect spatially and virtually to people, infrastructure and devices; (ii) monitoring capacity, which allows constant detection of the conditions of an urban system; (iii) knowledge, which enables information on events and processes to be developed; (iv) memory, which allows learning from past events to calculate possible future scenarios; (v) collaboration, which favours interactions and synergies between the different actors; (vi) participation, which allows people to be involved in decision-making processes. In addition, learning capacity and persistence are crucial to developing people and institutions' awareness of climate issues, improving the ability to anticipate future events that may threaten urban systems and, above all, rely on monitoring and knowledge to ensure effective management of the urban system over time (Papa et al. 2015) (see Fig. 11.3).

### 11.2.5 Methodological Approach

To methodological level, the application of complex systems and resilient city approaches to the Metropolitan Area of San Luis Potosí (MASLP) case has three phases. The first phase consists of an integrated diagnosis of the city as an object of study that shows its natural - geographical, economic, urban, socio – residential and institutional resources in order to detect its strategic components (elements, actors

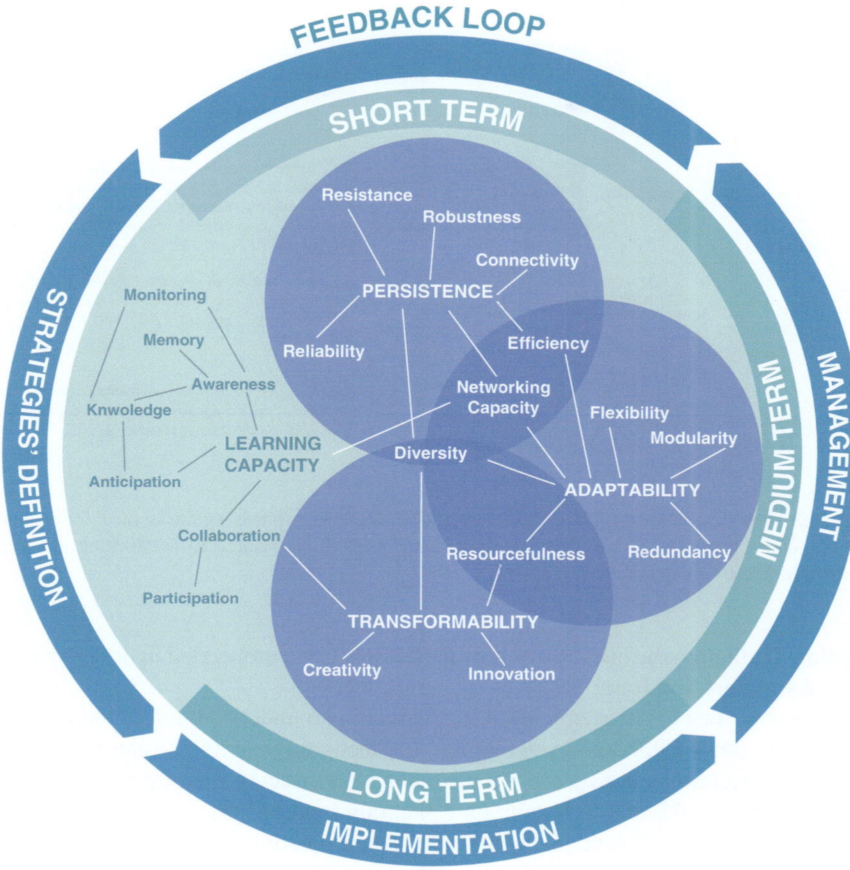


**Fig. 11.2** The Resilient City. Conceptual model: roles and linkages among the capacities in the face of climate change. Source: taken from Papa et al. 2015: 37)

and variables). The second phase consists in carrying out a systemic analysis that contemplates five dimensions or subsystems that integrate the mentioned resources that overlap in layers, creating value in the components.

The information obtained for each of these dimensions, subsystems and strategic components are used in the third phase, focused on to elaborate a proposal of strategic and participatory planning, with efficient measures that tries to solve the problems of development and sustainability that this metropolitan area presents. Both, the systemic analysis and the proposal are based on a territorial and multi/cross scalar approach, focused on anchoring local design in the wider geography, region or the watershed. When addressing the problem of management of water rain and flood risk inevitably city-wide and interrelated issues unfold: water management, urban climate, transportation networks, accessibility of jobs and services, social fragmentation and marginalization, densification, cultural heritage and landscapes,

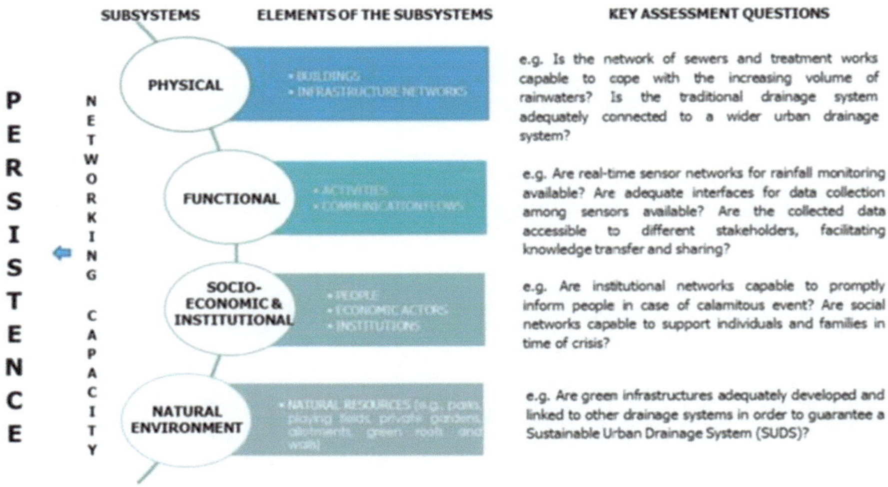


**Fig. 11.3** The Resilient City. Conceptual Model (Persistence): towards a guiding local for evaluating the characteristics of a resilient urban system (an example related to networking capacity. Source: taken from Papa et al. 2015: 39)

leading finally into the question of how these aspects are embedded in urban growth, spatial structure and form.

We subscribed to a systemic thinking, based on an understanding of the material and energy flows facilitated by ecological and technological infrastructures, engaging the interrelations between different realms, and tracing the effects of changes rippling through the urban system. Our planning methods are both context-based, carefully "excavating" data, on the one hand and projective on the other. They oscillate between conceptual strategies, the tactics of user perspectives and construction of future scenarios based on the idea that the city is a complex and dynamic system.

## 11.3 The Metropolitan Area of San Luis Potosí as a Complex System

### *11.3.1 Natural Characteristics: Climate; Water, Topography and Green Structure*

The metropolitan area of San Luis Potosí (MASLP) is one of the major urban concentrations in the Centre - West of Mexico region. It is located in la altiplanicie Mexicana in the micro-region San Luis and covers part of the area of the municipalities of San Luis Potosí and Soledad de Graciano Sánchez (see Fig. 11.4).

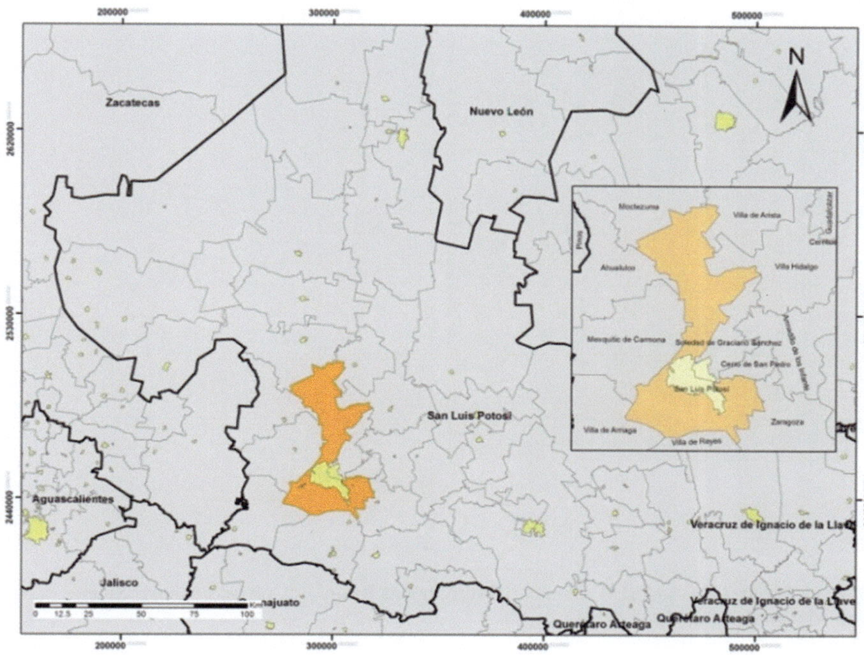


**Fig. 11.4** State of San Luis Potosí: principal cities, 2010. Source: taken from Moreno Mata y Villasís Keever 2015

This location influences the annual average temperature in the region varies between 12 ° C and 21 ° C therefore has a temperate and dry climate and semi dry temperate (see Fig. 11.5). Its rainfall is irregular and scarce, ranges from the 361 mm. at least and the 410.8 mm. at most and rainy season includes the months of May to September. The annual evaporation average reaches levels of 2038.7 mm (SWC 2005). The annual average relative humidity is 31%; in the course of the year is higher from June to December, reached 44% in June. Less humidity is recorded in the period from January to May; right in April is the minimum level of 18% (Stevens Vázquez 2012).

According to forecasts, the region where the MASLP will be affected by climate change, the process of urbanization, population growth and residential dynamics, changing its climate and vegetation, as well as hydrological cycles and the regime of rain - by what will tend will tend to rain more and more.

The MASLP is located in an endorheic basin, which is called San Jose - Los Pilares, whose extension is approximately 863,513.54 hectares, and is part of the hydrological region of the Salado, in an area with the presence of various bodies and streams. The hydrologic basin of San Luis Potosi has a North-South orientation and consists of numerous sub-basins (NWC 2002: 10). The main hydrographic flows of the valley of San Luis Potosi are the Santiago River, the Españita river, the Paisanos

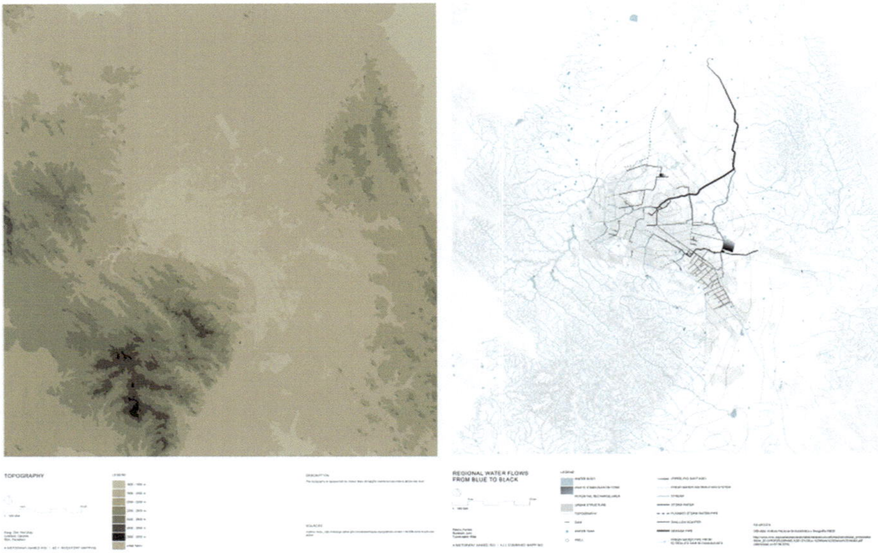


**Figs. 11.5 and 11.6** MASLP. Climate; Hydrological Basin and surface waters; Source: Taken from AHO/AUSLP 2016

river and the streams of Mexquitic San Antonio and San Carlos; all these flows originate to the west and south of the valley, and on them there is hydraulic infrastructure catchment and regulation of runoff, as well as natural, artificial channels and urbanized (see Figs. 11.6 and 11.7).

The Santiago River feeds on the basin corresponding to the dam San Jose, while the water catchment area of the river basin Españita corresponds to the dam Cañada del Lobo. Both dams control the flow of water runoff during the rainy season. All these flows are ephemeral, because the water only flows during a few days a year and always as a result of strong precipitation intensity hourly, downloading from the southern portion of the valley to the northern portion, due to their topographic features, in the areas referred to as gaps in the north and the area of the Ejido La Tinaja, located north of the MASLP, where disappear by infiltration and evaporation (see Fig. 11.7 again).

The slope of the relief is an important aspect in the volume and intensity of the flows that are observed in the study area. The elevations of the mountains that surround the valley where is located the MASLP range from 1584 to 2881 meters above the sea level. (m.o.s.l.). The altitude of the Sierra of San Miguelito to generate pending between moderate and pronounced (see Fig. 11.8). This Sierra presents a high tilt, detects slopes greater than 45° and a relief very abrupt and irregular. The Sierra San Pedro presented intermediate values (6 to 15°), counting on some occasions with values between 15 and 30° (Stevens Vázquez 2012). In the Sierra of Santa Maria, the inclination varies between 3 and 15°, while larger values to the steep escarpments close to the tops of the hills. Lower values of 3° inclination, presents the area of

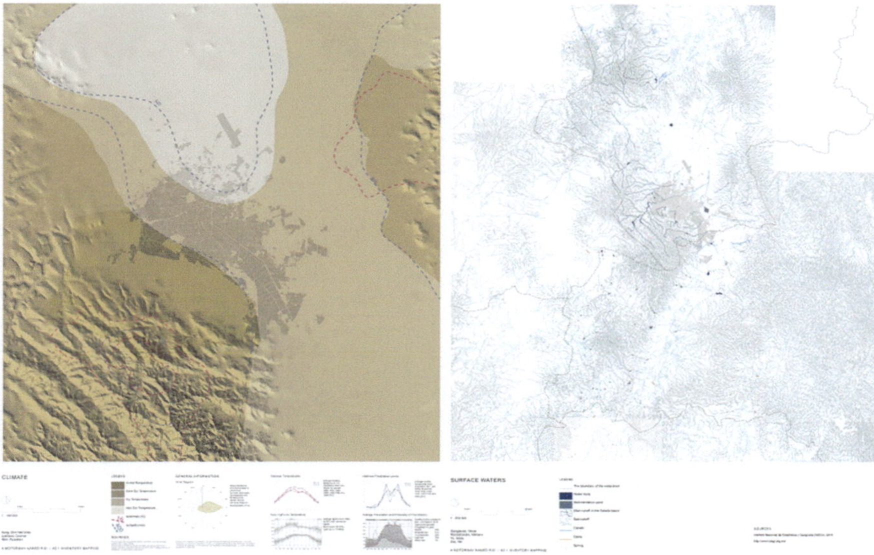

**Figs. 11.7 and 11.8** MASLP. Regional water flows: from blue to black; Topography and Pending. Source: Taken from AHO/AUSLP 2016

mount foot and close or equal to 0 per cent of tilt, the area of the plain. This aspect is very relevant in the assessment of the location of vulnerable settlements.

In the MASLP are flooding associated with rains of short duration and high currents and short times of concentration of runoff generated by the sierras San Miguelito and Alvarez, by lack of infrastructure for channeling and/or diversion of rainwater, by growth little controlled from the urban area and by a delay in the implementation of infrastructure previously raised to curb or control flooding (Candia Monsiváis 2015; MSGS 2010). The floods that occur not correspond to conventional floods; in the sense that it is hardly noted of rivers and Overflow is not present in large tracts of land (see Fig. 11.9).

San Luis Potosi Valley aquifer covers 1980 square kilometers and includes partially or totally the municipalities of San Luis Potosí, Soledad de Graciano Sánchez, Mexquitic de Carmona, Cerro de San Pedro and Villa de Zaragoza (see Fig. 11.4 again). This area is crossed by the sierras of San Miguelito to West and Alvarez to the East, which mark the contours of the plain where is the aquifer is located (Peña 2005). The MASLP is located in the central part of the aquifer and in the last two decades (1990–2010) urban growth has invaded considered recharge areas of the main aquifer, on the slopes of the sierra of San Miguelito, where also can be seen the loss of plant cover. The loss of plant cover representing 31.95% of the total of the Sub-basin; see Fig. 11.10 and Table 11.1.

Recent geohydrology studies (Carrillo-Rivera 1992) distinguish two bodies of the aquifer: a shallow aquifer and a deep aquifer. The brief has a thickness ranging from 5 to 40 meters. Due to its shallow, their behavior is very dynamic and recorded

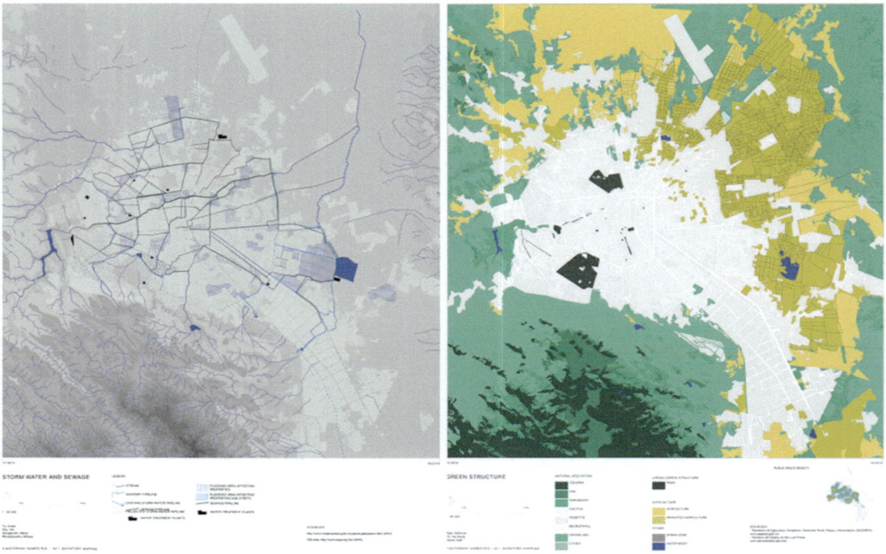

**Figs. 11.9 and 11.10** MASLP. Storm water and sewage systems 2010. Source: Taken from AHO/AUSLP 2016

**Table 11.1** MASLP. Demographic growth, urban surface, density and environmental impacts. Source: taken from Moreno Mata 2017

| Metropolitan Area of San Luis Potosi | | | |
|---|---|---|---|
| ***Total population*** | 2000 | | 850,828 |
| | 2010 | | 1,040,443 |
| | Demographic growth rate | | 1.7 |
| | Increase | | 1.22 |
| ***Urban surface (hectares)*** | 2000 | | 12,859 |
| | 2010 | | 23,445 |
| | Increase | | 1.82 |
| ***Average of density (inhabitants per hectares)*** | 2000 | | 66.16 |
| | 2010 | | 44.37 |
| | Increase | | 0.05 |
| ***Environmental impacts*** | Sub Basin | | San José – Los Pilares |
| | Surface | | 1′899,414.4 |
| | Geohidrological conditions | | Over exploitation |
| | **Loss of land cover** | Surface (Hectares) | 988,817.1 |
| | | % | 52.05 |

levels of pollution in its chemical composition. The deep aquifer, whose upper limit is located between 100 and 150 meters, is currently tapped through wells that reach up to 400 meters. This aquifer is confined to the center of the basin and the water that contains is considered antique water with more of one thousand years of antiquity studies(Carrillo-Rivera 1992; cited by Peña 2008: 26).

### *11.3.2 Economic, Urban and Socio – Residential Structure*

Economic characterization and residential partner of the conditions in which the aquifer, structure and the supply of water is located should be considered as a fundamental point the urban dynamics in the region, to understand the impacts affecting the aquifer and scenarios that can see. The situation that keeps the aquifer is the kind of urban development promoted.

In the State's urban hierarchy the Metropolitan Area of San Luis Potosí (MASLP) is the urban center of the first order and articulated at the municipal level to 485 locations. Recent studies located at the MASLP in the 14th place, within the classification of the cities more competitive at the national level, and the 207th at the global level (Moreno Mata and Villasís Keever 2015). The geographical location of the MASLP with respect to the cities of Mexico, Guadalajara and Monterrey, major cities of the country, is strategic because it is obligatory to refer to any of these cities, as well as the Gulf of Mexico and the border with the United States.

The MASLP is a pole of economic and industrial development; it has favored the growth of economic activities: health care, higher education (universities and research centers). The main industrial activities are concentrated in chemicals, metal and petroleum companies, processing of food, textile and automotive industries. The latter has had a significant growth in the last fifteen years, because its location, of the main automotive cluster in the region Centre - West, in the industrial zone of the municipality of Villa de Reyes, practically adjacent to the industrial area of MASLP (MPUDVDR 2015). The supply of water to the MASLP policy identifies as its main objective the economic development. This means to manage the supply of water to support the grow the capital of the State -economic and demographically- in the long run, under the argument that industrial production and services are the lines of greater economic profitability for the use of water (Peña 2008: 31).

The concentration of economic activities has favoured growth demographic and spatial, which has led to several consequences, such as an urban sprawl, speculation of urban land, social polarization, and urbanization in "protected" areas, problems of supply of water, flooding, pollution and increased traffic.

The city of San Luis Potosí was founded in 1592, during the colonial era. With an expansion space and population growing, for 2010 the MASLP reach a population of 1′ 040, 443 inhabitants. Between 1960 and 2000 the MASLP multiplied four times its population and almost eight times its surface. The acceleration of population growth together with urban sprawl and fragmentation (Alva Fuentes et al. 2015)

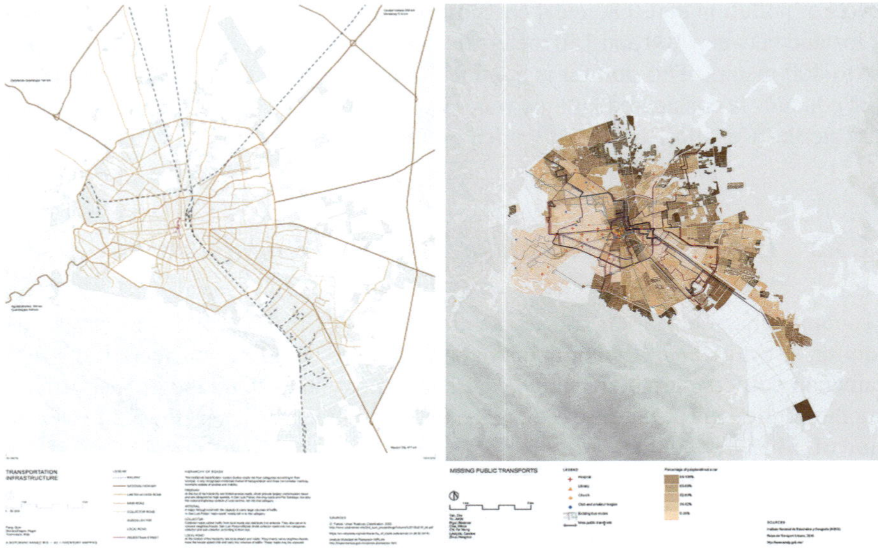

**Figs. 11.11 and 11.12** MASLP. Historical growth & Economic shift 2010. Source: Taken from AHO/AUSLP 2016

occurred predominantly so during the second half of the twentieth century, following the main axes of economic activity and new employment centers (see Figs. 11.11 and 11.12).

During the first three hundred years the city grew under an urban classic model of the New Spain cities, practically unchanged in all that time, and in which the old Center established the limits of it city space. From the twentieth century, this model began to turn slowly and in the middle of that century adopts a growth pattern of type concentric, combined with other sectorial radial pattern produced by the growing of the road system and the use of the automobile as the main way of transport (see Figs. 11.13 and 11.14). Since then, and particularly in the last three decades (1980–2010), this growth has become expansive (see Fig. 11.11 again).

The most intense urban growth occurs in the eastern sector of the city, in which the middle class and, mostly, the lower classes have made some acts of appropriation of the urban space; area that have been exposed to the displacement, while the middle class acts as a sort of transition by the rest of the city. The appropriation of space by purchasing high-level and upper middle class groups is observed to the southwest area of the city. The morphology of the city reveals a lack of planning (irregular layout). Although such urban planning has existed since 1993, the results obtained have been very few. There is no logic of zoning in the city, since there is no buffer green belts, the green areas are few and they are located in iniquitous form in the city (Moreno Mata et al. 2015).

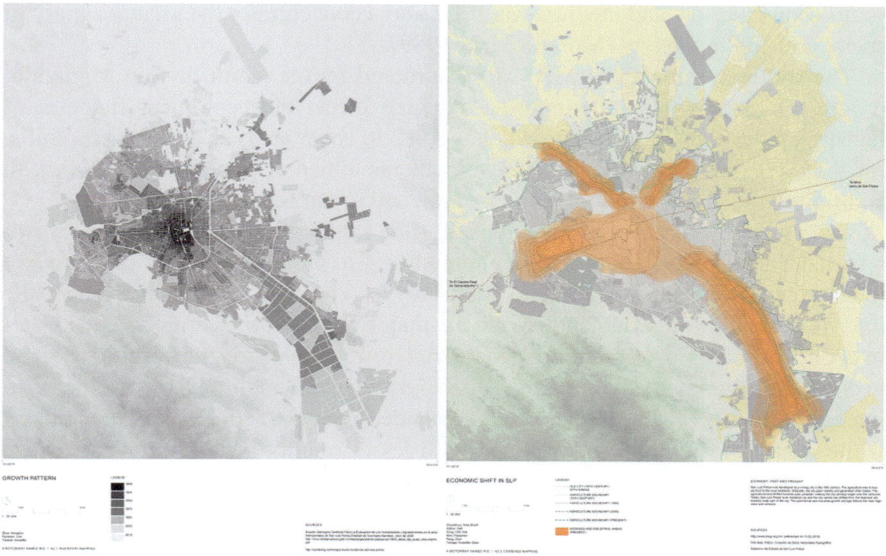


**Figs. 11.13 and 11.14** MASLP. Transportation infrastructure: Roads, Streets, Rails; Private cars use and public transports, 2010. Source: Taken from AHO/AUSLP 2016

### *11.3.3 Urban System and Water System*

From 2000 to 2010 the MASLP multiplied 1.2 times its population - under a rate of average annual growth of 1.7% and 1.82 times the urbanized surface, which reached 23,445 hectares in the period concerned (see Table 11.1 again). The average population density remained virtually unchanged from 66.16 to 44.37 inhabitants per hectare, this confirms a clear trend to Metropolitan sprawl, which is below the national average range to Metropolitan scale that was 73.05 (Alva Fuentes et al. 2015a). The fragmentation pattern indicates that 35% of the consolidated metropolitan area includes polygons fragmented (Alva Fuentes et al. 2016a). This reveals a strong speculation with the change of land use from rural to urban.

Derived from population growth, urban sprawl of the city of San Luis Potosí is accelerated and unplanned, that do not follow a regular path. Industrial development began in the 1970s, stimulated the production of housing under the system of subdivisions and housing developments of social housing, mainly for workers and blue collar employee (Moreno Mata 1989). During the decade of the nineties can be seen the emergence of new residential areas for upper and middle upper classes, mainly in the Metropolitan periphery South - West, in the form of gated communities (Moreno Mata and Cárdenas Nielsen 2015). Previously, the North-East of the city (municipality of Soledad de Graciano Sánchez) had large agricultural areas. Currently survive some of them, although this is the area of growing space because of the housing developments mainly targeted to groups purchasing level socio-economic low and medium-high. This area concentrates the greater supply of social housing in the city (Moreno Mata 1989).

The southern part of the MASLP is bordered by the sierra de San Miguelito, orographic barrier of vital importance which meets the ecological function of recharging the aquifer. Urban growth has come to cover areas of the sierra foothills, for the construction of real estate developments aimed at groups of high level, in addition to the construction of infrastructure and urban equipment.

The dynamics of changes in land use for the spatial growth drove urban growth on vacant land of the Ejido and communal origin. Other significant parties were moving peri-urban crops (Moreno Mata 1997). The land use change meant the intensification in water demand. While the majority of previous crops needed water in cyclic form, new uses of the soil (industrial, residential, and commercial) demand in constant volumes and with permanent supply water. In addition, household claiming better quality water and therefore demanded new extractions from the aquifer, while a part of agriculture developed with lesser quality water and even sewage (Cirelli 2004).

According with diverse studies (Moreno Mata 2012, 2010), the spatial growth of the MASLP adopts a residential pattern characterized by: (i) marked differences in patterns of urban accessibility (see Fig. 11.15), density of population and (in) efficiency in management and urban planning; (ii) the existence of a unsustainable model of intra-metropolitan mobility; (iii) the training of settlements popular in peripheral areas to MASLP, on land inadequate for urban growing incorporated to the use urban -sometimes through legal instruments allowed by the changes in the regulation of the use of soil, rural to urban within the modifications to the article 27 constitutional, made in 1993 and (iv) a complicated housing system with marked differences of marginalization index that are reflected from a sharp division of social space and in the deepening of inequalities in the living conditions of the population (see Fig. 11.16).

### *11.3.4 Relations Between Natural, Urban and Institutional Structures*

The most relevant effects of the model of urban growth - population of the MASLP is expressed by: (i) hydrologic balance sheet deficit; (ii) imbalance between supply and demand for water; (iii) a poor and uneven provision of drinking water, and (iv) serious environmental problems.

The deficit between the volume of extraction and recharge estimated (including the recharge, the induced and the underground flow) is 42.5 billion cubic meters a year (SWC 2005). The city isn't only growing in number of inhabitants, also on surface, which meant additional demands not only to supply domestic applications, but also to keep own supply system operation, above all to ensure that the pressure needed to reach more distant places. It also meant the accelerated infrastructure

**Figs. 11.15 and 11.16** MASLP. Marginalization Index; Transportation infrastructure: Roads | Streets | Rails, 2010. Source: Taken from AHO/AUSLP 2016

construction, became very complex in management and had repercussions on weaknesses in the supply system. The conditions in which these works were performed without planning and with few resources increased risk of leakage in the network by favouring investment in claimed extensions, and relegating to the background the rehabilitation of older sections. Under these conditions, plans to develop water supply service are overcome almost immediately. As a product of a speculative urban growth of MASLP has half a century of unregulated expansion of urban water demand, which has become the main threat for the stability of the aquifer.

From the environmental point of view, the growth of the population and the urban surface meant a radical change in the supply of water but also in the environmental conditions of the aquifer. The city went from the use of surface water and the shallow aquifer before 1950, accelerated and growing dependence of the deep aquifer. By 1960, each 100 liters of drinking water supply available, 59 came from 41 of the aquifer and surface waters. Currently 92 liters for every 100 urban networks are groundwater and only 8 come from surface water (Peña 2008).

Average spending, ranging from 6 to 95 liters per second, concerns the average total extraction of liters per second in a year. The trend of average spending has been to the upside. It is necessary to drill ever deeper to get water. The depth levels are in the range of 200 to 1185 meters. The depth of extraction wells indicates increasing pressure and demand for water and, at the same time, the abatement of the aquifers (MSGS 2010). All this is related to other processes such as depletion of the water table levels and subsidence caused by high levels of extraction, see Fig. 11.17. The

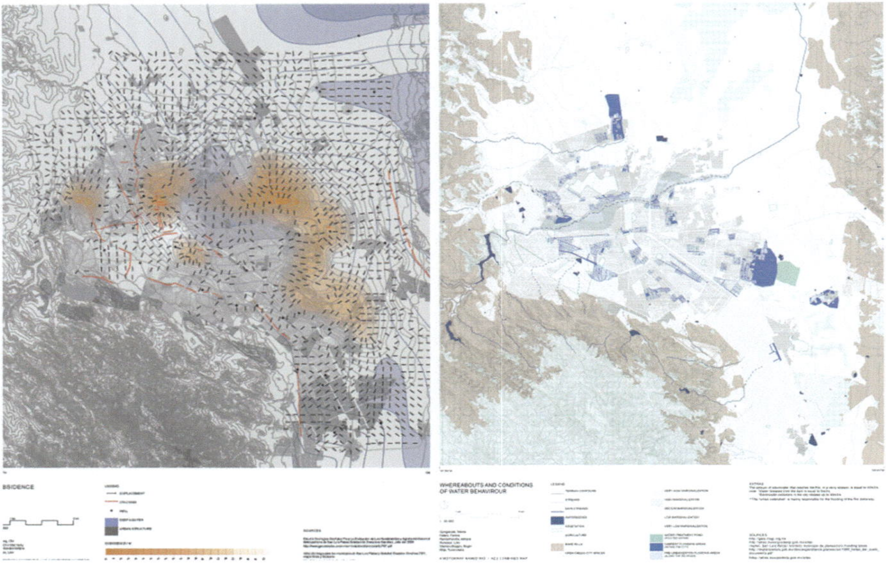

**Figs. 11.17 and 11.18** MASLP. Aquifer overdraw and subsidence risk; Whereabouts and Conditions of Water Behaviour. Source: Taken from AHO/AUSLP 2016

older than the thousand years of groundwater evidenced the lack of sustainability in the use of water in the city, indicating a low infiltration of rainwater (worse still when taking changes in the area of recharge (Stevens Vázquez 2012: 148).

As for the generation and treatment of waste water in the MASLP, there is an ambiguity in the official information concerning the volumes of treated water and the plants that are operating. According to Cirelli (2004), for the year 2000 were generated 1910 liters per second of wastewater: 87.0% was generated by the household and 12.3% for industrial use. According to projections of a company consultant (DEMM 1998), the generation of wastewater for the year 2018 could oscillate between 2119.6 liters per second (optimistic scenario) and 1611.8 (pessimistic scenario). Meanwhile, the State Water Commission reported that the residual water generated by the MASLP was 2572 liters per second. According to the annual report of the Interapas 2005, 8 treatment plants in operation in that year - municipal and private, altogether, have an installed capacity of 1882 liters per second (Interapas 2005). This implies that, in practice, the volume of water treated according to installed capacity approaches an intermediate, around scene of 1900 liters per second. Namely, that there is an important lap on the official estimates of this data, in the order of a 35.3% compared to that scenario (Santacruz De León 2008).

### 11.3.5 Urban Model, Water Structure and Risk to Floods

The urban flood represents a danger which has been gaining greater importance at the present time, mainly because the MASLP is in growth and because the processes of urbanization alter the components of the hydrological cycle, causing episodes of rain that previously were not significant, now have dangerous consequences or disastrous, causing widespread devastation, economic damage and even loss of human lives. If this is associated with the impact of climate change we are faced with a phenomenon whose consequences can reach disaster dimensions. In fact, natural climate variability coupled to the process of climate change is generally recognized as the main reason for the increase in the frequency and intensity of extreme events, which are one of the main factors of floods.

The spatial expression of the phenomenon of vulnerability or risk of flood disaster to the interior of the case studied reveals a marked polarization, biophysical and social origin. Although the risk factors, both natural and anthropogenic or social, are located in all directions of the Metropolitan spot, the concentration of poverty and inappropriate terms of location of human settlements tends to lead to a greater vulnerability and risk to flooding than in the rest of the city (see Fig. 11.18).

The analysis of the factors of the biophysical (hydraulic response, topography, runoff, prevalence) confirms the foregoing: social groups with higher incomes tend to be located in areas of lower risk biophysicist. This trend began in the 1990's, is consolidated at present occupying the highest areas of the valley of San Luis, which are the least prone to flooding, through the impetus on the part of the local real estate sector of an urban model known as gated communities. On the other hand, low-income groups (which include population strata media, media low and low or very low), observe clear trends of localization in the urban periphery, in clusters or agglomerations which are distinguished by inappropriate locations, prone to urban risk in general and in particular by flooding, according to biophysical indicators (in particular by its location in the lower areas of the valley).

The analysis of the factors of a social nature, in particular indicators of social backwardness and socioeconomic fragility, aggravates this differentiated situation with regard to the degree of vulnerability or risk to the disaster: in addition to the privileged conditions of location, higher socioeconomic strata are located in areas with better quality of life, living conditions, access to equipment and infrastructure of the first world and the best views of the city. On the contrary, in addition to inadequate spatial localization, the population of scarce resources are located in areas of low accessibility to basic equipment, inadequate coverage of water and sanitation services, serious problems of mobility and public transport, among other shortcomings.

It should be noted that in the phenomenon of differentiated residential location have a significant impact of the process of globalization, but also other factors that are located rather at the national and local levels. Among the latter, it may be noted: (i) changes in the use of soil, guided by the excessive growth of the urban sprawl and the expansion of the city mainly toward the peripheries; (ii) limitations and inefficiency of the urban planning model dating from institutional, 1976; (iii) the insecurity and growing violence, that "justify" the existence of products of the real estate market focus on the "needs" of live safer and with the best quality of life in the high socioeconomic strata, and (iv) the impact of urban zoning defined by the laws of the market, where are the private investors decide where the city grows, out of this decision-making the instance Governmental, which is restricted to orient the instruments of land-use.

### 11.3.6 Biophysical, Urban, Social and Institutional Relations

Several factors explain this complex interaction of biophysical, urban, social, and institutional factors that underlie the structure and dynamics of the urban water management: firstly, as regards the functioning of the system of potable water supply, it is important to note that: (i) an important area of the aquifer is located under the urban area, so that its condition is directly affected by the dynamics and growth of the city type: the invasion of their possible recharge areas, the risks of contamination and the location of the infrastructure of extraction and monitoring are some of the aspects that this involvement is manifested clearly; (ii) the largest volume of extractions is earmarked for urban uses (domestic consumption, operation of the municipal network, industrial applications, and supply shops and services), and (iii) as a political construction, the city through its agencies of administration and organizations of sectoral representation of its various actors, has become the most influential socio-political factor to define the type of use that is made of the aquifer and especially, has set the priority for use of the liquid (Peña 2008).

In the second hand, the runoff rainwater which dangerously down the slopes of the suburbs located in the southwest sector of the Metropolitan Area of San Luis Potosí (MASLP) have increased gradually over the last ten years, generating a vulnerability differential to flooding, according to residential location patterns. Factors associated with this phenomenon are: effects of climate change; intense rains of short duration; high intensities and short times of concentration of runoff generated by the mountains that surround the city; Metropolitan sprawl and unsustainable patterns of urbanization; lack of infrastructure for conduction and/or diversion of rainwater; backwardness in infrastructure to curb or control floods.

The analysis of the factors of the biophysical (hydraulic response, topography, runoff, prevalence) confirms the foregoing: social groups with higher incomes tend to be located in areas of lower risk biophysicist. This trend began in the 1990's, is consolidated at present occupying the highest areas of the valley of San Luis, which are the least prone to flooding, through the impetus on the part of the local real estate sector of an urban model known as gated communities. On the other hand, low-income groups (which include population strata media, media low and low or very low), observe clear trends of localization in the urban periphery, in clusters or agglomerations which are distinguished by inappropriate locations, prone to urban risk in general and in particular by flooding, according to biophysical indicators (in particular by its location in the lower areas of the valley).

The analysis of the factors of a social nature, in particular indicators of social backwardness and socioeconomic fragility, aggravates this differentiated situation with regard to the degree of vulnerability or risk to the disaster: in addition to the privileged conditions of location, higher socioeconomic strata are located in areas with better quality of life, living conditions, access to equipment and infrastructure of the first world and the best views of the city. On the contrary, in addition to inadequate spatial localization, the population of scarce resources is located in areas of low accessibility to basic equipment, inadequate coverage of water and sanitation services, serious problems of mobility and public transport, among other shortcomings.

It should be noted that in the phenomenon of differentiated residential location have a significant impact the impact of the process of globalization, but also other factors that are located rather at the national and local levels. Among the latter, it may be noted: (i) changes in the use of soil, guided by the excessive growth of the urban sprawl and the expansion of the city mainly toward the peripheries; (ii) limitations and inefficiency of the urban planning model dating from institutional, 1976; (iii) the insecurity and growing violence, that "justify" the existence of products of the real estate market focus on the "needs" of live safer and with the best quality of life in the high socioeconomic strata, and (iv) the impact of urban zoning defined by the laws of the market, where are the private investors deciding where the city grows, out of this decision-making the instance Governmental, which is restricted to orient the instruments of land-use planning and management of urban space under the criterion of private profitability.

### *11.3.7 Urban Planning and Water Management Model*

The current model of urban planning and management of the MASLP dates back to 1976, at that time was published the General Law of Human Settlements, and were created in Mexico organizations such as INFONAVIT and FOVISSSTE that drove the housing development based on massive sets of social housing. In the case of San Luis Potosí, in the same time was created the Secretariat of Urban Development and Ecology (now known as Ministry of Urban Development, Housing and Public Works), which is the regulatory entity in the field of urban development and territorial planning. There is also the city the Metropolitan Institute of planning, which recently replaced the Municipal Planning Institute, created in 1996. This management model, based on methods and techniques of obsolete planning, faces the challenge of modernizing, seriously behind schedule and the absence of participative and strategic approaches, fundamental to promote a holistic urban planning, inclusive and efficient. Much less there is an approach that conceives the MASLP as a complex system, which requires a considerable amount of economic and financial resources, and working in various sectors of society, such as Government, private companies, institutions and organizations, to promote the vision of a resilient city, sustainable and systemic.

The conventional management of water resources in the MASLP, both of clean water and sewage - domestic, industrial, and rain - has not had the capacity to face the challenges of key of this metropolitan area, as for example the growing competition for water, waste water management, sanitation, as well as, the protection of water resources. The traditional model of management of urban waters in the case study has failed to distinguish between the different qualities of water or to identify applications for them. As a result of this fact, good-quality water has been aimed at the indiscriminate needs of the metropolitan population, which has contributed to the scarcity of the resource (Peña 2008). In addition, in an uncoordinated manner, different authorities, each guided by various policies and laws, continue controlling at the level of city water subsectors. As urban governments become more complex and specialized, sectoral integration within Government and integration between the different levels of Government are becoming something increasingly important (Moreno Mata 1998).

In general, issues related to water resources in urban areas often remain disconnected from broader urban planning and management processes in the area of the basin. Urban master plans have failed to satisfactorily manage the various components infrastructural management of urban waters (supply of water, wastewater, dry sanitation, drainage of rainwater and solid waste treatment systems). Moreover, although the provision of water, sanitation and the planning of urban settlements could be included within the whole basin management plans, these plans neglect recognition of the interdependence between freshwater, wastewater, flood control and storm water (Stevens Vázquez 2012).

In this institutional context, planning, administration and management of the infrastructure for the supply of water and sanitation in the city of San Luis Potosí are facing great challenges. Some are problems of institutional, organizational type and technical product of previous processes, due to the coincidence between the bankruptcy of municipal public finances, which decreased investment in drinking water systems (and all public services) just in the years in which the city grew more (Moreno Mata 1998; Peña 2008). This circumstance explains partially the shortcomings accumulated in the supply infrastructure. In addition, supply and drainage infrastructures did not develop in parallel at the end of the nineteenth century. The infrastructure of the network, in addition to revealing a lack of planning and long-term strategic approach, had an expansive growth, according to the spatial pattern of Metropolitan growth. At present, the most obvious problems are the antiquity of these networks (which in some sections date back 60 years), leakage and collapses, the lack of a division of the network, problems related to the adequate control of the flow of water, in addition to low pressure in some areas of the city, rising costs of pumping and in general, poor technical and administrative coordination - organizational between municipal governments comprising the agency operator of water and sanitation of the MASLP (Interapas 2005; Peña 2008; Stevens Vázquez 2012).

Two questions arising from this systemic analysis are: do, in general, the supply management of water, sanitation and storm water management taken in a concerted way? Why, instead, the management of each of these elements is planned and carried out as isolated, consequently loses the interconnections between the problems and possible solutions. Of course, it is necessary this model moves away from other forms of urban planning and land - use territorial of multidimensional character and strategic, such as models of resilient cities, comprising the six basic lines for the design and management of cities: (i) living: (ii) people; (iii) economy; (iv) governance; (v) environment, and (vi) mobility (Serre et al. 2016).

## 11.4 Proposals to Tackle the Problem Studied

To address the alternative solutions to a complex problem such as the integrated management of water in the MASLP, from the conceptual and methodological framework proposed, as well as of the systemic analysis, it is essential to use the tools of strategic and participatory urban planning, and construction of future scenarios. As it is inferred from the systemic analysis developed in the previous sections, displayed in long-term scenarios are, necessarily, in the analysis of various subsystems and components that intersect: the urbanized watershed, i.e. the factor water; the network of mobility, roads and transport in its overall view, i.e. the factor mobility -understood as a structural network of other infrastructures (hydraulic, industrial, and domestic drainage of water from rain and sanitation)-, and the structural and functional dynamics of the city -including the accessibility to goods, services and own infrastructure networks indicated, land use planning, urban and landscape design, and the composition of the social fabric-, all these elements are condensed in the factor socio spatial equity. These three essential components of the city are articulated and interconnected among themselves, forming the vision of a complex system.

Construction of future scenarios based on the idea that the city is dynamic and can be transformed so. In this sense, the power of a scenario compared to the reality, is that it is a variant, option, vision, utopia or dystopia (Kosow 2008); It is the narration of a development path and why it is valuable to condense several narratives, representations and imaginaries as well (social, technical, political) values that become important elements in the discussion and the solution of the problem. A scene tells a story, it has a temporal dimension that starts in the present and unfolds forward, creating a coherent and complex image: a vision of the future that consists of a blend of facts, values, decision-making and partial logics (Kreibich et al. 2007).

Under the approach of scenarios as a tool for exploring possible futures, applied to the case study, the basic question raised is "what if"; i.e., that it could happen or what results could be achieved, according to three key aspects: political, economic and environmental techniques. You were analyzed and visualized as well as what might happen under conditions that include public intervention (political power), as opposed to the (local) Community intervention and the existence of resources (public or private) as opposed to the lack of them (local, not necessarily monetary resources).

**Table 11.2** Alternative solutions to problems. Exploring scenarios. Source: Moreno Mata and López Mares 2015

| Scenario | Conditioning | |
|---|---|---|
| | Stage power politician vs. community involvement | Financial resources vs. lack of resources |
| 1 | − | + |
| 2 | − | − |
| 3 | + | − |
| 4 | + | + |
| 5 | + | + |
| 6 | + | − |

## *11.4.1 Exploring Alternative Scenarios*

For the case study, the status quo that is derived from the characterization made in the preceding paragraphs is condensed, at the urban scale by the following problems: (i) Polarization emerging socio-spatial and a fragmented urban fabric which is related to the pattern of economic growth of contemporary MASLP; (ii) Fragmentation of the urban network and unequal access to the day-to-day basic services, especially evident in the functioning of the public transport network, unevenly distributed and exacerbated by the prevalence of a public road infrastructure focused on the use of the private car, and (iii) Environmental risk due to a linear problem, i.e., water, energy and matter flows, especially exemplified by a dialectic of scarcity of water, an overexploited aquifer and floods resulting from increased and more intense urban runoff.

Based on the mentioned conditions (political vs. community participation power and power economic vs. lack of resources) four possible combinations in six different scenarios were explored (see Table 11.2).

In relation to the management of urban runoff, these scenarios have in common the study of the MASLP to different scales, both Metropolitan - Suburban municipalities and hinterland, such as intra-urban and local or neighborhood, in order to understand what happens to water, for example, above, affects what happens downstream, with both urban, social and technical implications, as it would be the case of the quality of life. For all the scenarios we worked on three axes or dimensions: socio-spatial, mobility and water equity. For each of these strategies are prioritized, and built a box of tools and allows addressing each issue. Examples of these tools are the techniques to mitigate floods (such as reforestation, absorption and domestic rainwater collection pits).

Examples of tools in terms of mobility are different strategies to promote the use of non-motorized (as the public bicycle system) or to promote more democratic means of transport (including BTR, Metrobus or elevated light rail). In terms of equity, examples of tools are the conversion of bridges in places of equipment and conversion of urban sectors, colonies or eco districts.

On par with these techniques, was also proposed legal, fiscal and social tools that allow reaching the vision proposed by each scenario, according to the conditions given. For example, in the case of the stages with political power, but lack the resources, fiscal and planning tools are proposed to promote the densification of urban and residential; to promote the capture of capital gains resulting from public and private investment, or to regulate the uses of soil and the urban image. These tools can be implemented in specific locations in the MASLP, for which the corresponding partial master plans are to be compiled.

Objectives and shared values of all scenarios should subscribe to the premises of sustainability and, therefore, to the perspective of the right to the city [livable city], with layout and greater access to opportunities for work [efficient city] and a strong commitment to environmental justice [inclusive and sustainable city]. The elements of this position are condensed in a city resilient to the MASLP proposal:

A livable city. With an urban fabric permeable and safe, healthy climate and a universal access and multi-scale to public goods and services.

An efficient city. To facilitate mobility and connectivity, both physical and virtual. That integrates transportation options alternative to the private car, both motorized transport and non-motorized.

An inclusive and sustainable city. To ensure access to public goods and services, and networks of urban infrastructure that promote the mitigation of environmental risks and a circular management of resources; focused on infrastructure services and green ecosystems, which include the recycling of wastewater, integrated waste management and the management of storm water, as well as adaptation and mitigation of the impacts on the environment and the use of renewable energy.

Under these premises, resulting from the combination of constraint scenarios mentioned earlier propose alternate solutions to improve the studied problem, under an approach that contemplates the natural context as part of a water and urban system.

### *11.4.2 Proposed Intervention*

Given the natural, economic, social and spatial limitations of the MASLP were the questions that were raised at this stage of the research: what are the opportunities to define the synergies between the different dimensions and subsystems of the city? What are the scales appropriate to the challenges? Are there any alternatives for the solution of various problems facing the city from approaches to complex systems, urban resilience and environmental sustainability?

The right to the city (the socio-spatial equity) is promoted through infrastructure, public services and access to services and job opportunities, which is closely linked to redressing the marginalization of the poor and the fragmentation of the city. Social segregation and physical and perceptual barriers that entrench the residential location of different social groups, are part of the analytical elements of the studied problem.

The problem of urban mobility is associated with a strong demand for improvements in the functioning of the road networks, public transport system and reducing the distances or transfer times. The study of the possibilities of incorporating alternative systems to urban mobility as pedestrian mode and the construction of bike paths, cycling with public transport connections, could be also main avenues reforestation strategies, improvement of urban micro-climate and reduction of the heat islands in the city (Lárraga Lara and Moreno Mata 2015).

The issue of management of flood and rain water should be located within the global and regional scales, and at the same time relate to the possibilities of facilitating new sources of water improved in underserved areas. Thus, reducing urban risks caused by the floods, affecting the life and health of marginalized populations, could also be considered as a matter of Justice and right to the city, related to the environment and widespread access to the urban satisfiers.

This is also linked with the issue of urban fragmentation, supported in part by the character of the road structure that cleaves and divides the city; a problem that could be reversed or at least alleviated through updating and retraining in an ecological green corridors-based infrastructure.

On the other hand, it is the question of urban and residential densification, control of the uses of the soil and the impulse to a model of a compact city, distinguishing the current urban functions; distinguish the socioeconomic profile and the demographic dynamics, but also the urban landscape of the different sectors of the MASLP.

In summary, proposals addressing the multidimensional problem associated with the analyzed problems require compromise with obvious issues, urban, but which are situated at the intersection (necessarily trans disciplinary and multi-scale) between region, landscape, urbanism and architecture.

## 11.5 Conclusions and Recommendations: Lessons Learned from the Case Study

The problems of the cities whether large, medium or even small, are derived in many cases inadequate planning and urban design, the lack of vision in the long term and forecast the behavior of the various factors that make up the life of a city. This is the case of the MASLP, whose trends of urbanization and population growth, coupled with climate change, have important impacts of territorial, urban and socio – environmental conditions that require priority attention by various disciplines, but also of the authorities and citizens in general.

From the case study has been able to recognize the existence of important links between geography, climate change and the urban development trends. This explains, for example, the impact in terms of vulnerability of cities - in this case the MASLP- that can generate meteorological phenomena, such as heavy rains, floods of runways and unexpected avenues of water.

Considering all these aspects, as well as the possible involvement in the technical solutions and enhancement or mitigation of identified urban and environmental effects, then is an inescapable part of urban designers and landscape, and other related disciplines, such as architecture, road and transportation engineering, water management. This you should add an appropriate urban management policy and municipal, in many cases of metropolitan character, whose effectiveness and efficiency can play in favour of the living conditions of the population of the cities and the characteristics which adopts the urban habitat.

Study the MASLP objectives were to rethink its use, functional and dysfunctional, as a key element not only for urban mobility in that sector of the metropolitan area studied, but also to detonate a socially and environmentally sustainable development, from a broader perspective, of metropolitan character. As we have seen the growth of the MASLP, as well as climate change imposes serious challenges to the results of an obsolete urban policy and with serious limitations.

In conclusion, in the context described, the problems described are prototypical (metaphorically) of contemporary cities: either from the perspective of environmental toxicology, from the field of the vulnerability of the city to the effects of climate change, or social, urban and road conflict. This metropolitan model currently generates multiple negative impacts on the functioning and efficiency of the MASLP and also affects the degree of socio-environmental and urban vulnerability.

When you are tackling urban problems from the perspective complex systems, sustainability and the urban resilience, other metropolitan scale factors are involved inevitably: sustainable management of water and the urban climate, networks of mobility and transport, accessibility to services and jobs, the spatial fragmentation, social marginalization and socio - residential polarization.

Altogether, these impacts or multidimensional effects constitute an adverse situation for a sustainable future of the MASLP, and a formidable challenge for the city, with important connotations of instrumental and methodological type for urban planning and design, which could become, the centerpiece of an urban agenda that environmental systems are integrated, strategic and comprehensive, way space and residential mobility, and at the same time pressures that the physical – geographical environment, the urbanization process and a culture of social participation. In this regard, the results of this approach to the case study could provide conceptual categories, analytical tools and intervention strategies to address and understand the complex problems currently facing the MASLP, which interact with different layers and interrelate with urban sustainability.

## References

AHO/AUSLP. (2016). A Motorway Named Rio, Publication resulting to international workshop «A Motorway named Rio» organized in March 2016 by The Oslo School of Architecture and Design (AHO) and Master in Urban Landscape.

Alberti, M., Marzluff, J. M., Shulenberger, E., Bradley, G., Ryan, C., & Zumbrunnen, C. (2003). Integrating humans into ecology: Opportunities and challenges for studying urban ecosystems. *Bioscience, 53*, 1169–1179.

Alva Fuentes, B., Moreno Mata, A., & Zavala Ojeda, G. (2015a). Chapter II: Metropolitan systems and urban sprawl in Mexico: Theory and reality. In A. Moreno Mata (Cord.), Urban environment, sustainability and territory in mexican cities (Vol. II, pp. 54–77). Mexico: Autonomous University of San Luis Potosí.

Alva Fuentes, B., Lopez Mares, M., & Durán, G. (2015b). Chapter VI: Urban design and metropolitan fragmentation in San Luis Potosí. In Moreno Mata, A. (Cord.), Urban environment, sustainability and territory in Mexican cities (Vol. II, pp. 139–155, Mexico: Autonomous University of San Luis Potosí.

Balsells, M., Barroca, B., Amdal, J. R., Diab, Y., Becue, V., & Serre, D. (2013). Analyzing urban resilience through the alternative storm water management options: Application of the conceptual DS3 model at the neighborhood scale. *Water Science and Technology, 68*(11), 2448–2457.

Batty, M. (2008). Cities as complex systems. Scaling, interactions, networks, dynamics and urban morphologies. Working Papers Series, Paper 131, Centre for Advanced Spatial Analysis, University College London, ISSN 1467–1298.

Boccolini, S. M. (2016). The urban event. The city as a complex system away from equilibrium. *Quid 16, Journal of Area Urban Studies of the Institute of Research Gino Germani from the Faculty of Social Sciences (UBA),* 220–251.

Candia Monsiváis, M. A. (2015). *Risk analysis by flooding in the metropolitan area of San Luis Potosi.* Thesis of Master in Applied Geosciences. San Luis Potosi: Potosino Institute of Scientific and Technological Research.

Carrillo-Rivera, J. (1992). *The hydrogeology of the San Luis Potosí area.* Ph. D. Thesis. London: University of London.

Cirelli, C. (2004). *Discarded water, used water. Growing on the margins of the city.* San Luis Potosí: El Colegio de San Luis.

Conte, E., & Monno, V. (2016). The regenerative approach to model an integrated urban – Building evaluation method. *International Journal of Sustainable Built Environment, 5*, 12–22.

Cutter, S., Barnes, L., Berry, M., Burton, C., Evans, E., Tate, E., & Webb, J. (2008). A place-based model for understanding community resilience to natural disasters. *Global Environmental Change, 18*, 598–606.

DEMM. (1998). *Master plan for the improvement of the services of drinking water, sewerage and sanitation in San Luis Potosí.* Soledad de Graciano Sánchez and Cerro de San Pedro, Mexico: DEMM Consultores, S.A. de C.V., Mexico.

Folke, C., Carpenter, S. R., Walker, B., Scheffer, M., Chapin, T., & Rockstrom, J. (2010). Resilience thinking: Integrating resilience, adaptability and transformability. *Ecology and Society, 15*(4), 20.

Fusco Girard, L., Baycan, T., & Nijkamp, P. (2012). The creative city in a multidimensional perspective. In *Chapter of sustainable city and creativity: Promoting creative urban initiatives.* London: Ashgate Publishing, Ltd.

García, R. (2006). *Complex systems. Concepts, method and epistemological foundation of interdisciplinary research.* Barcelona: Gedisa.

Godschalk, D. R. (2003). Urban hazard mitigation: Creating resilient cities. *Natural Hazards Review, 4*, 136–143.

Harvey, D. (1977). *Urbanismo y desigualdad social.* México: Siglo XXI.

Holling, C. S. (1996). Engineering resilience versus ecological resilience. In P. Schulze (Ed.), *Engineering with ecological constraints.* Washington, DC: National Academy.

Interapas. (2005). *Annual report 2005, San Luis Potosí.* San Luis Potosí: Body Intermunicipal Metropolitan water, sewerage, sanitation and services related of the municipalities of Cerro de San Pedro, San Luis Potosí and Soledad de Graciano Sánchez (Interapas).

IPCC. (2012). *Special report on managing the risks of extreme events and disasters to advance climate change adaptation.* Cambridge/New York: Cambridge University Press.

Jacobs, J. (1971). *The economy of cities.* Peninsula: Barcelona.

Johnson, S. (2003). *Sistemas Emergentes: O Qué Tienen en Común Hormigas, Neuronas, Ciudades y Software.* México: FCE.

Kosow, H. (2008). Methods of future and scenario analysis: Overview, assessment, and selection criteria. In H. Kosow & R. Gaßner (Eds.), *Development policy: Questions for the future.* Bonn: DIE Research Project.

Kreibich, R., Oertel, B. & Wölk, M. (2007). *Future-oriented technology analysis principles, methodology and research questions*. Paper prepared for the 1st Berlin Symposium on Internet and Society, Oct. 25–27, 2011.

Larraga Lara, R., & Moreno Mata, A. (2015). Chapter III: Urban sprawl, environmental justice and equity in access to green spaces in the metropolitan area of San Luis Potosí. In A. Moreno Mata (Coord.), *Urban environment, sustainability and territory in Mexican cities* (Vol. II, pp. 78–107). Mexico: Autonomous University of San Luis Potosí.

Lhomme, S. (2012). *Technical networks as a vector of spread of risks in urban areas - a theoretical and practical contribution to the urban resilience the analysis*. Thesis of doctorate, University of Paris Diderot, 2012, 365.

Luhmann, N. (1998). *Social systems: Guidelines for a general theory*. Barcelona: Anthropos Editorial.

Moreno Mata, A. (1989). *The production process of the urban space in the metropolitan area of San Luis Potosí: The problem of housing*. Master's thesis in urban development. Mexico: The Colegio de Mexico, A.C.

Moreno Mata, A. (1997). The occupation of municipal land for urban use in the metropolitan area of San Luis Potosí. Notes. *Journal of Information and Analysis, 1998*, 9–17. Aguascalientes: National Institute of Statistics, Geography and Informatics.

Moreno Mata, A. (1998, January–June). Local government, planning and management of the public service in the middle of Mexico cities. The case of the metropolitan area of San Luis Potosí. *Economy, Society and Territory, 1*(3), 519–545.

Moreno Mata, A. (2010). Utopia and reconceptualization of the Mexican urban model. Alternative approaches to the study of the drift and the new centralities: The case of the metropolitan area of San Luis Potosí. Journal of architecture. *Design and Urban Planning, 7*, 180–195.

Moreno Mata, A. (2012). *Residential segregation in the metropolitan area of San Luis Potosí: Causes and manifestations*. Memoirs of 12th Meeting of the Mexican Society of Demography. México: Mexican Society of Demography.

Moreno Mata A. (2017a). Low density urban expansion, access to urban land and territorial policies in intermediate cities of Mexico. Paper presented at the Third Ibero-American Urban Land Congress, 21–26 October 2017, Curitiba, Brazil.

Moreno Mata, A. (2017b). *Dispersion, suburbanization and sustainability of cities. Evaluation of their interactions and effects in the metropolitan area of San Luis Potosí, México (1990–2010)*. Work Paper, UASLP.

Moreno Mata, A., & Cárdenas Nielsen, A. (2015). Peri – urbanity, inequality and segregation in San Luis Potosi: 1990–2010. Journal Development Observatory, 4(14), 63–70, Mexico: Autonomous University of Zacatecas.

Moreno Mata A. & López Mares L. M. (2015) Chapter VIII: Vulnerability,dispersion and fragmentation in the San Luis Potosí Metropolitan Area. The case of a road called Rio. In Moreno Mata, A. (Cord.), Urban environment, sustainability and territory in Mexican cities (Vol. II, pp. 139–155, Mexico: Autonomous University of San Luis Potosí.

Moreno Mata, A., & Villasís Keever, R. (2015). Urban fragmentation and segregation socio-residential in the cities of the center region - western México. The cases of Aguascalientes, Leon, San Luis Potosi and Zacatecas. In À. Layuno & A. Acosta (Eds.), *Heritage, territory and landscape*. Alcalá de Henares: University of Alcalá de Henares/Autonomous University of Aguascalientes.

MPUDVDR. (2015). *Municipal plan of urban development of Villa de Reyes, 2015–2035, Municipality of Villa de Reyes/State Government of San Luis Potosí*. San Luis Potosí: Official Review.

MSGS. (2010). Municipal Atlas of risk. Municipality of Soledad de Graciano Sanchez, S.L.P.

Nacopo (National Council of Population) (2010). Delimitation of metropolitan areas of Mexico, 2010. Mexico: National Council of Population.

NCW (National Commission of Water). (2002). *Determination of the availability of water in the aquifer San Luis Potosi, state of San Luis Potosi*. Mexico: National Commission of Water.

Newman P., Beatley T., & Boyer H. (2009). *Resilient cities: Responding to Peak oil and climate change*. Island Press, ISBN: 1597268631.
Papa, R., Galderisi, A., Vigo Majello, M. C., & Saretta, E. (2015). Smart and resilient cities. A systemic approach for developing cross-sectoral strategies in the face of climate change. *TeMA, Journal of Land Use, Mobility and Envoronment, Dossier: Cities, Energy and Climate Change, 1*, 19–50.
Peña, F. (2005). The supply of water to the city of San Luis Potosí. In D. Barkin (Coord.), *The supply of urban water in Mexico*. Guadalajara: Universidad de Guadalajara.
Peña, F. (2008). Protection of the aquifer and urban growth in San Luis Potosí. *Bulletin of the File Historical Water, 2008*, 21–32.
Prigogine, I. Y., & Stengers, I. (1994). *Between time and eternity*. Madrid: Alianza.
Rockefeller Foundation. (2015). Available at: http://www.100resilientcities.org/cities#/-_/
Rueda, S. (2005). *A new urban planning for a more sustainable city*. Presented at meeting of networks of sustainable development and combat climate change, Vitoria-Gasteiz, Spain.
SADSN. (2014). *Urban flood and climate change*. Buenos Aires: Recommendations for the management, Secretariat of Environment and Sustainable Development for the Nation.
Salingaros, N. (2005). *Principles of urban structure*. Amsterdam: Design Science Planning.
Sanders, T. I. (2003). *What is complexity? Catalogue essay for fine arts exhibit entitled, Complexity* (pp. 1–2). Washington, DC: Board of Governors of the Federal Reserve Board.
Santacruz De León, G. (2008). Generation and treatment of waste water in the metropolitan area of the city of San Luis Potosí. *Bulletin of the File Historical Water, 2008*, 33–37.
Serre, D., Lhomme, S., Peyras, L., Laganier, R., & Diab, Y. (2012). Analyzing the civil engineering infrastructures to prioritize urban flood resilient actions. In *7th International Conference on Water Sensitive Urban Design*, Melbourne, Australia, 21–23 February 2012.
Serre, D., Barroca, B., & Laganier, R. (2013). *Resilience and urban risk management*. London: CRC Press, Taylor & Francis Group.
Serre, D., Barroca, B., Balsells, M., & Becue, V. (2016). Contributing to urban resilience to floods with neighbourhood design: The case of am Sandtorkai/Dalmannkai in Hamburg. *Journal of Flood Risk Management, Browse Early View Articles, Online Version of Record published before inclusion in an issue*. https://doi.org/10.1111/jfr3.12253.
Shelfaut, K., Pannemans, B., Craats, I. V. D., Krywkow, J., Mysiak, J., & Cools, J. (2011). Bringing flood resilience into practice: The FREEMAN project. *Environmental Science Policy*. 2011, *14*(7), 825–833.
Soja, E. (2008). *Postmetropolis: Critical studies of cities and region*. Buenos Aires: Dreams Dealer Editions.
Stevens Vázquez, S. (2012). The water vulnerability in the city of San Luis Potosí. *Journal of the College of San Luis, New Era, 11*(4), 130–159.
SWC (State Water Commission). (2005). *Annual report 2005*. San Luis Potosí: Body of information of the State Water Commission.
Toubin, M., Laganier, R., Diab, Y., & Serre, D. (2014). Improving the conditions for urban resilience through collaborative learning of Parisian urbanservices. *Journal of Urban Plannning Development, 141*(4).
Tucci, C. E. M., & Bertoni, J. C. (2006). *Urban flood management*. Washington, DC: World Meteorological Organization.
UNISDR. (2012). How to make cities more resilient – A handbook for local government leaders. Available at: http://www.unisdr.org/we/inform/publications/26462
Vale, L., & Campanella, T. H. (2005). *The resilient city: How modern cities recover from disaster*. New York: Oxford University Press.
Vergara, R. A., & Zureck Varela, E. (2013). *Model of sustainable urban management*. Barranquilla: Universidad del Norte.
World Bank Group. (2011). *Guide to climate change adaptation in cities*. Available at: http://siteresources.worldbank.org/INTURBANDEVELOPM-ENT/Resources/3363871318995974398/GuideClimChangeAdaptCities.pdf

# Chapter 12
# Urban Resilience and Landslide Risk Management: The Case of Santos (Brazil)

**Karolyne Ferreira and Alex Abiko**

**Abstract** The concept of Urban Resilience has entered discussions on how cities can prepare for, adapt to, and cope with acute natural disasters. Urban resilience is understood as a process that involves enhancing capabilities for learning about changing conditions and for adapting to such conditions in order to reduce the risk of disasters, restore essential urban services and improve quality of life. In the past, landslides were the major cause of deaths from natural disasters in Santos, a large port city on the coast of São Paulo state, Brazil. The number of fatalities however fell significantly with the introduction of the Civil Defence Prevention Plan (PPDC) which has been in annual operation since 1989. The aim of this paper was to identify the landslide risk management tools employed by the Santos Civil Defence Department (responsible for disaster management), and to analyse the role of these tools in promoting urban resilience. Field, documentary and qualitative research (including literature reviews) revealed the Department's use of the following instruments: weather monitoring; a map to chart susceptibility to gravitational mass movements and floods; a Risk Reduction Plan; and a geotechnical map to illustrate aptitude for urbanization. The paper concludes that promotion of urban resilience by the Santos Civil Defence has been only partially achieved to date. The above-mentioned tools are still being implemented only gradually and challenges remain, such as the need to improve coordination between local government departments in favour of risk reduction.

**Keywords** Landslides · Civil defence · Risk management · Disaster risk reduction · Brazil

K. Ferreira (✉) · A. Abiko
University of Sao Paulo, Sao Paulo, Brazil
e-mail: karolyneaf@gmail.com; alex.abiko@poli.usp.br

 
G. Brunetta et al. (eds.), *Urban Resilience for Risk and Adaptation Governance*, Resilient Cities, https://doi.org/10.1007/978-3-319-76944-8_12

## Case Study Fiche/Resume

| | |
|---|---|
| **Image or graphic** |  |
| **Location** | [Region: South America] [Brazil}{Santos] |
| **Duration** | [total duration of the project: 39 months] |
| **Time Frame** | [Start date: 03/2013] [End date: 06/2016] |
| **Consortium** | None |
| **Cost** | – |
| **Resources** | Civil Defence and Protection agents, computers, camera, voice recorder, car. |
| **People affected** | [35.000] |
| **Donors** | None |
| **Urban Sectors** | Building, housing, sanitation, planning, management. |
| **Resilience dimension** | Adaptation, mitigation, preparedness, learning, disaster risk reduction, vulnerability, bounce forward. |
| **Strengths** | Dialogue with the local community, regular training for adults and children, uninterrupted operation of the landslides prevention plan since 1989. |
| **Weaknesses** | Only partial implementation of risk reduction plan; poor coordination between municipal departments for promoting risk reduction; need to improve the Irregular Settlements and Substandard Housing Control Group. |
| **Impact** | Urban resilience in terms of landslide management. |
| **Stakeholders List** | None |

| | |
|---|---|
| **Challenge** | In the past, landslides were the major cause of deaths from natural disasters in Santos, a large port city on the coast of São Paulo state. The number of fatalities however fell significantly after the implementation of the Civil Defence Prevention Plan (PPDC), operated annually since 1989. The aim of this research study was to identify the landslide risk management tools employed by the Santos Civil Defence Department (responsible for disaster management), and to analyse their role in terms of improving urban resilience. |
| **Project description** | Santos, Latin America's largest port, occupies a strategic position in the State of São Paulo. In view of its severe housing shortage, settlements inhabited by the low-income population proliferate on the city's hills. The tragic history of landslides has mobilized the State and local government in favour of risk reduction. |
| | The concept of urban resilience reflects management and planning practices designed to reduce the risk of natural disasters. This study explored the concept in terms of natural disaster (especially landslides) risk management from an institutional point of view. We chose to concentrate on the Civil Defence (Defesa Civil) which has a long history of coping with disasters in Brazil, including landslides, particularly at the local level where the municipal authorities are responsible for disaster risk management. |
| | Our qualitative research study involved unstructured data collection including: (i) documentary research - laws, decrees and other materials related to civil protection and natural disaster risk management; and (ii) field research, consisting of interviews, visits to - and an internship at - the Civil Defence headquarters in Santos, and participation in 2013–15 in official government events related to civil defence and protection. Data analysis and interpretation focused on the interface between urban resilience, natural disasters, vulnerability and adaptive capacity. |
| | The purpose of risk management tools is to support disaster prevention, preparedness and mitigation measures in the areas of urban planning and management, land tenure regularization and redevelopment, and risk reduction works. An overarching goal is of course to save lives. The various tools, introduced in response to having to cope with highly precarious situations or major tragedies, form part of a learning process. But the tools alone cannot boost urban resilience. Local government managers need to better coordinate efforts in order to address risk reduction issues more convincingly. |
| **Urban resilience elements** | Laws, tools and maps for promoting preventive measures in areas such as land management, land tenure regularization, urban redevelopment and training (adults and children). |
| | Public awareness-raising dialogue between Civil Defence and the population living in risk areas. |
| | Cooperation between Federal, State and Municipal governments through monitoring, early warning systems and risk reduction centres. |
| **List of innovation elements.** | The National Civil Protection and Defence Policy, instituted in 2012, highlights risk reduction and the development of resilient cities as sectorial policy and land management priorities. |
| | The Irregular Settlements and Substandard Housing Control Group, designed to prevent new settlements being installed in previously-mapped risk areas, is an example of joint action in the areas of security, public services, urban development, social welfare, the environment and housing. |
| | The Landslide Prevention Monitoring Project: sensors installed on the hills can detect evidence of minor ground movements and send out early warning signals. |

| | |
|---|---|
| **Lessons learned** | Disaster risk management is a responsibility of government and society as a whole. Technological solutions and raising public awareness are keys to implementing effective risk reduction measures. Discussions on urban resilience often focus on the concept of vulnerability, given that disasters are not confined to natural phenomena but can also involve physical and social aspects and socio-economic development dynamics within a particular sample of territory. Fostering urban resilience thus calls for well-structured institutions with coordination capacities and good legal frameworks, and which have earned public confidence in their ability to guide the planning and implementation of adaptation measures. |
| **Upscaling and replication** | In view of the political, administrative and financial autonomy enjoyed by Brazil's local (municipal) governments, this research was confined to a one-off case study. For a broader view we suggest that future research might involve conducting a comparative case study focused on two or more municipalities. A further approach to evaluate progress in urban resilience would be to develop landslide assessment metrics or indicators. |
| **Web page** | – |
| **Reference** | – |

## 12.1 Introduction

The climate of south-eastern Brazil, marked by heavy rainfall in summer, together with the existence of substantial mountain ranges such as the Serra do Mar, makes the area highly susceptible to landslides. It is now widely accepted that the social and economic damage caused by landslides, despite their being of natural origin, can be partly blamed on human activities (Fernandes and Amaral 2000; Guimarães et al. 2008).

The areas most vulnerable to landslides are predominantly occupied by low-income groups. These settlements consist of low-value, poorly constructed properties, lacking basic infrastructure and adequate public services (IPT 1979; Nogueira 2002).

It is not difficult to see that, given the highly precarious human occupation of the steep slopes, disasters are not exclusively "natural" and that management/planning practices can in fact increase or decrease disaster risks (UNISDR 2012). The concept of urban resilience takes the latter into account.

It is widely believed that customary political culture involves largely reacting to hazardous events after they have taken place. This makes it more difficult to learn from errors and to seize opportunities for seeking and introducing risk reduction alternatives.

From an institutional point of view the aim of this research was to identify and evaluate the performance of landslide risk reduction legal instruments used by the Civil Defence authorities to foster urban resilience.

We chose the city of Santos on the coast of São Paulo State (Map 12.1). Santos is included in the national register compiled by the National Civil Defence Secretariat

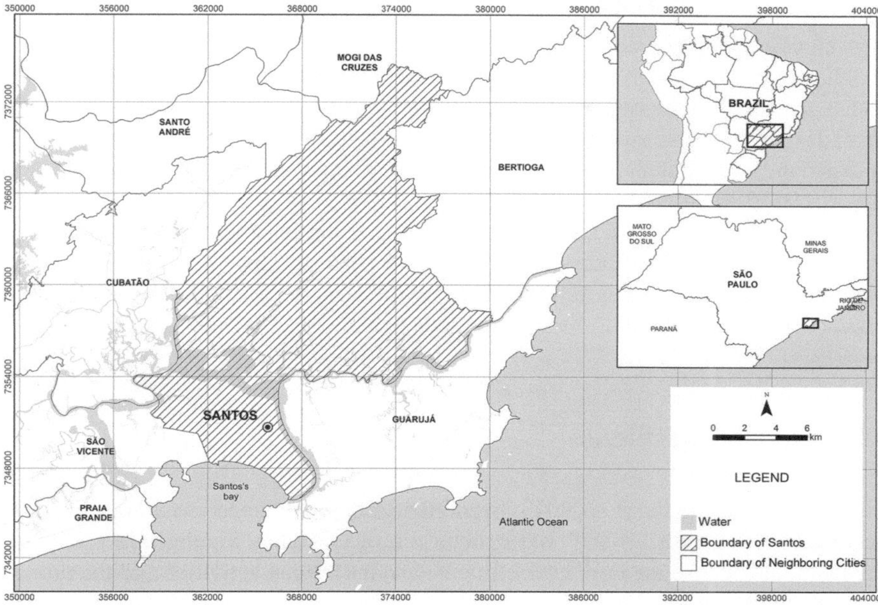


**Map 12.1** Location of Santos (São Paulo state). (Prepared by André Luiz Ferreira (2016))

(SEDEC) of municipalities that are most susceptible to natural disasters, based on "number of fatalities", "frequency of major destructive events" and "people directly involved or affected."

This register lists various events that can be classified as the possible causes of disasters in the municipalities such as erosion, heavy rainfall, storms, flooding, landslides, etc., all of which call for the continuous scaling up and improvement of landslide risk management by the local authority.

As part of our qualitative research according to which "a phenomenon can be best understood in the context in which it occurs and of which it forms part, and which must be comprehensively analysed" (Godoy 1995, p.21), data collection included a documentary survey (laws, decrees and other material related to Civil Defence, natural disaster risk management, etc.) and field research (fieldwork, interviews and participatory observation).

The analysis was based on the premise that urban resilience is a process that includes learning and adaptation aimed at reducing the risks arising from natural disasters, supporting a city's return to normality and ensuring better quality of life for the population (Cutter et al. 2008; Silva 2014; UNISDR 2009).

Natural disasters result from the interaction of the Earth's natural processes with the dynamics of economic and social development (Freitas et al. 2012; Lavell 1998, 2000). This dynamic is related to the vulnerability distributed unequally throughout society. It is also related to physical (geological, geomorphological, etc.) and social (social, economic, cultural and political) variables, all of which are important issues to be considered in the realm of applied sciences (Adger 2006; Gallopín 2006; Hufschmidt and Glade 2010; Susman et al. 1983).

The concept of vulnerability goes together with that of resilience (Miller et al. 2010), especially with regard to the adaptive capacity of individuals or groups that plan, design, facilitate and deploy adaptation measures (Klein et al. 2003), against a background of heightened risks of disasters occurring as a result of climate change (IPCC 2007; PBMC 2014).

Our study also took into account that fostering urban resilience primarily calls for a well-structured institutional and administrative structure with excellent coordination capacity and a good legal framework as prerequisites for undertaking disaster risk reduction initiatives (Malalgoda et al. 2013; UNISDR 2012).

## 12.2 Urban Resilience

Malalgoda (Malalgoda et al. 2013) argue that urban resilience to natural disasters needs to bring together social and structural projects. They highlight in particular the fact that because the built environment sustains human activities and the proper functioning of society, local governments need to commit themselves to mitigating disasters by focusing on urban planning, conforming to building codes and providing support to the civil construction industry.

The United Nations International Strategy for Disaster Risk Reduction - UNISDR posits the view that a city that is resilient to natural hazards can be built providing that a committed and inclusive local government is determined to minimize the effects of a disaster by encouraging housing to be built in safe areas, by informing and educating its public about natural hazards, by placing value on local knowledge, by anticipating and mitigating hazards affecting infrastructure, housing, environmental and cultural heritage sites through monitoring and early warning systems, and by defining post-disaster reconstruction strategies and the reestablishment of basic services (UNISDR 2012).

Urban resilience to natural disasters gained strength as a guided process following the 2004 Indian Ocean tsunami and the 2005 World Conference on Disaster Reduction - WCDR in Kobe, Hyogo, Japan, which raised the question of building resilient communities and societies (Cutter et al. 2008; Manyena 2006; Manyena et al. 2011 Sulaiman 2014.). Incorporating the concept of resilience into legal instruments was considered to be a way of re-discussing and highlighting disaster risks reduction by means of participatory processes and respect for the knowledge held by traditional populations (UNISDR 2012; Khailani and Perera 2013; Saavedra and Budd 2009).

Examples of this can be found in Brazil. The National Civil Defence Protection Policy (PNPDEC) - Law n°12.608/2012 - listed the development of resilient cities as one of its objectives, although it failed to clarify what was meant by this. The City of Santos Master Plan (Supplementary Law 821/2013) refers in Chapter IV to a strategy for building a resilient society, embracing a series of legal measures and public sector and society initiatives based on technical studies and coordinated by the local government civil defence and protection agency.

The campaign Building Resilient Cities: My City is Getting Ready (from 2010 to 2015), promoted by the UNISDR in partnership with Brazil's National Secretariat for Protection and Civil Defence (SEDEC), was an awareness-raising initiative to encourage local municipal governments to incorporate disaster risk reduction guidelines in their planning and land management activities and to strengthen municipal protection and civil defence agencies.

A prerequisite for boosting resilience is to have a well-structured institutional and administrative framework with good coordination capacity and a legal framework for taking disaster risk reduction initiatives (Malalgoda et al. 2013; UNISDR 2012). Suassuna (2014) put forward a proposal for institutional urban resilience indicators for dealing with floods: governance capacity and investment in disaster management and reduction, organizational capacities to reduce and manage hazards, capacity for understanding risks, and, finally, territorial planning capacity. In addition, according to this author, institutional factors such as the need to gain the confidence and engagement of the population in the institutions are seen as fundamental for ensuring the resilience of cities.

In the planning and management areas, resilience has been associated with issues concerned with the reduction in mitigation of natural disasters (Pizzo 2014), given that it is a concept viewed positively and appropriately by large sectors of academia (Meerow et al. 2016). The concept has also become widely accepted in the political arena mainly because it does not raise issues requiring radical transformation or change (Leichenko 2011; Pizzo 2014).

Clarifying and delimiting prospective research involving the concept of the urban resilience is a way of avoiding misunderstandings while providing opportunities for genuine contributions to be made to the debate on natural disasters. In this way questions such as 'resilience for whom, why, what, when, where?' need to be seriously broached (Carpenter et al. 2001; Leichenko 2011; Meerow et al. 2016; Pizzo 2014).

Research on urban resilience and natural disasters has been conducted on the following: (a) cities where disasters have primarily affected the population (Ainuddin and Routray 2012; Carpenter 2015; Joerin et al. 2012; Tang et al. 2015.); (b) the impact of climate change (Brown et al. 2012; Jabareen 2013; Khailani and Perera 2013; Klein et al. 2003); and (c) specific natural disasters such as floods (Gupta 2007; Lhomme et al. 2013; Liao 2012; Suassuna 2014; Sudmeier et al. 2013).

During the revision of the literature focused on proposals we unearthed no studies to relate urban resilience to landslide risk from the institutional standpoint. It is hoped therefore that the present study will contribute to filling this particular gap.

Based on definitions by Silva (2014), UNISDR (2009) and Cutter et al. (2008), this study considers urban resilience to be a process that involves learning and adaptation capacities for reducing natural disaster risks, recovering normal functioning, and improving people's quality of life.

As for the questions raised in the literature, we sought to confine our responses to the following:

- For whom and where? For people who live, work or frequent areas that are knowingly susceptible to natural hazards (according to historical records, maps, etc.);
- To what? Natural disasters and especially landslides;
- When? Before, during and after a disaster;
- Why? Because the concept encourages integrated thinking. While this research study deals with the political/institutional dimension, disasters can of course affect infrastructure and people's social and economic lives.

### *12.2.1 Vulnerability, Adaptation and Climate Change*

Resilience refers primarily to recovery. It can however also be thought of as a "bounce forward" (Manyena et al. 2011; Revi et al. 2014.). Adopting structural and non-structural measures to reduce the impact of a disaster can contribute to a more efficient and less painful recovery for those affected. Vulnerability, climate change and adaptability are all part of the debate on the urban resilience process.

The concept of vulnerability in general is related to the potential losses or damage caused to an individual or group (Cutter 1996), expressed in terms of structural damage or human, economic, cultural and environmental losses (Crozier and Glade 2010). The concept is applicable to different areas, and as result there is no consensus on the different definitions, methodologies and recommendations (Cutter 1996; Gallopín 2006; Mcentire 2005; Miller et al. 2010).

The vulnerability debate is rooted in two main themes. The core idea of the human ecology approach is that a disaster occurs as the result of man-made decisions to occupy risk areas (Burton et al. 1978; Hewitt 1980). The human adjustment approach considers that by adapting to natural hazards the negative effects of a disaster could be reduced by monitoring, structural works and land use planning and occupation (Hufschmidt 2011).

Vulnerability is the degree to which different social classes are clearly at risk, both in terms of the probable occurrence of an extreme event, and from the coping and resilience angles (Susman et al. 1983).

The literature contains other approaches related to vulnerability to natural disasters. Examples are: (a) exposure to risk; (b) social status and response capacity; and (c) a combination of these two approaches (Adger 2006; Cutter 1996; Gallopín 2006; Sherbinin et al. 2007). The third approach is a combination of physical and social aspects. This is regarded as the most appropriate approach by the applied sciences (Adger 2006; Gallopín 2006; Hufschmidt and Glade 2010).

The temporal and territorial sample selected is an important question since it may highlight certain elements while simultaneously camouflaging others, as well as relating global and regional phenomena to local impacts. Both cases call for careful consideration and clear delimitation (Cutter 1996; Cutter et al. 2008; Sherbinin et al. 2007).

Folke et al. (2002) affirmed that vulnerability was the opposite or the antonym of resilience (although they failed to clarify this appropriately). A resilient system is less vulnerable than a non-resilient system, although this relationship is not necessarily symmetrical (Gallopín 2006). Cutter et al. (2008) accepted resilience as a broader concept containing the element of vulnerability typified as the situation existing prior to a given disaster. Hufschmidt (2011) considered adaptation as the core element of resilience as well as a crucial way of anticipating and reducing vulnerability.

Adger (2006) stressed how vulnerability and resilience converge as a result of pressures suffered by the socio-ecological system (SES) and how the system adapts and responds to such pressures. In the opinion of Miller et al. (2010), understanding the political and social processes, the costs, risks and benefits distribution of environmental changes espoused in the vulnerability concept is also of substantial importance for understanding resilience.

Adaptability, the main component of resilience, was defined by Carpenter et al. (2001) as an element following exposure to post-disaster system behaviour. Meanwhile, Klein, Nicholls and Thomalla (Klein et al. 2003) considered adaptability as the ability to plan, prepare, facilitate and deploy adaptation measures to confront natural hazards intensified by climate change. Folke (2006) defined the adaptation process as the ability to tolerate and cope with external system changes.

Lorenz (2010) concurred with Walker et al. (2004) that adaptability is primarily the capacity of the social component - individuals and groups involved in management of the system - to influence resilience intentionally and unintentionally. Gallopín (2006) defined adaptability as emanating from human activity (individual or humankind) aimed at maintaining or increasing quality of life in a specific environment or a range of different environments.

Smithers and Smit (1997) pointed out that people possess an ability to plan and manage adaptation measures, since social groups are able to successfully perceive environmental risk and evaluation of such risks as vital elements germane to adaptation strategies.

It is worth noting that the purpose of adaptation could be understood in terms of maintaining existing socioeconomic imbalances. It could also be argued that adaptation is conservatively biased since the strategies involved do not necessarily get to the root of problems such as land tenure regularization, social inequality, corruption and governance (Pizzo 2014; Sudmeier-Rieux 2014).

The existence of political will can however favour adaptation measures. Adger (2006) for example argues that if resources and goodwill exist, adaptation capacity can be used to increase resilience and reduce the vulnerability of marginalized sectors of society.

Adaptation, or the capacity for adaptation, are clearly associated (although indirectly) with the concepts of resilience and vulnerability, and also with the subject of climate change.

Climate change is a global concern, but its effects can be experienced locally, such as floods and heat waves, particularly in the urban centres as a result of carbon emissions, and heavy concentrations of buildings, economic activities and people (Romero-Lankao and Dodman 2011).

According to the 2014 Intergovernmental Panel on Climate Change report on urban areas, the highest number of deaths caused by natural hazards related to extreme weather events specifically affected low and medium income population groups. It follows that adaptation initiatives need to be systemic and contribute to the well-being and security of such groups, as well as ensuring that basic services are provided for them.

The frequency of extreme climatic events is a major concern, and the difficulty of forecasting such events and estimating the magnitude of their impact in different areas of the globe often means that the poorer and developing countries suffer the greatest damage (IPCC 2014; Mechler and Bouwer 2014; Thomalla et al. 2006).

This study concurs with Adger (2006), Gallopín (2006), Hufschmidt and Glade (2010) and Susma, O'Keefe and Wisner (Susman et al. 1983) that vulnerability is not evenly distributed in societies and possesses inherent physical (i.e. geological, geomorphological, etc.) and social (social, economic, cultural and political) characteristics that call for serious consideration by the applied sciences. This approach dovetails with the concept of resilience (Miller et al. 2010), especially in terms of the ability to adapt against a background of increased disaster risk resulting from climate change (IPCC 2007; PBMC 2014).

## 12.3 Natural Disaster Management in Brazil

Natural disaster risk management is a multidisciplinary activity involving a range of public sector and social practitioners operating in different areas with responsibilities for generating prevention, mitigation, preparedness, and response measures, as well as for undertaking reconstruction work. All these functions follow a cyclic sequence (Cardona 1996; Faria and Santoro 2011; Kobiyama et al. 2006; Sapountzaki et al. 2011).

Law No. 12,608, of 10 April 2012 (BRASIL 2012), which established Brazil's National Protection and Civil Defence Policy (PNPDEC), introduced a series of new approaches and guidelines for disaster risk management throughout Brazil. The PNPDEC created guidelines and commitments for the National Protection and Civil Defence System (SINPDEC) and the National Protection and Civil Defence Council (CONPDEC) and authorized the creation of an information and disaster monitoring system. These new arrangements gave new meaning to natural disaster and risk management, which henceforth would be applied to prevention, mitigation, preparation, response and recovery actions, in addition to being integrated into inter alia land management and sectorial policy planning for health, infrastructure, education, climate change and the environment.

The new Law also focused on the adoption of preventive measures to minimize the impacts of disasters, encouraged the development of resilient cities, and established a national public database containing information about hazardous occurrences. The Law also outlined the kind of responses to be jointly taken by the federal, state and municipal governments (including the Federal District) to reduce disasters and assist communities affected by them.

By prioritizing disaster risk reduction activities in land use and other sectorial policies, and acknowledging the multifarious nature of disasters, the PNPDEC signalled the beginning of a transition towards a more systemic approach to actions that needed to be taken (Almeida 2015; MI and SEDEC 2014; Nogueira et al. 2014). This approach focuses on a number of challenging areas such as resource planning, professional training and upgrading for government agents involved in civil defence and protection, as well as reorganizing the three levels of government to work in a more coordinated manner in disaster prevention and relief. This is particularly important at the municipal level where most disasters occur.

The main remit of the Municipal Civil Defence Agencies is to: (a) understand and identify risks; (b) prepare on a permanent basis to cope with disasters; (c) focus on prevention at every level of their operations; (d) provide institutional visibility and continuing education for coping with risks; and (e) ensure integrated operations with other agencies and related bodies.

The recommended PNPDEC instruments for the municipalities include:

- A Master Plan.
- A Map displaying susceptibility to gravitational mass movements and floods;
- A Geotechnical Map displaying fitness for urbanization;
- A Civil Defence and Protection Plan (PPDC);
- A Municipal Risk Reduction Plan (developed in 2003 by the Ministry of Cities).

The Community Civil Defence Centre (NUDEC) is channel for the Municipal Civil Defence Agency to communicate with the population. This is responsible for informing, organizing and preparing the local community about what to do in the event of a disaster, and how to prevent and respond promptly to and minimize the damage caused (Calheiros et al. 2009, p.6).

## 12.4 The Case Study: Prevention and Coping with Risk in Santos

The City of Santos has a land area of 280.67 km$^2$ (IBGE 2015), with 14.5% corresponding to the island and 85.4% to the mainland. The city, with 99.93% built-up area (SEADE 2015), is home to a population of 433,966 (IBGE 2015) concentrated on São Vicente Island. The mainland portion contains a small urbanized area and four environmental conservation units. The principal economic activities are tourism and the Port of Santos, the largest in Latin America.

The hills around Santos were occupied in two distinct phases. The first influx occurred in the second half of the nineteenth century when immigrants from the Atlantic Islands, Madeira and Azores, attracted by the opportunities presented by the coffee-generated economic cycle, ended up being employed in construction work on the port and the São Paulo Railway. These immigrants were experienced in construction techniques on rugged terrain and built their chalet-type homes of timber mounted on stone pillars to suit the contours of the hilly areas.

The second phase, from the twentieth century onwards, involved migrants mainly from the Northeast of Brazil, who were attracted by job opportunities in industry, construction and the port. These built their houses in topographically difficult areas with no knowledge of the type of construction techniques needed for this kind of terrain. The cut and fill used to build homes and open access roads are the root cause of substantial ground instability to this day.

According to the city records (Table 12.1) of Santos, landslides are a recurrent phenomenon on the hills around the city and led to the preparation of the first urban environmental geotechnical map in Brazil.

The Civil Defence Preventive Plan (PPDC) has operated uninterruptedly since 1989 and has drastically reduced the number of deaths associated with landslides, thus demonstrating the importance of developing and applying risk reduction tools. Since 1989 only four deaths have been recorded from landslides, the last being in year 2000.

The instruments analysed (Table 12.2) are in line with PNPDEC and have common goals: to support prevention, preparedness and mitigation in the areas of planning and urban management, land regularization, urban development, risk reduction works and also to save lives.

Despite the susceptibility to landslides in the hills (Map 12.2), the Geotechnical Map did not classify the hills as unsuitable for human occupation providing (a) basic sanitation infrastructure was correctly installed, (b) areas with original vegetation (primary and/or secondary) were preserved, and (c) housing was built in less geotechnically problematic areas.

The Santos Community Civil Defence Centre (NUDEC) promotes a range of activities for adults and children. According to interviews conducted during our study, adult training consists of three courses: firefighting, first aid, and notions of civil defence.

In the event of an emergency, trained NUDEC volunteers work together with civil defence and protection agents. It is clearly accepted that the idea of joining the NUDEC is to empower local citizens to assist voluntarily with community safety and risk prevention. In the course of our field visits we identified no evacuation training programs. Around 500 volunteers were trained in the period 2008–2014.

A community civil defence centre normally consists of eight or nine people from the same neighbourhood. A total of 17 centres has been formed in the Santos hills. Training sessions are run throughout the year either at Santos Civil Defence (COMPDEC – Santos) headquarters or in the actual neighbourhoods (particularly in those that are more remote from the HQ). While training sessions were initially

**Table 12.1** Landslides recorded in santos

| Month/Year | Mean rainfall recorded | Outcomes |
|---|---|---|
| 03/1928 | 649.5 mm (January) | 80 fatalities |
| | 564.1 mm (February) | Destruction of part of the Santa Casa de Misericórdia Hospital |
| 03/1956 | 178.7 mm (January) | 21 fatalities |
| | 312.5 mm (February) | Over 40 people injured |
| | 954 mm (March) | Destruction of approx. 50 houses |
| | | Not recorded |
| | | Obstruction of the Anchieta-São Vicente highway |
| | | 20 victims |
| | | Destruction of 7 dwellings |
| | | Destruction of 1 house |
| | | Material damage (not detailed) |
| | | Indeterminate number of fatalities |
| 1959 | Not recorded | Destruction of houses and many fatalities |
| 01 e 03/1978 | 212 mm (January)[a] | 13 fatalities[b] |
| | | 20 injured[b] |
| | | Approx. 500 made homeless[b] |
| | | Obstruction of roads[b] |

Source: IPT 1979;
[a, b]Folha de São Paulo Records 1978.
[a]number recorded on Sunday 15/01/1979, at 21:00 hrs.

open to all citizens, including those not living in risk areas, training has been restricted more recently (no date specified) to residents of the risk areas.

The 'Civil Defence for Schools Project' targets children in the fourth year of elementary school, with lectures, booklets and guided visits to the Fire Department and the Civil Defence headquarters. 11,526 children participated in this project from 2007–2014.

On 14 October 2014 - the United Nations International Day for Natural Disaster Reduction - the Civil Defence Knowledge Olympics, devoted to issues related to droughts, landslides, flooding, storms, lightning, etc., attracted schoolchildren from local authority schools in nine municipalities of the Baixada Santista Metropolitan Region.

In the course of fieldwork for this study three cases of preventive actions were followed up:

- inspection of an area with an imminent risk of landslides;
- an action undertaken by the Irregular Settlements and Substandard Housing Control Group; and
- a visit to the installations of CEMADEN's Robotic Total Station (RTS).

The first case involved an 'imminent risk survey' where a landslide had taken place in an area mapped and classified as 'at risk' in view of its previous history of

**Table 12.2** Legal instruments used in landslide risk management

| Legal Instrument | Description and use |
|---|---|
| The Master Plan | It is a basic instrument of the local government's development and urban expansion policy improving the social functions of the city and the welfare of its inhabitants. |
| Map of Susceptibility to Gravitational Mass Movements and Floods | It is one of the Master Plan's supporting documents drawn on a scale of 1:25,000 suitable for urban planning. |
| Geotechnical Map of the Santos and São Vicente Hills | The first of its kind to be prepared for urban areas in Brazil, is a landmark document. |
| | The map has several functions: |
| | relates the occurrence of landslides to specific hillside characteristics and positive slope breaks; |
| | relates rainfall to landslides; and |
| | provides a clearer understanding of the behaviour of the Serra do Mar Mountains. |
| | The process of land tenure regularization is based upon this Map. |
| The Civil Defence Preventive Plan (PPDC) | It is based on three elements: |
| | weather forecasting; |
| | rainfall; and |
| | field surveys of risk areas. |
| | Prior to officially undertaking the field surveys, Civil Defence agents visit the risk areas detailed in the PMRR in order to deliver notices to householders containing instructions on danger signs and how to proceed. |
| | Residents are required to sign a receipt for the notices and provide a telephone number to enable the Civil Defence authorities to send them early warnings and make other announcements. |
| Municipal Risk Reduction Plan (PMRR) | It is a planning tool aimed at diagnosing risk areas and proposing structural measures to reduce such risks. |
| | The plan also contains cost estimates and prioritization and financial support criteria. |
| | The first PMRR prepared in 2005 identified 19 risk areas in the hills. |
| | The updated PMRR prepared in 2012 identified 22 risk areas, divided into 104 risk sectors: |
| | 22 sectors with R4 rating - Very High; |
| | 44 sectors with R3 rating - High; |
| | 37 sectors with R2 rating - Medium; and |
| | 01 sector with R1 rating - Low. |
| | The proposed risk reduction measures related to: |
| | removing trash and rubble; |
| | clearing inappropriate or unstable vegetation; |
| | carrying out structural works; and |
| | relocating houses. |

Source: Alheiros 2006; Braga 2001; IPT 2014a, b; Sobreira and Souza 2012; IPT 1979; IPT 2012; Santos 2004; Canil 2010; PREFEITURA MUNICIPAL DE SANTOS 2013

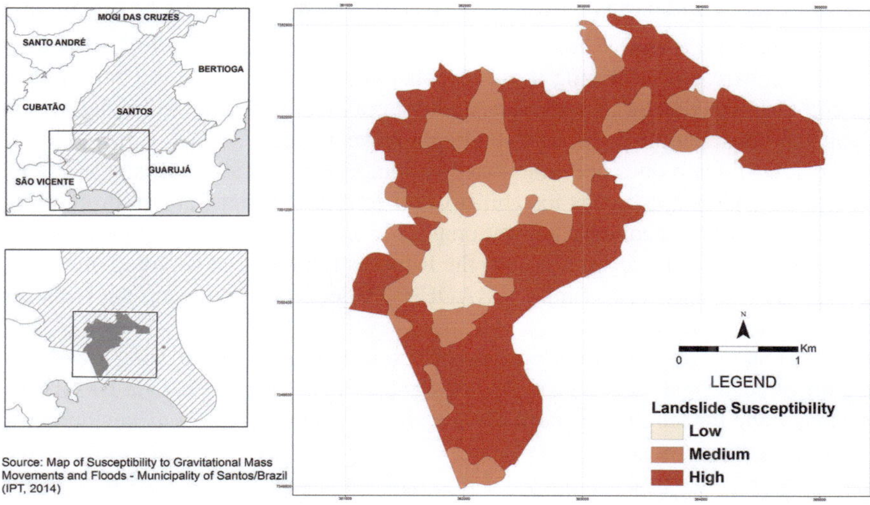


**Map 12.2** Landslide susceptibly map of Santos Hills (Prepared by Karolyne Ferreira (2016))

landslides. This event resulted in the destruction of a housing block but left no victims. The first survey conducted by geologists assessed the conditions of the terrain and the material (rocks, earth and trees) involved. The initial visit to the place where the family had lived was carried out by the Civil Defence and Protection agents who explained the situation to the previous occupants in an amicable manner. The family had been fully aware of the risks. The Civil Defence Agency requested the Municipal Social Assistance Department to relocate this family.

However, in view of the negative response and the unlikely chance of the family obtaining emergency rent assistance, it was suggested that the family should be housed in the municipal communal shelter. The family refused and went to live with friends in a house situated in a flood risk area. According to the social worker involved, this kind of situation is commonplace: families refuse to go to the municipal shelter because of the lack of privacy there.

The Civil Defence and protection agents expressed dissatisfaction with the social worker on account of the reduction in rent assistance (for financial or policy-related reasons) for residents of risk areas.

The second case involved a survey undertaken with the Irregular Settlements and Substandard Housing Control Group. This group is responsible for promoting and facilitating intersetorial cooperation between the municipal, state and federal agencies and civil society. While the Civil Defence is responsible for carrying out surveys to detect imminent risk, this group undertakes regular activities throughout the year aimed at preventing the installation of irregular settlements.

The Group, coordinated by the Civil Defence, consists of 13 employees who deal with safety, public services, urban development, social assistance, housing and the environment.

During the field visits we were able to verify the presence of half-constructed houses. Once the irregular constructions have been identified, the following step normally consists of informing those responsible to stop work on the building - a difficult task, since inspections are carried out only from Monday to Friday, which means that most people continue to build at the weekends.

The control team does not undertake the demolition of brick-built constructions since that could lead to further instability of the terrain. The main aim of prohibiting building is to allow secondary vegetation to grow and cover the unoccupied spaces.

The third survey involved a visit to the Robotic Total Station installed on one of the hills and forming part of the CEMADEN 'Project for Monitoring the Hills for the Prevention of Landslides'.

The RTSs are devices fitted with sensors installed at a central point within an area surrounded by slopes and which emit signals which are reflected via 100 prisms installed along the length of a particular monitored slope that can detect evidence of even the smallest movements.

CEMADEN has also installed 11 automatic rain gauges in Santos to monitor rainfall and issue early warning signals. This type of initiative shows that some of the SINPDEC coordinated risk reduction activities are being put into practice.

## 12.5 Discussion and Conclusions: Building Urban Resilience in Santos

Given that resilience is being fostered by the public administration, it is obvious that institutions and instruments are required for regulating the entire process. The institution mentioned in this study COMPDEC - Santos in general undertakes activities at all stages of the civil defence management cycle. Its Coordinating Unit has identified and mapped the risk areas, developed training activities for preparing for and coping with risks through the NUDEC and PPDC, carries out prevention (liaising with other agencies) and enjoys a good image among the population.

During the delivery of the PPDC notices the agents made every attempt to avoid visiting the hills where police activities were being undertaken, so the residents were not given the mistaken impression that the two public institutions were acting in collusion. Furthermore, they avoided putting flashing lights on the roofs of their official vehicles to avoid being confused with the police. It is clear that this careful approach, together with the polite treatment of residents, helps to sustain the good image that COMPDEC - Santos has with the residents of risk areas, even including notoriously violent areas. Trust in institutions is essential for building resilience (Suassuna 2014).

COMPDEC - Santos has played an effective prevention role, as can be seen from the reduction of fatalities caused by landslides, but its role needs to be expanded.

We understand that the NUDEC training restricted to residents of risk areas is a priority for training people who live with risk on a day-to-day basis. However, this places constraints on building a 'disaster prevention culture' and developing further awareness of disaster risk among the entire population, as provided for in the PNPDEC.

The Irregular Settlements and Subnormal Housing Control Group is not staffed by a large team. The group, tasked with providing a service for the entire city, consists of a total of only 13 members. It is obvious that this can undermine the capillarity of the Group's actions in the risk areas of Santos regardless of the Group's willingness to fulfil its mission correctly.

The various interviews revealed that the 2012 PMRR had not been implemented because the works suggested in the 2005 PMRR needed completing. The bidding processes for works to be carried out in the hills faces resistance by some local government sectors, although certain emergency works have in fact been approved and undertaken.

The agents argued that budgetary restrictions arose from the fact that the public maintenance services did not serve equally the different areas of the city. The agents claimed that maintenance priority was given to the outer rim of the hills or to the north-western area of the city that suffers from flooding, as if these areas were not Santos.

The various instruments introduced in response to having to cope with highly precarious situations or major tragedies form part of a learning process which is one of the aspects of urban resilience. Although well formulated, except for the PPDC they have not been used or implemented fully for adapting to and minimizing risks. In the course of our research survey it became obvious that risk management is not restricted to saving lives but also involves improving quality of life.

As for COMPDEC - Santos's areas of responsibility, we found that resilience promotion is partially carried out by this body, based on the requirement to "demonstrate efficient intersetorial coordination, involving the public and private spheres, in order to reduce or mitigate the effects of frequent threatening factors and, in essence, to promote appropriate interaction with civil society by encouraging resilience processes in people's daily lives" (Valencio and Valencio 2011, p.148).

Urban resilience to natural disasters, specifically to landslides, is only one of the many ways of approaching a concept that reflects the overall complexity of an urban system. This approach makes it possible to envisage landslides as a vehicle for encouraging joint efforts to minimize their effects. The occurrence of a disaster, and the damage and losses arising from it, reveal how a particular society is able to organize itself. It also demonstrates the potential and limitations of a society when faced by such disasters.

Our study examined urban resilience from the standpoint of a process, and not as an occasional or static activity. The process involves learning and adaptation skills aimed at reducing the risk of natural disasters, for recovering a city's necessary functionalities and improving people's quality of life. The study focused on urban resilience promoted by public managers through legally-supported institutions committed to undertaking risk reduction activities in conjunction with society itself.

With the establishment of the National Protection and Civil Defence Policy (PNPDEC), risk reduction and the minimization of the impact of disasters was acknowledged in law as a priority in the context of sectorial policies, land management, and the development of resilient cities.

In view of the importance of the port and its related activities, the City of Santos occupies a strategic position in the state of São Paulo. Tourism and associated activities also provide essential government income. The hills of Santos are vital for the city from an economic standpoint since they contain quarries and other areas for extracting building materials. They also provide low-income groups with housing alternatives in view of the current low supply.

Human occupation of the hillsides has however generated a natural phenomenon in the dynamics of the hillsides: landslides. The severity of such occurrences in a city as strategic as Santos led to the introduction of Brazil's first Geotechnical Map on the urban environment. This was followed up with regular Civil Defence efforts that have preserved countless lives.

The Santos Civil Defence continues to undertake landslide risk management activities. The promotion of urban resilience calls for well-structured institutions with good coordination capacity and an appropriate legal framework, and which enjoy the confidence of the population to guide the process of planning and implementation of adaptation measures.

Urban resilience in Santos is partly promoted by COMPDEC - Santos. On the one hand there is potential for expanding interface with the population and the local media, for example by conducting regular training sessions for adults in NUDEC, as well as for schoolchildren. The uninterrupted operation of the PPDC since 1989 is also important. On the other hand, challenges remain to be overcome, such as fully implementing PMRR, strengthening the Irregular Settlements and Subnormal Housing Control Group, and the need for better inter-agency coordination to reduce risks and improve citizens' quality of life. Legal instruments alone do not promote resilience but, provided they are incorporated into the activities of the various agencies, they can prove to be extremely effective.

## References

Adger, W. N. (2006). Vulnerability. *Global Environmental Change, 16*(3), 268–281.

Ainuddin, S., & Routray, J. K. (2012). Earthquake hazards and community resilience in Baluchistan. *Natural Hazards, 63*(2), 909–937. 26 abr.

Alheiros, M. M. (2006). O Plano Municipal de Redução de Risco. In C. S. Carvalho, T. Galvão (org.) (Ed.), *MINISTÉRIO DAS CIDADES; CITIES ALLIANCE. Prevenção de Riscos de Deslizamentos em Encostas: Guia para Elaboração de Políticas Municipais*. Brasília: Ministério das Cidades; Cities Alliance.

Almeida, P. E. (2015). *A Política Nacional de Proteção e Defesa Civil: os desastres como problema político. Anais do I Seminário Internacional de Ciência Política*. Anais...Porto Alegre: UFRGS.

Braga, R. (2001). Política urbana e gestão ambiental: considerações sobre o plano diretor e o zoneamento urbano. In P. F. Carvalho & R. Braga (Eds.), *Perspectiva de gestão ambiental em cidades médias*. Rio Claro: LPM-UNESP.

BRASIL. Lei n° 12.608, de 10 de abril de 2012. Institui a Política Nacional de Proteção e Defesa Civil - PNPDEC; dispõe sobre o Sistema Nacional de Proteção e Defesa Civil - SINPDEC e o Conselho Nacional de Proteção e Defesa Civil - CONPDEC; autoriza a criação de sistema de informações e monitoramento de desastres; altera as Leis nos 12.340, de 1o de dezembro de 2010, 10.257, de 10 de julho de 2001, 6.766, de 19 de dezembro de 1979, 8.239, de 4 de outubro de 1991, e 9.394, de 20 de dezembro de 1996; e dá outras providências.

Brown, A., Dayal, A., & Rio, C. R. D. (2012). From practice to theory: emerging lessons from Asia for building urban climate change resilience. *Environment and Urbanization, 24*(2), 531–556.

Burton, I., Kates, R. W., & White, G. F. (Eds.). (1978). *The environment as hazard* (1st ed.). New York: Oxford University Press.

Calheiros, L. B., Castro, A. L. C. D. E., & Dantas, M. C. (2009). *Apostila sobre Implantação e Operacionalização de COMDEC.* (4th ed.). Brasília: Ministério da Integração Nacional; Secretaria Nacional de Proteção e Defesa Civil.

Canil, K. (2010). *Áreas de Risco aos Processos de Escorregamentos no Município de Santos, SP: Análise e Indicadores de Vulnerabilidade. Anais XVI Encontro Nacional dos Geógrafos.* Anais...Porto Alegre - RS: jul.

Cardona, O. D. A. (1996). El manejo de riesgos y los preparativos para desastres: compromiso institucional para mejorar la calidad de vida. In E. Mansilla (Ed.), *Desastres: modelo para armar. Colección de piezaz de un rompecabezas social.* [s.l.] LA RED.

Carpenter, A. (2015). Resilience in the social and physical realms: Lessons from the Gulf Coast. *International Journal of Disaster Risk Reduction, Risking Disaster – The role of private investment and public regulation in disaster risk management, 14*(3), 290–301.

Carpenter, S., et al. (2001). From metaphor to measurement: resilience of what to what? *Ecosystems, 4*(8), 765–781.

Crozier, M., & Glade, T. (2010). Hazard assessment for risk analysis and risk management. In I. Alcántara-Ayala & A. S. Goudie (Eds.), *Geomorphological hazards and disaster prevention.* Cambridge: Cambridge University Press.

Cutter, S. L., et al. (2008). A place-based model for understanding community resilience to natural disasters. *Global Environmental Change, 18*(4), 598–606.

Cutter, S. L. (1996). Vulnerability to environmental hazards. *Progress in Human Geography, 20*(4), 529–539.

Folha de São Paulo. (1978, 11 mar.). Local, Primeiro Caderno, p. 13. Deslizamento mata 9 em Santos. Acervo Folha. Disponível em:http://goo.gl/CLLmVG. Acesso em: 15/12/2015.

Faria, D. G. M., & Santoro, J. (2011). Gerenciamento de desastres naturais. In L. K. Tominaga, J. Santoro, & R. do Amaral (Eds.), *Desastres naturais: conhecer para prevenir* (1st ed.). Instituto Geológico: São Paulo.

Fernandes, N. F., & Amaral, C. P. (2000). Movimentos de massa: uma abordagem geológico-geomorfológica. In A. J. T. Guerra & S. B. Cunha (Eds.), *Geomorfologia e meio ambiente* (3rd ed.). Bertrand: Rio de Janeiro.

Folke, C., et al. (2002). Resilience and Sustainable Development: Building Adaptive Capacity in a World of Transformations. *AMBIO: A Journal of the Human Environment, 31*(5), 437–440.

Folke. (2006). Resilience: the emergence of a perspective for social–ecological systems analyses. *Global Environmental Change, 16*(3), 253–267.

Freitas, C. M. D. E., et al. (2012). Vulnerabilidade socioambiental, redução de riscos de desastres e construção da resiliência: lições do terremoto no Haiti e das chuvas fortes na Região Serrana, Brasil. *Ciência & Saúde Coletiva, 17*(6), 1577–1586. jun.

Gallopín, G. C. (2006). Linkages between vulnerability, resilience, and adaptive capacity. *Global Environmental Change, 16*(3), 293–303.

Godoy, A. S. (1995). Pesquisa qualitativa: tipos fundamentais. *Revista de Administração de Empresas, 35*(3), 20–29.

Guimarães, R. F., et al. (2008). Movimentos de massas. In T. G. Florenzano (Ed.), *Geomorfologia: conceitos e técnicas* (2nd ed.). Oficina de Textos: São Paulo.

Gupta, K. (2007). Urban flood resilience planning and management and lessons for the future: A case study of Mumbai, India. *Urban Water Journal, 4*(3), 183–194.
Hewitt, K. (1980). Review of the environment as hazard. *Annals of the Association of American Geographers, 70*(2), 306–311.
Hufschmidt, G. (2011). A comparative analysis of several vulnerability concepts. *Natural Hazards, 58*(2), 621–643.
Hufschmidt, G., & Glade, T. (2010). Vulnerability analysis in geomorphic risk assessment. In I. Alcántara-Ayala & A. S. Goudie (Eds.), *Geomorphological hazards and disaster prevention*. Cambridge: Cambridge University Press.
IBGE. (2015). IBGE Cidades - Santos - SP. Disponível em: http://cod.ibge.gov.br/2B8. Acesso em: 23/11/2015.
IPCC – Intergovernmental Panel on Climate Change. (2014). Clicate Change 2014: Synthesis report. In Core Writing Team, R. K. Pachauri & L. A. Meyer (Ed.), *Contribution of Working Groups I, II and III to the Fifth Assessment Report of the Intergovernmental Panel on Climate Change*. Geneva: IPCC.
IPCC. (2007). Summary for policymakers. In M. L. Parry, O. F. Canziani, J. P. Palutikof, P. J. Van Der Linden, & C. E. Hanson (Eds.), *Climate change 2007: impacts, adaptation and vulnerability. Contribution of Working Group II to the Fourth Assessment Report of the Intergovernmental Panel on Climate Change* (pp. 7–22). Cambridge: Cambridge University Press, UK.
IPT – INSTITUTO DE PESQUISAS TECNOLÓGICAS DO ESTADO DE SÃO PAULO. (1979). *Relatório da Carta Geotécnica dos Morros de Santos e São Vicente.*
IPT – INSTITUTO DE PESQUISAS TECNOLÓGICAS DO ESTADO DE SÃO PAULO. (2012). *Atualização do Plano Municipal de Redução de Riscos para o Município de Santos, SP, 2012.*
IPT – INSTITUTO DE PESQUISAS TECNOLÓGICAS DO ESTADO DE SÃO PAULO. (2014a). *Cartas de suscetibilidade a movimentos gravitacionais de massa e inundações do município de Santos-SP, 2014. Escala*: 1:25.000.
IPT – INSTITUTO DE PESQUISAS TECNOLÓGICAS DO ESTADO DE SÃO PAULO. (2014b). *Você sabia? Que. o IPT elaborou a primeira Carta Geotécnica do País? [s.d.]*. Disponível em: http://goo.gl/S6QXpB. Acesso em: 14/12/2014.
Jabareen, Y. (2013). Planning the resilient city: Concepts and strategies for coping with climate change and environmental risk. *Cities, 31*, 220–229.
Joerin, J., et al. (2012). Assessing community resilience to climate-related disasters in Chennai, India. *International Journal of Disaster Risk Reduction, 1*, 44–54.
Khailani, D. K., & Perera, R. (2013). Mainstreaming disaster resilience attributes in local development plans for the adaptation to climate change induced flooding: A study based on the local plan of Shah Alam City, Malaysia. *Land Use Policy, 30*(1), 615–627.
Klein, R. J. T., Nicholls, R. J., & Thomalla, F. (2003). Resilience to natural hazards: how useful is this concept? *Environmental Hazards, 5*(1–2), 35–45.
Kobiyama, M., et al. (2006). *Prevenção de desastres naturais conceitos básicos* (1st ed.). Organic Trading: Curitiba.
Lavell, A. (2000). *Desastres y Desarrollo: Hacia un entendimiento de las formas de construcción social de un desastre: el caso del huracán Mitch en Centroamérica.* San José: BID, CIDHS.
Lavell, A. (1998). Un encontro con la verdad: los desastres en América Latina durante 1998. In *Anuario social y político de América Latina y El Caribe* (Vol. 2). San José: FLACSO, Nueva Sociedad.
Leichenko, R. (2011). Climate change and urban resilience. *Current Opinion in Environmental Sustainability, 3*(3), 164–168.
Lhomme, S., et al. (2013). Analyzing resilience of urban networks: a preliminary step towards more flood resilient cities. *Natural Hazards and Earth System Sciences, 13*(2), 221–230.
Liao, K.-H. (2012). A theory on urban resilience to floods-a basis for alternative planning practices. *Ecology and Society, 17*(4), 48.
Lorenz, D. F. (2010). The diversity of resilience: contributions from a social science perspective. *Natural Hazards, 67*(1), 7–24.

Malalgoda, C., Amaratunga, D., & Haigh, R. (2013). Creating a disaster resilient built environment in urban cities: The role of local governments in Sri Lanka. *International Journal of Disaster Resilience in the Built Environment, 4*(1), 72–94.

Manyena, S. B., et al. (2011). Disaster resilience: a bounce back or bounce forward ability? *Local Environment, 16*(5), 417–424.

Manyena, S. B. (2006). The concept of resilience revisited. *Disasters, 30*(4), 434–450.

Mcentire, D. A. (2005). Why vulnerability matters: Exploring the merit of an inclusive disaster reduction concept. *Disaster Prevention and Management, 14*(2), 206–222.

Mechler, R., & Bouwer, L. M. (2014). Understanding trends and projections of disaster losses and climate change: is vulnerability the missing link? *Climatic Change, 133*(1), 23–35.

Meerow, S., Newell, J. P., & Stults, M. (2016). Defining urban resilience: A review. *Landscape and Urban Planning, 147*, 38–49.

MI - MINISTÉRIO DA INTEGRAÇÃO NACIONAL; SEDEC - SECRETARIA NACIONAL DE PROTEÇÃO E DEFESA CIVIL. (2014). *Proteção e Defesa Civil: novos paradigmas para o Sistema Nacional.* Texto de referência para a II Conferência Nacional de Proteção e Defesa Civil. Brasília.

Miller, F., et al. (2010). Resilience and vulnerability: Complementary or conflicting concepts? *Ecology and Society, 15*(3), 11.

Nogueira, F. R. (2002). *Gerenciamento de riscos ambientais associados a escorregamentos: contribuição às políticas públicas municipais para áreas de ocupação subnormal. 2002* (p. 268). Rio Claro: Tese (Doutorado) - Instituto de Geociências e Ciências Exatas, Universidade Estadual Paulista.

Nogueira, F. R., Oliveira, V. E. D. E., & Canil, K. (2014). Políticas públicas regionais para gestão de riscos: o processo de implementação no ABC. *SP. Ambiente & Sociedade, 17*(4), 177–194.

PBMC - Painel Brasileiro de Mudanças Climáticas. (2014). Impactos, vulnerabilidades e adaptação às mudanças climáticas. In E. D. Assad & A. R. Magalhães (Eds.), *Contribuição do Grupo de Trabalho 2 do Painel Brasileiro de Mudanças Climáticas ao Primeiro Relatório da Avaliação Nacional sobre Mudanças Climáticas* (Vol. 2). Rio de Janeiro: COPPE. Universidade Federal do Rio de Janeiro.

Pizzo, B. (2014). Problematizing resilience: Implications for planning theory and practice. *Cities, 43*, 133–140.

PREFEITURA MUNICIPAL DE SANTOS. (2013) *Lei complementar n° 821, de 27 de dezembro de 2013.* Institui o plano diretor de desenvolvimento e expansão urbana do município de Santos, e dá outras providências.

Revi, A., et al. (2014). *Urban areas: Climate change 2014: Impacts, adaptation, and vulnerability. Part A: Global and sectoral aspects. Contribution of working group II to the fifth assessment report of the intergovernmental panel on climate change.* Cambridge/New York: IPCC.

Romero-Lankao, P., & Dodman, D. (2011). Cities in transition: transforming urban centres from hotbeds of GHG emissions and vulnerability to seedbeds of sustainability and resilience: Introduction and Editorial overview. *Current Opinion in Environmental Sustainability, 3*(3), 113–120.

Saavedra, C., & Budd, W. W. (2009). Climate change and environmental planning: Working to build community resilience and adaptive capacity in Washington State, USA. *Habitat International, Climate Change and Human Settlements, 33*(3), 246–252.

Santos, A. R. (2004). *Dos. A grande barreira da Serra do Mar: da trilha dos Tupiniquins à Rodovia dos Imigrantes* (1st ed.). São Paulo: O Nome da Rosa.

Sapountzaki, K., et al. (2011). Disconnected policies and actors and the missing role of spatial planning throughout the risk management cycle. *Natural Hazards, 59*(3), 1445–1474.

SEADE. (2015). *Perfil Municipal. Disponível em.* http://goo.gl/OCG5jA. Acesso em: 05 jan.

Sherbinin, A. D., Schiller, A., & Pulsipher, A. (2007). The vulnerability of global cities to climate hazards. *Environment and Urbanization, 19*(1), 39–64.

da Silva, C. A. M. (2014). *Em busca de resiliência? Urbanização, ambiente e riscos em Santos (SP). 2014* (p. 253). Campinas: Tese (Doutorado) - Campinas, SP: Instituto de Filosofia e Ciências Humanas da Universidade Estadual de Campinas.

Smithers, J., & Smit, B. (1997). Human adaptation to climatic variability and change. *Global Environmental Change, 7*(2), 129–146.

Sobreira, F. G., & Souza, L. A. (2012). Cartografia geotécnica aplicada ao planejamento urbano. *Revista Brasileira de Geologia de Engenharia e Ambiental, 2*(1), 79–89.

Suassuna, C. C. DE A. (2014). *Cidade resiliente: sistema de indicadores dos aspectos institucionais. 2014.* (p. 292). Recife: Tese (Doutorado) - Centro de Artes e Comunicação, Desenvolvimento Urbano Universidade Federal de Pernambuco.

Sudmeier, K. I., Jaboyedoff, M., & Jaquet, S. (2013). Operationalizing "resilience" for disaster risk reduction in mountainous Nepal. *Disaster Prevention and Management, 22*(4), 366–377.

Sudmeier-Rieux, K. I. (2014). Resilience – an emerging paradigm of danger or of hope? *Disaster Prevention and Management: An International Journal, 23*(1), 67–80.

Sulaiman, S. N. (2014). *De que adianta? O papel da educação para a prevenção de desastres naturais. 2014* (p. 291). São Paulo: Tese (Doutorado) – Faculdade de Educação, Universidade de São Paulo.

Susman, P., O'keefe, P., & Wisner, B. (1983). Global disaster, a radical interpretation. In K. Hewitt (Ed.), *Interpretations of calamity from the viewpoint of human ecology*. Boston: Allen & Unwin Inc.

Tang, W., et al. (2015). Creating social–physical resilience to natural disasters: lessons from the Wenchuan earthquake. *Natural Hazards, 79*(2), 1111–1132.

Thomalla, F., et al. (2006). Reducing hazard vulnerability: towards a common approach between disaster risk reduction and climate adaptation. *Disasters, 30*(1), 39–48.

UNISDR – United Nations International Strategy for Disaster Reduction. (2012). *How to make cities more resilient - a handbook for local government leaders. A contribution to the global campaign 2010–2015*. Geneva: UNISDR.

UNISDR – United Nations International Strategy for Disaster Reduction. (2009). *2009 UNISDR Terminology on disaster risk reduction*. Geneva: UNISDR.

Valencio, N., & Valencio, A. (2011). Os desastres como indícios da vulnerabilidade do Sistema Nacional de Defesa Civil: o caso brasileiro. *Territorium: Revista Portuguesa de riscos, prevenção e segurança, 18*, 147–156.

Walker, B., et al. (2004). Resilience, adaptability and transformability in social-ecological systems. *Ecology and Society, 9*(2), 5.

# Chapter 13 Probabilistic Assessment of Seismic Risk of Dwelling Buildings of Barcelona. Implication for the City Resilience

**Armando Aguilar-Meléndez, Lluis G. Pujades, Josep De la Puente, Alex H. Barbat, Mario G. Ordaz S., Sergio Natan González-Rocha, Carlos M. Welsh-Rodríguez, Héctor E. Rodríguez-Lozoya, Nieves Lantada, Luis Ibarra, Alejandro García-Elias, and Amelia Campos-Rios**

**Abstract** The knowledge of seismic risk of buildings can contribute to increase the resilience of cities. In the present work a new assessment of the seismic risk of dwelling buildings of Barcelona was done. This assessment was performed according to a probabilistic methodology, which is summarized in the following steps: (1) performing a probabilistic seismic hazard assessment (PSHA) to obtain exceedance rates of macroseismic intensities; (2) performing a probabilistic seismic vulnerability assessment (PSVA) of each building in order to determine probability density functions that describe the variation of a vulnerability index; and (3) performing a probabilistic seismic risk assessment (PSRA) to generate seismic risk curves in terms of frequencies of exceedance of damage states. In the present work 69,982 dwelling buildings of Barcelona were assessed. According to the results the percentage of dwelling buildings of Barcelona that have a probability equal or greater than 1% of suffer partial collapse in the next 50 years is a value between 0% and 34.29%. A value of 0% corresponds to the results of seismic risk obtained for the case where regional vulnerability modifiers were not considered during the procedure to assess the seismic vulnerability of buildings and 34.29% correspond to the case where regional vulnerability modifiers were considered. For the same two options, the losses due to the physical damage of the dwelling buildings of Barcelona assessed for an exposure time of 50 years, could vary from 807.3 to 1739.4 millions of euros, respectively. Finally, possible uses of the seismic risk results computed in the present work are mentioned.

**Keywords** Seismic risk · Urban zones · Vulnerability index · Barcelona

---

A. Aguilar-Meléndez (✉)
Barcelona Supercomputing Center (BSC), Barcelona, Spain & Faculty of Civil Engineering, Universidad Veracruzana, Poza Rica, Mexico
e-mail: armaguilar@uv.mx



G. Brunetta et al. (eds.), *Urban Resilience for Risk and Adaptation Governance*, Resilient Cities, https://doi.org/10.1007/978-3-319-76944-8_13

## 13.1 Resilience in Cities

### *13.1.1 Importance of the Assessment of Risks in Cities*

In order to reduce the disasters that occur every year in the world different efforts are done. One of these efforts is promoted by United Nations (UN), who generated the Hyogo Framework for Action 2005–2015: Building the Resilience of Nations and Communities to Disasters (ISDR-UN 2005), where the expected outcome for the next 10 years was: "The substantial reduction of disaster losses, in lives and in

L. G. Pujades · N. Lantada
Division of Geotechnical Engineering and Geosciences, Department of Civil and Environmental Engineering, Universitat Politècnica de Catalunya. BarcelonaTECH, Barcelona, Spain
e-mail: lluis.pujades@upc.edu; nieves.lantada@upc.edu

J. De la Puente
Barcelona Supercomputing Center (BSC), Barcelona, Spain
e-mail: josep.delapuente@bsc.es

A. H. Barbat
Division of Mechanics, Continuous Media and structures, Department of Civil and Environmental Engineering, Universitat Politècnica de Catalunya. BarcelonaTECH, Barcelona, Spain
e-mail: alex.barbat@upc.edu

M. G. Ordaz S.
Engineering Institute, Universidad Nacional Autónoma de México, Ciudad de México, Mexico
e-mail: mordazs@iinen.unam.mx

S. N. González-Rocha
Barcelona Supercomputing Center (BSC), Barcelona, Spain & Faculty of Chemical Sciences, Universidad Veracruzana, Poza Rica, Mexico
e-mail: ngonzalez@uv.mx

C. M. Welsh-Rodríguez
Earth Sciences Center, Universidad Veracruzana, Xalapa, Mexico
e-mail: cwelsh@uv.mx

H. E. Rodríguez-Lozoya
Faculty of Engineering, Universidad Autónoma de Sinaloa, Los Mochis, Mexico

L. Ibarra
Department of Civil & Environmental Engineering, The University of Utah, Salt Lake City, USA
e-mail: luis.ibarra@utah.edu

A. García-Elias
Faculty of Civil Engineering, Universidad Veracruzana, Poza Rica, Mexico
e-mail: alejagarcia@uv.mx

A. Campos-Rios
Services of Engineering, Tuxpan, Mexico

the social, economic and environmental assets of communities and countries (ISDR-UN 2005)". As a consequence of the Hyogo Framework specific programs were developed. Some of these programs were oriented to increase the resilience of cities. For instance, it was created the "City Resilience Profiling Programme (CRPP)" (UN-Habitat 2016). In this last program the resilience is defined as: "the ability of a system to reduce, prevent, anticipate, absorb and adapt, or recover from the effects of a hazardous event in a timely and efficient manner, including through ensuring the preservation, restoration, or improvement of its essential basic structures and functions" (United Nations 2013).

Another product of the UN initiatives about resilience is the handbook for local government leaders called "How to make Cities more Resilient" (United Nations 2012), where the third of ten essentials to increase the resilience of cities, states: "Maintain up to date data on hazard and vulnerabilities. Prepare risk assessments and use these as the basis for urban development plans and decisions, ensure that this information and the plans for your city's resilience are readily available to the public and fully discussed with them" (United Nations 2012). Therefore, the knowledge of the whole hazards and risks that exist in a city is an essential part to do an appropriate management of these risks in order to contribute to increase the level of resilience of the city (United Nations 2012).

Additionally, since 2015 the UN programs mentioned previously have a new support called the Sendai Framework for Disaster Risk Reduction 2015–2030, which was adopted at the Third UN World Conference in Sendai, Japan, on March 18, 2015 and it is considered as the successor instrument to the Hyogo Framework (UNISDR 2015). The expected outcome of the Sendai Framework over the next 15 years is the following: "The substantial reduction of disaster risk and losses in lives, livelihoods and health and in the economic, physical, social, cultural and environmental assets of persons, businesses, communities and countries" (UNISDR 2015). At the same time the Sendai Framework states that this outcome will be obtained base on the following goal: "Prevent new and reduce existing disaster risk through the implementation of integrated and inclusive economic, structural, legal, social, health, cultural, educational, environmental, technological, political and institutional measures that prevent and reduce hazard exposure and vulnerability to disaster, increase preparedness for response and recovery, and thus strengthen resilience" (UNISDR 2015).

The Sendai Framework states the importance of develop multi-hazard approach in the disaster risk reduction, but this Framework not included a definition of the term multi-hazard. However, according to Gardoni and LaFave (2016) "multi-hazards can be classified as *concurrent* (e. g., wind and surge), *cascading* (e.g., fire following earthquake), or independent and likely to occur a different times (e.g, wind and earthquake)". In the present document the multi-hazard composed word will be considered as a concept that has the objective of reminding that all potential hazards must be appropriately considered to reduce the occurrence of disasters. In any way, it is important to have in mind that both single-hazard and multi-hazards considerations must contribute to reach the outcome of the Sendai Framework: "the substantial reduction of disaster risk and losses …" (UNISDR 2015).

In the last decades the major development has occurred in the methodologies where the single-hazards are assessed. However, in general, nowadays it is recognized that the new approaches to consider multi-hazard must be considered as complementary approaches to the existing single-hazard approaches. In fact, some multi-hazard approaches require the results of single-hazard approaches (Zschau 2017). Therefore, in general, both single-hazard and multi-hazard procedures are relevant to assess both hazards and risks that are present in cities.

In this context was developed the present work, where the seismic risk of the dwelling buildings of Barcelona was assessed. The probabilistic methodology applied in the present work was oriented to both single-hazard and single-risk analysis. However, the seismic hazard computed in the present work can be considered as a starting point to continue with a multi-hazard assessment, for instance it is possible to use some of the computed results to assess the hazard of fire due to earthquakes or the hazard due to tsunamis. Similarly, some of the seismic risk results computed in the present work are expressed in a risk curve in terms of economic losses versus return period. Therefore it is possible to aggregate these last results to other single-risk curves, in order to build a multi-risk curve that incorporates the risks due to several hazards. In the next sections details about the essentially probabilistic methodology applied to assess the seismic risk of the dwelling buildings of Barcelona are included. At the same time, main data and fundamental steps that were applied to obtain the seismic risk of the dwelling buildings of Barcelona are mentioned in the subsequent sections of this document.

## 13.2 Assessment of Seismic Risk in Urban Areas

### *13.2.1 Antecedents*

The partial or total collapse of buildings is the main source of deaths that occur during certain earthquakes (Lizarralde et al. 2009). For instance, Table 13.1 shows data from earthquakes occurred in 2010 and 2011 that caused a significant number of deaths. The death of persons related to an earthquake is the main negative aspect about these events. However, an additional negative aspect related to some earthquakes are the significant economic losses that they eventually produce.

More recently, in 2016, some earthquakes also have triggered significant damage, for instance, it is possible to highlight the case of Ecuador earthquake that occurred in April 16, and the Italy earthquake that occurred in August 24. In the first case the number of death people due to the earthquake was of 673, and in the second case 298 persons died. Then it is clear that earthquakes are a natural hazard that can trigger disasters. Therefore, it is necessary to increase our knowledge about the levels of seismic risk that exist in the cities. For this purpose, nowadays there are different methodologies to assess the seismic risk in urban areas. Table 13.2 high-

**Table 13.1** Earthquakes with 10 or more deaths during 2010 and 2011[a]

| Deaths | Magnitude | Location | Year | Date |
|---|---|---|---|---|
| 222,570 | 7.0 | Haiti Léogâne Haiti | 2010 | January 12 |
| 2,968 | 6.9 | China Qinghai, China | | April 14 |
| 711 | 7.7 | Indonesia Sumatra, Indonesia[b] | | October 25 |
| 562 | 8.8 | Chile Maule region, Chile | | February 27 |
| 51 | 6.1 | Turkey Elâzığ province, Turkey | | March 8 |
| 17 | 7.0 | Indonesia Papua, Indonesia | | June 16 |
| 19,846 | 9.0 | Japan Tōhoku, Japan[b] | 2011 | March 11 |
| 604 | 7.1 | Turkey Van, Turkey | | October 23 |
| 181 | 6.1 | New Zealand Christchurch, NZ | | February 22 |
| 112 | 6.9 | India Nepal India-Nepal border | | September 18 |
| 74 | 6.9 | Myanmar Thailand Shan State, Myanmar | | March 24 |
| 25 | 5.5 | China Myanmar-China border region | | March 10 |
| 13 | 6.1 | Uzbekistan Kyrgyzstan Tajikistan Fergana Valley | | July 19 |
| 10 | 6.6 | Indonesia Aceh, Indonesia | | September 5 |
| 10 | 5.1 | Spain Lorca, Spain | | May 11 |

[a]Data obtained from USGS (2016a, b) and EM-DAT (2015).
[b]Earthquake and Tsunami.

light some examples of relevant methodologies that have been applied to assess the seismic risk of buildings in urban areas.

As can be inferred from previous information, nowadays, there is not a standard methodology to assess the seismic risk of buildings in urban areas in the whole world. But there are important coincidences between some of these methodologies. For instance, it is widely accepted that the assessment of the seismic hazard is a basic step in the assessment of seismic risk. Additionally, the assessment of seismic vulnerability is another basic step in the evaluation of the seismic risk. However, this last step can be considered explicitly or implicitly. Therefore, the seismic risk is computed considering both the seismic hazard and the seismic vulnerability. In the next section, the main aspects of the probabilistic methodology applied in the present work to assess the seismic risk of the dwelling buildings of Barcelona are described.

## 13.3 The RISKBUA-E Methodology to Assess Seismic Risk of Buildings in Urban Areas

### *13.3.1 The RISKBUA-E Methodology*

Since 2008 Aguilar-Meléndez and collaborators started to propose the basic elements of a probabilistic methodology to assess seismic risk of buildings (Aguilar-Meléndez et al. 2008; Aguilar-Meléndez et al. 2010; Aguilar-Meléndez et al. 2011;

**Table 13.2** Main characteristics of relevant methodologies to assess seismic risk in urban areas

| Methodology | Seismic | | |
|---|---|---|---|
| | a) Hazard | b) Vulnerability | c) Risk |
| ATC-13 (1985) | Determination of a ground motion in terms of an MMI grade. | The vulnerability is implicitly considered in the damage probability matrix defined for each facility type. | Probabilities of damage states based on damage probability matrices for each facility type. Damage degree (0 to 100%). |
| HAZUS (FEMA 2015a, b) | a) Deterministic ground motion analysis; b) United States Geological Survey (USGS) probabilistic ground motion maps; c) other probabilistic or deterministic ground motion maps. | Capacity curves for specific structural typologies. | Probabilities of damage states based on fragility curves for specific structural typologies. |
| GNDT II (Benedetti et al. 1988) | Value of peak ground acceleration (PGA). | Vulnerability index with scores and weights, according to characteristics in buildings. | Damage degree (0–1). |
| Risk-UE Vulnerability index method (VIM) or LM1 method (Milutinovic and Trendafiloski 2003) | A seismic scenario in terms of a macroseismic intensity obtained by means of: a) probabilistic method; b) deterministic method. | Vulnerability index for each building classified into a structural typology. | Probabilities of damage grade (5 no-null damage states), for each building, based on a damage function depending of a vulnerability index and a macroseismic intensity. |
| Risk-UE LM2 method (Milutinovic and Trendafiloski, 2003) | Demand spectrum | Capacity curves for structural typologies. | Probabilities of damage states (4 no-null damage states), for each building of a group of buildings; based on fragility curves of structural typologies. |
| CAPRA (Cardona et al. 2012; ERN-AL 2010) | Seismic hazard scenarios obtained by means of a probabilistic assessment | Vulnerability functions for structural typologies. | Damage states probabilities based on fragility curves, for each structural typology. |

Aguilar-Meléndez et al. 2012). This probabilistic methodology is called VIM_P and it was derived from the Vulnerability Index Method (VIM) that was proposed in the Risk-UE project (Milutinovic and Trendafiloski 2003). The name VIM_P is due mainly to the fact that in this methodology more probabilistic elements than in the VIM method are considered to assess the seismic risk of buildings (Aguilar-Meléndez 2011). In the present work the VIM_P was applied to assess the seismic risk of the dwelling buildings of Barcelona. According to this methodology (Aguilar-Meléndez et al. 2010) the main steps to compute seismic risk of buildings are the following: a) a probabilistic seismic hazard assessment (PSHA); b) a probabilistic seismic vulnerability assessment (PSVA), and; c) a probabilistic seismic risk assessment (PSRA).

The application of the VIM_P can be done with the support of two computer codes: (a) CRISIS2015 (Ordaz et al. 2015) in order to perform the PSHA and; (b) USERISK2015 (Aguilar-Meléndez et al. 2015a) in order to perform the PSVA and the PSRA. The code USERISK2015 also allows to compute PSRA with the seismic hazard results computed in a code different of CRISIS. However, we strongly recommend the use of CRISIS2015 to perform the PSHA. Therefore, we called RISKBUA-E (Risk of Buildings in Urban Areas due to Earthquakes) to the methodology that allows to compute seismic risk of buildings in urban areas with the application of two computer codes: CRISIS2015 and USERISK2015. In the present work the seismic risk of the dwelling buildings of Barcelona was assessed applying the RISKBUA-E methodology (Fig. 13.1).

CRISIS2015 is a recent version of the computer code CRISIS that has been continuously updated and widely validated in the last 15 years (Ordaz et al. 2013; Aguilar-Meléndez et al. 2017; Faccioli 2006; Faccioli et al. 2003; Villani et al. 2010). On the other hand, the code USERISK2015 is a recent code, which first version was developed by Aguilar-Meléndez et al. (2011). The USERISK2015 code allows to compute seismic vulnerability and seismic risk of buildings according to the VIM_P method.

In the present work the methodology RISKBUA-E was applied to assess the seismic risk of a 69,982 dwelling buildings of Barcelona. However, in order to high-

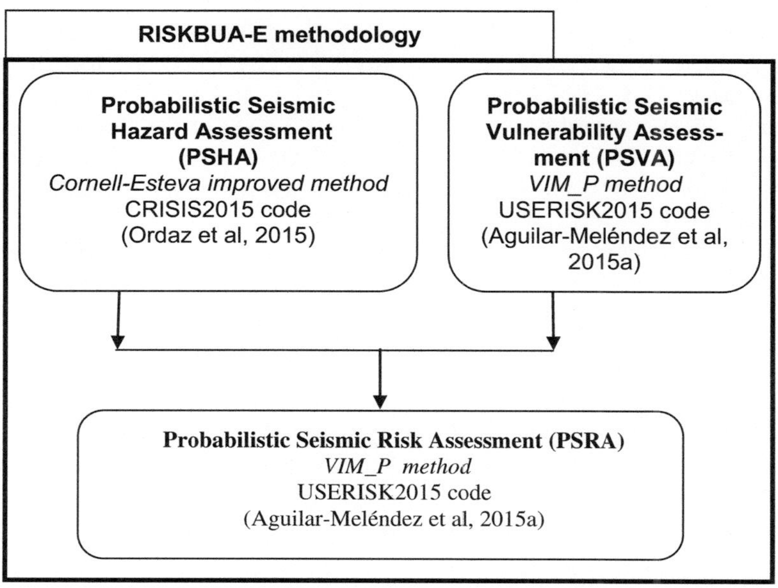


**Fig. 13.1** Diagram with the main steps and the principal software to apply the methodology RISKBUA-E, to perform a PSRA

**Table 13.3** Data of two buildings located at Barcelona, Spain

| Data | Building A | Building B |
|---|---|---|
| 1. Structural Typology. | *Unreinforced masonry bearing walls with composite steel and masonry slabs.* | *Irregular concrete frames with unreinforced masonry infill walls.* |
| 2. Reliability factor in the assignment of the structural typology. | 8 | 8 |
| 3. Conservation state. | Good | Good |
| 4. Number of levels. | 3 | 3 |
| 5. Construction date. | 1960 | 1970 |
| 6. Type of terrain. | Rock | Rock |

light the main steps of the RISKBUA-E methodology an example of the assessment of the seismic risk of two buildings of Barcelona is included in the present document. The main data of these two buildings are shown in Table 13.3. It is important to highlight that the purpose of the RISKBUA-E methodology is to assess the seismic risk of buildings in an urban scale, and not to assess a single building.

### *13.3.2 Seismic Hazard in the RISKBUA-E Methodology*

According to the RISKBUA-E methodology (Fig. 13.1), the first step to assess seismic risk is to perform a PSHA. Consequently, to assess the seismic risk of buildings A and B (Table 13.3), it is necessary to perform a PSHA in the site where the buildings are located (Fig. 13.2). For this purpose, we apply the CRISIS2015 code, which computes seismic hazard based on an improved version of the classical Cornell-Esteva approach (Cornell 1968; Esteva 1970; McGuire 2008). More details about the theoretical background of CRISIS can be found in Ordaz et al. (2015) and in Villani (2010).

The curve of Figure 13.2 represents the seismic hazard curve for a rock of the site of Barcelona where the buildings A and B are located. This curve was computed by means of the CRISIS2015 code (Ordaz et al. 2015), which allows assessing seismic hazard using Ground Motion Prediction Equations (GMPE) in terms of macroseismic intensities.

According to the curve of Fig. 13.2, the macroseismic intensity that has a return period of 475 years is equal to VI. However, in the cases where the buildings are not located in rock, it is also necessary to evaluate local effects that usually increase the size of the seismic ground motions. More details about the data considered to perform the PSHA for Barcelona are included in an upcoming section.

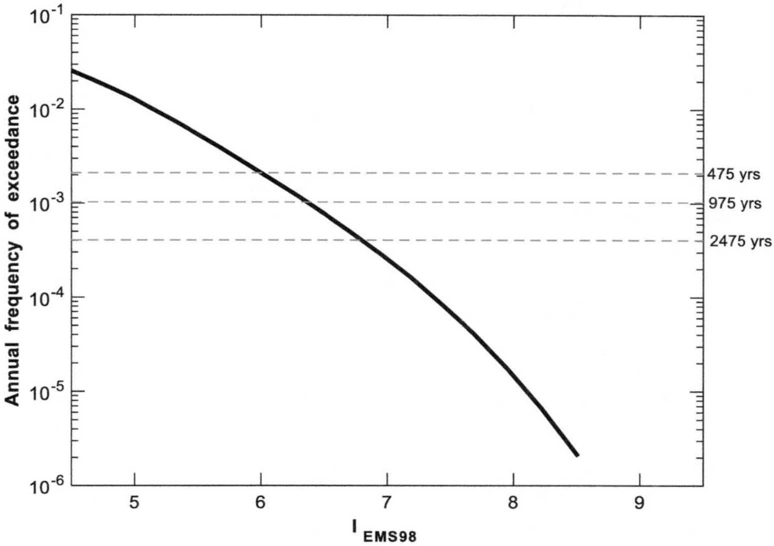


**Fig. 13.2** Seismic hazard curve for a rock site of Barcelona

## *13.3.3 Seismic Vulnerability in the RISKBUA-E Methodology*

The seismic vulnerability is a measure of the expected seismic behavior of a building. For example, a building with a high value of seismic vulnerability is a building with a high probability of suffering significant damage due to the occurrence of strong ground motions. In the opposite case, a building with a low seismic vulnerability is a building with a high probability of withstanding without damage the occurrence of strong ground motions. According to RISKBUA-E methodology, it is necessary to perform a PSVA where the main objective is to assess the seismic vulnerability of the buildings. For this purpose it is necessary to apply the VIM_P. Particularly, in the VIM_P the purpose of the assessment of the seismic vulnerability is to determine probability density functions beta type that represent the seismic vulnerability of each building. These functions describe the variation of a vulnerability index that usually varies in a range close to the one defined by 0 and 1. Values close to cero mean low vulnerability and values close to one mean high vulnerability. In the original VIM method of the Risk-UE project (Milutinovic and Trendafiloski 2003; Lantada et al. 2009a), the seismic vulnerability of a building is represented by a scalar called total vulnerability index, $\overline{V_I}$, which is computed according to Eq. 1. However, in the VIM method $\overline{V_I}$ is only the mean value of the function beta type that represents the seismic vulnerability of a building.

$$\overline{V_I} = V_I^* + \Delta V_R + \Delta V_m \quad (13.1)$$

In Eq. 13.1 $V_I^*$ is the vulnerability index by structural typology (Table 13.7), $\Delta V_R$, means regional vulnerability index, and $\Delta V_m$ is the sum of the scores related to all the factors that modify the vulnerability of each building, whereas $\Delta V_R$ and $\Delta V_m$ are optional values. However $V_I^*$ is a basic parameter, because this term represent the mean value of the seismic vulnerability of a building according to its structural typology (Barbat et al. 2006). Therefore, at least the value of $V_I^*$ must be assigned. For this reason, in the VIM_P method is fundamental to know the structural typology of each building that will be assessed.

The $\Delta V_{\mathrm{R}}$ term can be used if it is considered that the mean value of vulnerability assigned to a structural typology does not correspond to the mean seismic vulnerability of the local buildings that are classified into a structural typology. In other words, $\Delta V_{\mathrm{R}}$ can be used when exists evidence that highlights the fact that, in the region of study, the buildings of a specific structural typology are not appropriately represented by the values that were considered by Milutinovic and Trendafiloski (2003). The $\Delta V_{\mathrm{R}}$ values must be determined by expert opinion. On the other hand, $\Delta V_{\mathrm{m}}$ modifiers can be used to include additional characteristics of the buildings that increase or reduce its seismic vulnerability (Milutinovic and Trendafiloski 2003). These modifiers can be related to features of the building or can represent a condition between the building and its surrounding environment. For instance, the conservation state is a feature of the buildings, but the position of the building respect to its neighbouring buildings is a condition between the building and its environment.

The data of the buildings A and B (Table 13.3) were used to perform the PSVA with the software USERISK2015. The detailed theoretical background of the VIM_P method is described in Aguilar-Meléndez (2011). Table 13.4 shows the main seismic vulnerability results of building A that were used to generate the seismic vulnerability curves of Fig. 13.3. Similarly, Table 13.5 shows the main seismic vulnerability results of building B, which were used to generate the seismic vulnerability curves of building B (Fig. 13.4).

If the seismic vulnerability curves of buildings A and B (Tables 13.3 and 13.4; Figs. 13.3 and 13.4) are compared, then it is possible to mention that building B is a building with higher uncertainty in its seismic vulnerability than that of building A. At the same time it is possible to affirm that building A corresponds to a building with higher levels of seismic vulnerability than those of building B. For instance, if

**Table 13.4** Results of seismic vulnerability of building A (Table 13.3). Values of $\alpha_m$ and $\beta_m$ that define the three seismic vulnerability curves that represent the seismic vulnerability of building A (Fig. 13.3)

| *Seismic vulnerability curve for a range of values of V = (−0.04,1.04)* | $\alpha_m$ | $\beta_m$ | *Mean* $\bar{V}$ | *Standard deviation* |
|---|---|---|---|---|
| Lower | 4.26 | 1.41 | 0.77 | 0.18 |
| Best | 4.74 | 1.11 | 0.84 | 0.16 |
| Upper | 5.37 | 0.81 | 0.90 | 0.14 |

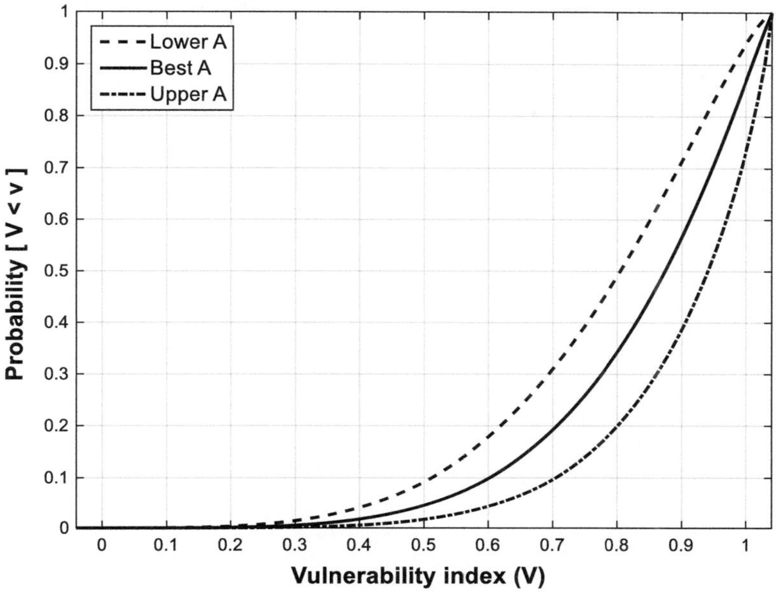


**Fig. 13.3** Seismic vulnerability curves of building A (Table 13.3), corresponding to values of Table 13.4

**Table 13.5** Results of seismic vulnerability of building B (Table 13.3). Values of $\alpha_m$ and $\beta_m$ that define the three seismic vulnerability curves that represent the seismic vulnerability of building B (Fig. 13.4)

| *Seismic vulnerability curve for a range of values of V = (−0.04,1.04)* | $\alpha_m$ | $\beta_m$ | *Mean* $\bar{V}$ | *Standard deviation* |
|---|---|---|---|---|
| Lower | 0.76 | 1.01 | 0.42 | 0.32 |
| Best | 1.08 | 0.91 | 0.55 | 0.31 |
| Upper | 1.54 | 0.81 | 0.67 | 0.28 |

only the best curve of seismic vulnerability of both buildings is considered, then it is possible to determine the values of Table 13.6; and according to these values the probability that the value of the vulnerability index could exceeds the value of 0.8 is 66% in building A and only of 27% in building B (Table 13.6). Therefore, it is possible to confirm that building A is more vulnerable than building B.

## 13.3.4 Seismic Risk in the RISKBUA-E Methodology

According to the RISKBUA-E methodology the PSRA of buildings can be computed applying the VIM_P procedure (Aguilar-Meléndez et al. 2010), which is included in the code USERISK2015. This last software compute the seismic risk of

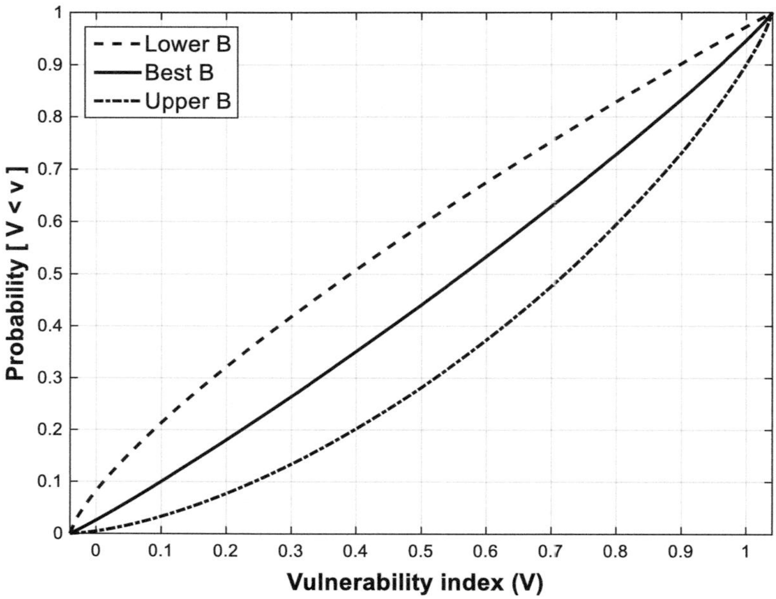


**Fig. 13.4** Seismic vulnerability curves of building B (Table 13.3), corresponding to values of Table 13.5

**Table 13.6** Values of probability that the vulnerability index *V* exceeds a specific value

| Building | P(*V* > 0.2) | P(*V* > 0.5) | P(*V* > 0.8) |
|---|---|---|---|
| A | 0.99 | 0.96 | 0.66 |
| B | 0.82 | 0.56 | 0.27 |

buildings considering three main elements: the seismic hazard where each building is located (PSHA), the seismic vulnerability of each building (PSVA), and a semi-empirical damage function that takes into account this seismic hazard and this seismic vulnerability in order to determine frequencies of exceedance of damage states of each building (Aguilar-Meléndez et al. 2010, 2012; Aguilar-Meléndez 2011). The way in that those elements are taken into account to estimate seismic risk is summarized in Eq. 13.2. This last equation adapted from McGuire (2004) is applied to compute the annual frequencies of exceedance of the damage *D*.

$$\nu[D > D_k] \approx \sum_{I}\sum_{V} P[D > D_k \mid V, I] P[V] \gamma'[I] \tag{13.2}$$

where $\nu[D > D_k]$ is the annual frequency of exceedance of the damage *D* and $D_k$ are the damage states (Table 13.8). In Eq. 13.2 the approximation is due to the fact that the annual frequency of occurrence of the intensity, $\gamma'[I]$, is considered as a value equivalent to $P[I]$. This last consideration is in agreement to McGuire (2004) that

**Table 13.7** Structural typologies and representatives values of their vulnerability in terms of the vulnerability index (Milutinovic and Trendafiloski 2003)

| Group | Tipology | Description | Representative values of the vulnerability[a] | | | | |
|---|---|---|---|---|---|---|---|
| | | | $V_I^{min}$ | $V_I^-$ | $V_I^*$ | $V_I^+$ | $V_I^{max}$ |
| Masonry | M31 | *Unreinforced masonry bearing walls with wooden slabs* | 0.460 | 0.650 | 0.740 | 0.830 | 1.020 |
| | M32 | *Unreinforced masonry bearing walls with masonry vaults* | 0.460 | 0.650 | 0.776 | 0.953 | 1.020 |
| | M33 | *Unreinforced masonry bearing walls with composite steel and masonry slabs* | 0.460 | 0.527 | 0.704 | 0.830 | 1.020 |
| | M34 | *Unreinforced masonry bearing walls with reinforced concrete slabs* | 0.300 | 0.490 | 0.616 | 0.793 | 0.860 |
| Reinforced concrete | RC32 | *Irregular concrete frames with unreinforced masonry infill walls* | 0.060 | 0.127 | 0.522 | 0.880 | 1.020 |
| Steel | S3 | *Steel frames with unreinforced masonry infill walls* | 0.140 | 0.330 | 0.484 | 0.640 | 0.860 |
| | S5 | *Steel and RC composite systems* | −0.020 | 0.257 | 0.402 | 0.720 | 1.020 |
| Wood | W | *Wood* | 0.140 | 0.207 | 0.447 | 0.640 | 0.860 |

[a]$V_I^*$ is the more likely value of the vulnerability index for the corresponding typology. $V_I^-$ and $V_I^+$ delimit the range of the probable values of the vulnerability index for the corresponding typology. $V_I^{min}$ and $V_I^{max}$ increase the range of the likely values of the vulnerability index in order to include the less likely values of the vulnerability index for the same typology.

addresses that the value of $\gamma'[I]$ can be considered as a "very close estimator of the probability", ($P[I]$), for values of $\gamma'[I] < 0.1$. At the same time Ellingwood (2006) considers a similar criteria according to the following phrase: "the annual probability and annual mean rate of occurrence are numerically interchangeable for randomly occurring events with probabilities less than 0.01/year". In the VIM_P method the seismic intensity corresponds to a macroseismic intensity. On the other hand, $P[V]$ is the probability of occurrence of the vulnerability index $V$. This last probability is computed from the respective curve of seismic vulnerability of each building. Also, $P\ [D > D_k \mid V, I]$ is the probability that damage $D$ will be exceeded, given that a seismic intensity $I$ and a seismic vulnerability $V$ have occurred. This last probability is assessed applying an earthquake damage function, which was proposed in the LM1 method of the Risk-UE project (Milutinovic and Trendafiloski 2003). Table 13.1 shows the description of the five damage states that are considered in the RISKBUA-E methodology. In Eq. 13.2 the total probability theorem is applied and it is considered that the intensity $I$ and the vulnerability $V$ are independent random variables (Aguilar-Meléndez et al. 2010). On the other hand, the damage function considered in the present methodology can be summarized by the semi-empirical function of Eq. 13.3.

**Table 13.8** Classification of damage to both masonry and reinforced concrete buildings (EMS-98) (Grünthal 1998)

| Masonry buildings | Reinforced concrete buildings |
|---|---|
| Grade 1. Negligible to slight damage (no structural damage, slight non-structural damage) | |
| Hair-line cracks in very few walls.<br>Fall of small pieces of plaster only.<br> | Fine cracks in plaster over frame members or in walls at the base.<br>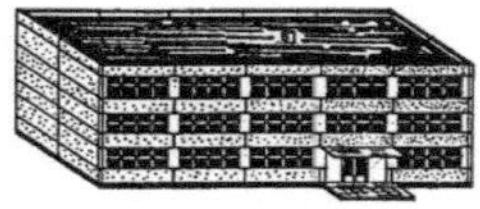 |
| Grade 2. Moderate damage (slight structural damage, moderate non-structural) | |
| Cracks in many walls.<br>Fall of fairly large pieces of plaster.<br> | Cracks in columns and beams of frames and in structural walls.<br>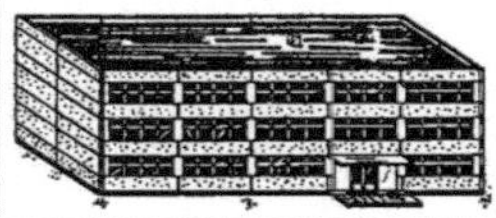 |
| Grade 3. Substancial to heavy damage (moderate structural damage, heavy non-structural damage). | |
| Large and extensive cracks in most walls. Roof tiles detach.<br> | Cracks in columns and bean column joints of frames at the base and at joints of coupled walls.<br>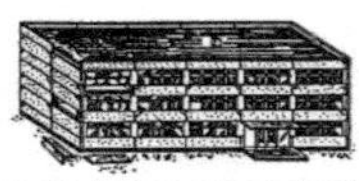 |
| Grade 4. Very heavy damage (heavy structural damage, very heavy non-structural damage) | |
| Serious failure of walls; partial structural failure of roofs and floors.<br> | Large cracks in structural elements with compression failure of concrete and fracture of rebars. Collapse of a few columns or of a single upper floor.<br> |
| Grade 5. Destruction (very heavy structural damage) | |
| Total or near total collapse.<br> | Collapse of ground floor or parts (e. g. wings) of buildings.<br> |

$$\mu_D = 2.5\left[1+\tanh\left(\frac{I+6.25\overline{V_I}-13.1}{2.3}\right)\right] \tag{13.3}$$

The Eq. 13.3 allows estimating only a mean damage grade, for this reason, in order to completely define the damage probability matrices, it can be assumed that the damage probability follows a beta probability density function (*pdf*) (Lantada et al. 2009a). The *pdf* beta type can be represented by Eq. 13.4.

$$PDF : p_\beta(x) = \frac{\Gamma(t)}{\Gamma(r)\Gamma(t-r)} \frac{(x-a)^{r-1}(b-x)^{t-r-1}}{(b-a)^{t-1}} \quad a \le x \le b; t, r > 0 \qquad (13.4)$$

where *a*, *b*, and *r* are parameters of the distribution, and Γ is the gamma function (Abramowitz and Stegun 1964). In the case of the VIM method *a* is set to *0* (no damage state) and *b* is *6* (destruction damage state) (Lantada et al. 2009a). On the other hand the parameter *t* affects the scatter of the distribution; therefore it can take different values. However, due to the fact that the damage distribution in the EMS98 scale (Grünthal 1998) is considered as a binomial distribution (Giovinazzi 2005) it was determined that *8* was an appropriate value for *t*, because with this value the beta distribution is similar to the binomial one (Lantada et al. 2009a). The parameter *r* is defined as a function of $\mu_D$ according to Eq. 13.5.

$$r = t\left(0.007\mu_D^3 - 0.0525\mu_D^2 + 0.2875\mu_D\right) \qquad (13.5)$$

Then, it is possible to compute the probability that the damage will be less or equal to a damage grade $P_\beta(x)$integrating this value in Eq. 13.4. between *0* and the *k*-damage grade (Lantada et al. 2009a). With that result it is possible to compute the probability of occurrence of each damage grade, $p_k$ using Eq. 13.6.

$$p_k = P_\beta(k+1) - P_\beta(k) \qquad (13.6)$$

The procedure summarized in Eq. 13.2 was applied by USERISK2015 to compute the seismic risk of buildings A and B. The computed results are represented in the seismic risk curves of Figs.13.5 and 13.6, respectively.

In summary, the seismic risk results of Fig. 13.5 were computed considering the seismic vulnerability curves of Fig. 13.3 (Building A) and the seismic hazard curve of Fig. 13.2 (truncated to a return period of 475 years). Similarly, the seismic risk results of Fig. 13.6 were computed considering the seismic vulnerability curves of Fig. 13.4 (Building B) and the seismic hazard curve of Fig.13.2 (truncated to a return period of 475 years).

According to the results of seismic risk, in the building A the annual frequency of exceedance of the damage state 4 is a value between $2.88 \times 10^{-5}$ and $7.77 \times 10^{-5}$, with a mean value of $4.77 \times 10^{-5}$. Similarly, in the building B the annual frequency of exceedance of the damage state 4 is a value between $1.09 \times 10^{-5}$ and $3.24 \times 10^{-5}$, with a mean value of $1.92 \times 10^{-5}$. It is important to remember that in the damage state 4 the partial collapse of the building can occur (Table 13.8). This is an example of the type of seismic risk results that can be obtained with the RISKBUA-E methodology. In the next section the main data and the main results about the assessment of the seismic risk of the 69,982 dwelling buildings of Barcelona with the RISKBUA-E methodology are described.

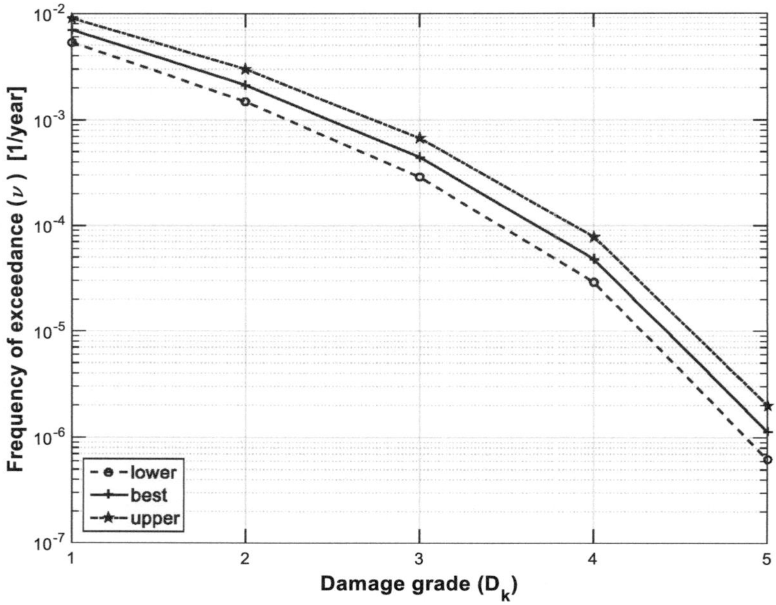


**Fig. 13.5** Seismic risk curves for building A located in a rock site of Barcelona (Table 13.3), for a seismic hazard truncated to 475 years

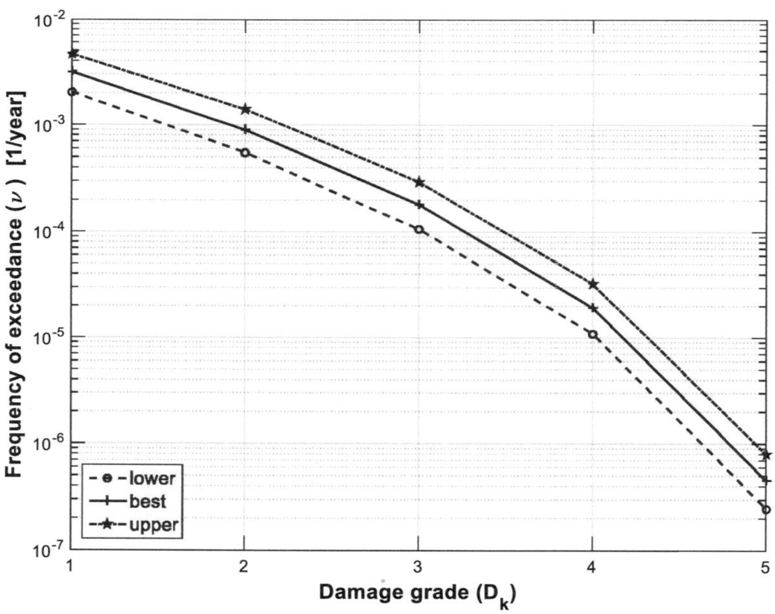


**Fig. 13.6** Seismic risk curves for building B located in a rock site of Barcelona (Table 13.3), for a seismic hazard truncated to 475 years

## 13.4 Application of the RISKBUA-E Methodology to Assess the Seismic Risk of Barcelona

### *13.4.1 Previous Studies of Seismic Risk of Barcelona*

The city of Barcelona (Fig. 13.7) has been part of different relevant international projects about seismic risk. For example, Table 13.9 shows data about recent projects where seismic risk results of Barcelona have been published. On the other hand Barcelona also has been involved in significant projects related to the resilience of cities. For instance, Barcelona has been included as a study site in the following projects: a) City Resilience Profiling Programme (CRPP) (UN-Habitat 2016) and; b) 100 Resilient Cities, Rockefeller Foundation (2017).

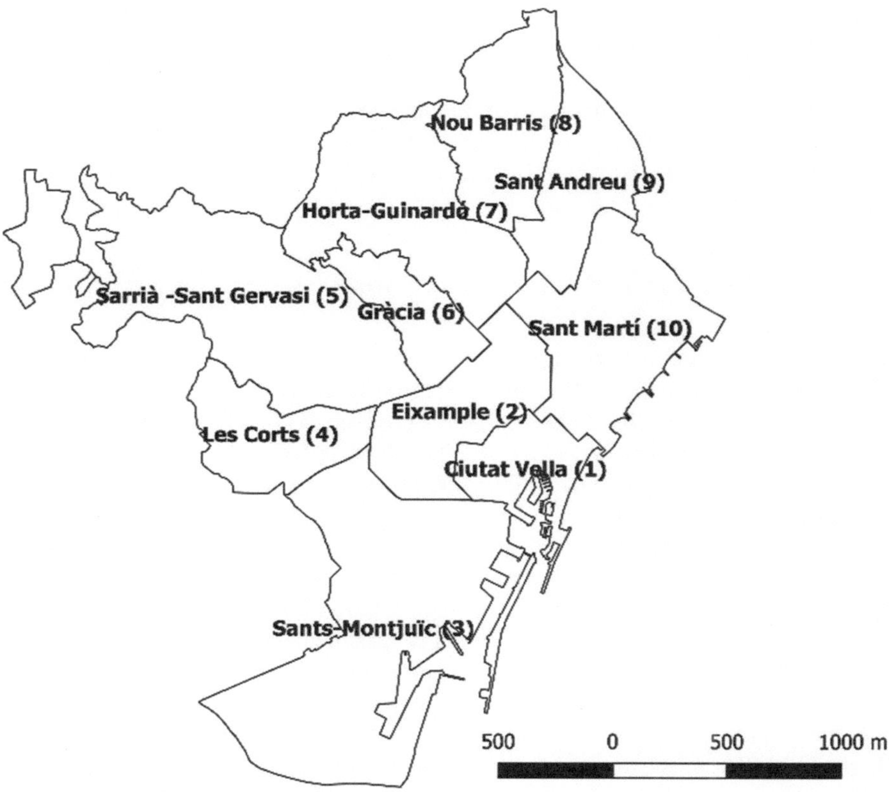


**Fig. 13.7** Map of the ten districts of Barcelona

**Table 13.9** Examples of recent projects where the seismic risk of dwelling buildings of Barcelona has been assessed

| No | Main results | Main methodology applied | Author (s) and date |
|---|---|---|---|
| 1 | Seismic risk of Barcelona in terms of mean damage grade for a specific seismic hazard scenario. | Risk-UE methods: (a) Vulnerability index and; (b) capacity spectrum. | Lantada et al. (2010) |
| 2 | Seismic risk of Barcelona in terms of probabilities of damage states. | A probabilistic version of the vulnerability index method of the Risk-UE project | Aguilar-Meléndez (2011); Aguilar-Meléndez et al. (2015b). |
| 3 | Seismic risk of Barcelona in terms of economic losses. | CAPRA | Marulanda et al. (2013) |

### *13.4.2 Basic Data of Barcelona*

The population of Barcelona in 2015 was of 1,604,555 inhabitants (Department d'Estadística 2016). However, the city hosts a significant number of persons that visit Barcelona for tourism or business. For instance according to tourism statistics 8,303,649 tourists stayed in a hotel during its visit to the city of Barcelona in 2015 (Ajuntament de Barcelona 2016). Figure 13.7 shows the main regions (called Districts), which were defined in Barcelona for administrative purposes.

### *13.4.3 The Seismic Hazard of Barcelona*

Barcelona is located in a region with low-to-moderate seismicity. The more active region near to Barcelona is the Pyrenees region to the northwest of that city. As a reference of the seismicity in the region it is possible to observe the epicentres map in Fig. 13.8. In this map it is also located Barcelona and it is possible to observe that according to Susagna and Goula (1999), earthquakes with a moderate macroseismic intensity have occurred below Barcelona and also near to Barcelona.

Additionally, Olivera et al. (2006) published intensity values that probably occurred in the city of Barcelona during the centuries XIV and XV. These values are shown in Table 13.10. Therefore, according to these values if a deterministic study to assess the seismic hazard is done, then it is possible to consider as reasonable for Barcelona a seismic scenario where a macroseismic intensity of VI-VII can occur. On the other hand according to Cid et al. (1999), the city of Barcelona can be divided into the seismic zones that are shown in Fig. 13.9 and that they are also described in Table 13.11. These seismic zones were defined according to the types of soils that were identified by Cid et al. (1999), and they are the basic reference to consider local site effects.

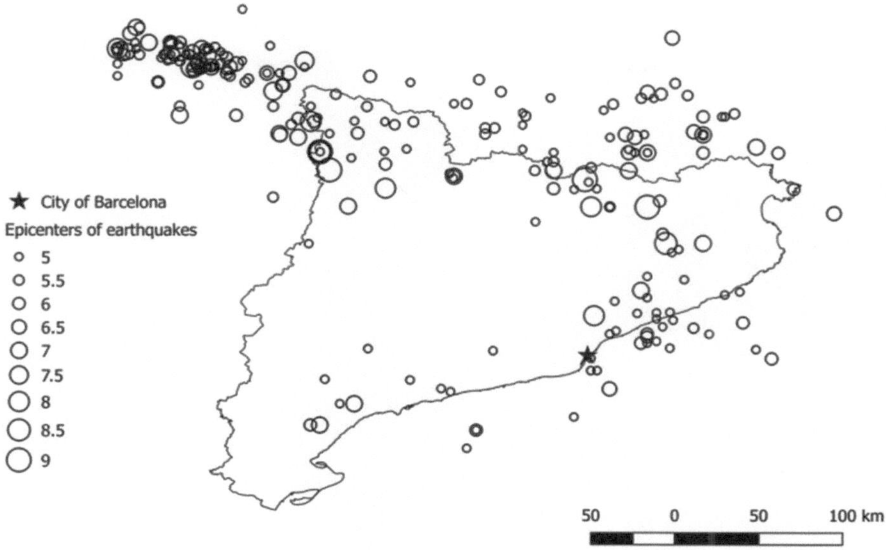


**Fig. 13.8** Epicentres of earthquakes with macroseismic intensities greater or equal to V, that occurred from 1152 to 1998 in the Catalonia region according to Susagna and Goula (1999)

**Table 13.10** Values of macroseismic intensities of earthquakes that according to Susagna and Goula (1999) occurred in Barcelona

| Date | Intensity |
|---|---|
| 03/03/1373 | V-VI |
| 02/02/1428 | VI-VII |
| 05/25/1448 | V-VI |

#### 13.4.3.1 Seismic Sources

Figure 13.10 shows the geometry of the seismic sources that were considered in CRISIS2015 to perform the PSHA for Barcelona. These seismic sources have been applied in previous studies of seismic hazard of Barcelona and Catalonia (Irizarry et al. 2010; Secanell et al. 2004; Irizarry et al. 2003).

The seismicity of each seismic source can be defined mainly with the following parameters: the minimum epicentral intensity considered ($I_{min}$), the maximum epicentral intensity possible in each seismic source ($I_{max}$), the annual frequency of exceedance of intensities greater or equal to $I_{min}$ ($\alpha$), and the slope ($\beta$) associated to the Guttenberg-Richter relation (Goula et al. 1997; Ordaz et al. 2015). These values related to the seismic source of Fig. 13.10 are shown in Table 13.12. These last parameters have been used in recent works about the seismic hazard of Barcelona and other regions of Catalonia (Irizarry et al. 2010; Secanell et al. 2004; Irizarry 2004).

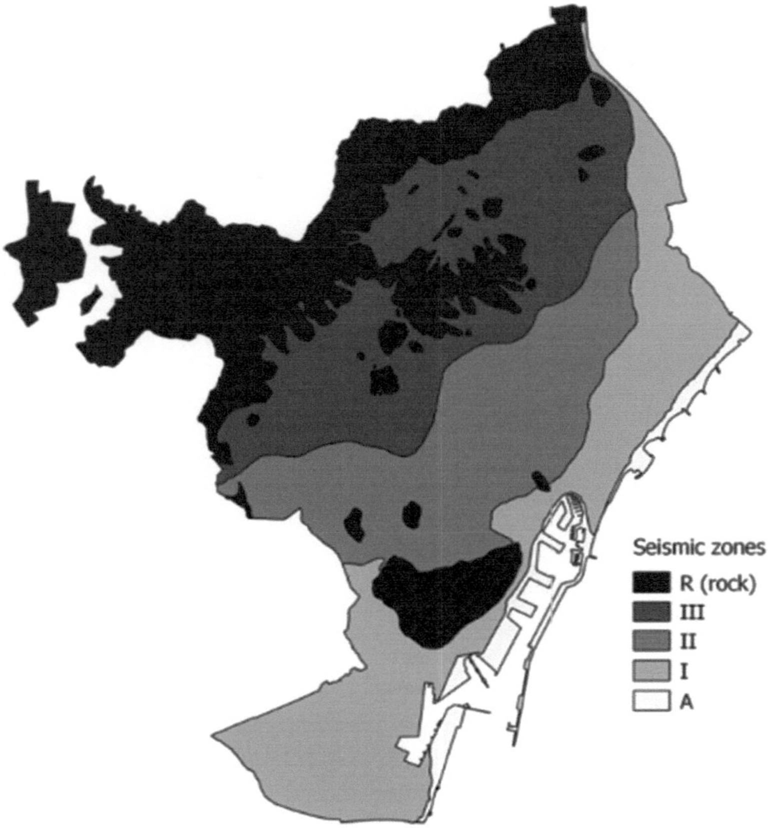


**Fig. 13.9** Seismic zones of Barcelona (Cid et al. 1999)

**Table 13.11** Description of the characteristics of the soil in each seismic zone of Barcelona (Cid et al. 1999)

| Zone | Description |
|---|---|
| R (0) | Rocky outcrops. |
| I | Holocene outcrops. |
| II | Pleistocene outcrops with Tertiary substrate, thick enough to influence the response. |
| III | Pleistocene outcrops without Tertiary substrate, thick enough to influence the response. |
| A | Artificial terrain. |

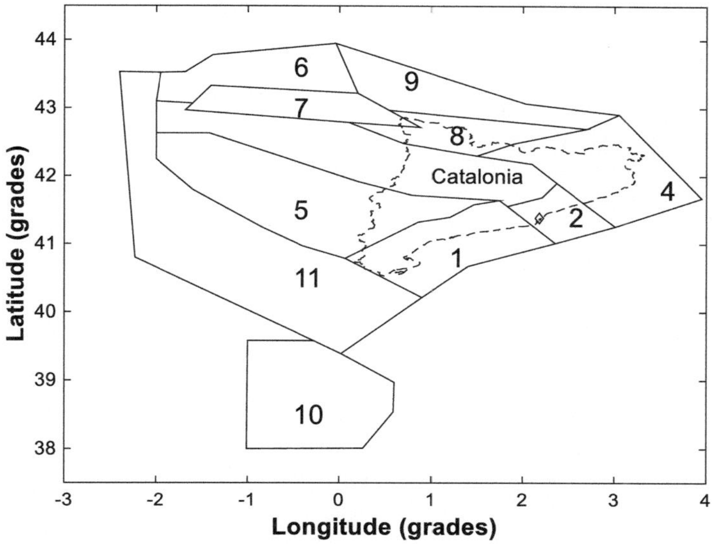


**Fig. 13.10** Geometry of seismic sources considered to assess the seismic hazard of Barcelona

#### 13.4.3.2 Attenuation Relationships

Another basic data to perform a PSHA with CRISIS2015 are the GMPE. For this work two attenuation relationships defined by López Casado et al. (2000) were chosen, one of them is called "Attenuation Relationship for High Attenuation" and the other one is called "Attenuation Relationship for Low Attenuation". These attenuation relationships were mainly determined with catalogues of map of isoscists of the Iberian Peninsula (López Casado et al. 2000). Both attenuation relationships are represented by the same general Eq. 13.7, but they are differenced according to the values of $f\,(I_{epic})$, $a_2$, $a_3$, $R_0$ and $\sigma$ that are showed in Table 13.13.

$$I = f\left(I_{epic}\right) - a_2 \ln \Delta - a_3 \Delta \tag{13.7}$$

where $I$ is the macroseismic intensity to a focal distance $\Delta = \left(R^2 + R_0^2\right)^{1/2}$ with $R$ equal to the epicentral distance in km, and $R_0$ a value used to improve the fitting and it means focal depth in km; $I_{epic}$ is the epicentral macroseismic intensity MSK; $f\,(I_{epic})$ is the value according to Table 13.13; $a_2$ and $a_3$ are coefficients with the values shown in Table 13.13. According to the specification of López Casado et al. (2000) only the attenuation relationship for low attenuation must be applied to the seismic source that represents the seismicity in the Pirineus (Table 13.14).

According to the seismic hazard curve of Barcelona (Fig. 13.11) that was computed by CRISIS2015 for a rock site, the macroseismic intensity that has a return period of 475 years corresponds to a value equal to VI. The seismic hazard results that were obtained in the present work have important coincidences with the results

**Table 13.12** Seismic parameters of the seismic sources that were considered to assess seismic hazard in Barcelona (Secanell et al. 2004)

| Seismic source | $\alpha$ | $\beta$ | $Cv\ (\beta)$* | $I_{min}$[a] | $E(I_{max})$ | Uncertainty interval of $I_{max}$ |
|---|---|---|---|---|---|---|
| 1 | 0.100 | 1.864 | 0.3 | V | VII | 1 |
| 2 | 0.128 | 1.608 | 0.202 | V | VIII | 1 |
| 4 | 0.157 | 1.256 | 0.148 | V | IX | 1 |
| 5 | 0.040 | 1.319 | 0.283 | V | VIII | 1 |
| 6 | 0.099 | 1.977 | 0.324 | V | VI | 1 |
| 7 | 0.957 | 1.420 | 0.082 | V | VIII | 2 |
| 8 | 0.218 | 1.716 | 0.143 | V | VIII | 1 |
| 9 | 0.070 | 1.737 | 0.123 | V | VII | 1 |
| 10 | 0.635 | 1.201 | 0.069 | V | X | 1 |
| 11 | 0.060 | 0.886 | 0.273 | V | VIII | 1 |

[a]$I_{min}$ is the minimum macroseismic intensity assigned to the seismic source; $Cv(\beta)$ is the variation coefficient of $\beta$; $E(I_{max})$ is the expected value of the maximum macroseismic intensity that in this case was considered equal to the $I_{max}$ observed.

**Table 13.13** Values of the two attenuations relationships that were determined by López Casado et al. for the Iberian Peninsula in terms of macroseismic intensities (2000)

| Attenuation relationship | $f(I_{epic})$ | $a2$ | $a3$ | $R_0$ | $\sigma$ |
|---|---|---|---|---|---|
| 1) For high attenuation (AR-HA) | $6.016+0.090 \cdot I_{epic}+0.069 \cdot I_{epic}^2$ | 1.477 | 0.01035 | 4 | 0.46 |
| 2) For low attenuation (AR-LA) | $5.557+0.902 \cdot I_{epic}+0.014 \cdot I_{epic}^2$ | 1.762 | 0.00207 | 2 | 0.59 |

**Table 13.14** Attenuation relationships assigned to each seismic source

| Seismic source | Attenuation relationships assigned |
|---|---|
| 1,2,4,5,6,8,9,10,11 | AR-HA (López Casado et al. 2000) |
| 7 | AR-LA (López Casado et al. 2000) |

that were obtained in previous studies. For instance, Secanell et al. (2004) obtained a mean value of 6.5 for the macroseismic intensity related to a return period of 475 years. A similar value was estimated by Goula et al. (1997).

In order to assess local effects the criterion that was applied by Lantada (2007) and Aguilar-Meléndez (2011) it was considered still valid for the present work. According to this criterion the local effects for any type of ground except rock can be increased in a half degree of macroseismic intensity, with respect to the macroseismic intensity that can occur in rock. According to this criterion was possible to obtain the seismic hazard curve con local effects for the seismic zones I, II, II and A that is shown in Fig. 13.11.

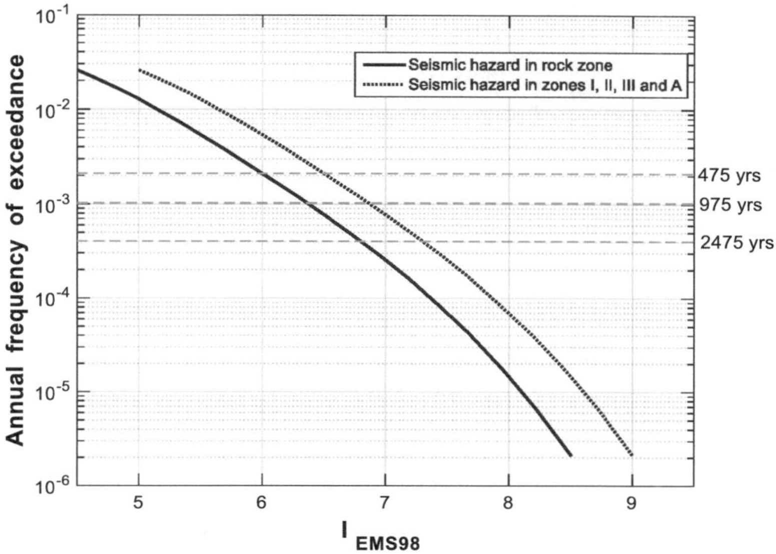


**Fig. 13.11** Seismic hazard of Barcelona in the following seismic zones: (**a**) R (rock) [continuous line]; (**b**) I, II, III and A [dashed line]

## *13.4.4 The Seismic Vulnerability of Barcelona*

### 13.4.4.1 The Dwelling Buildings of Barcelona

The data of the buildings is essential in order to perform a PSVA, according to the RISKBUA-E methodology. In the present work a valuable data base of the main data of buildings was used to compute the seismic vulnerability of the dwelling buildings of Barcelona. This data base has been generated and improved in the last 30 years, and it contains valuable information of each building of the city. The origin of this data base was a work from the mid '80s, which was done with the purpose of determining the economic value of the buildings of the city in order to determine a tax related to each property (Aguilar-Meléndez 2011). Unfortunately, the new procedures that are applied nowadays in Barcelona to determine the economic value of a property, in order to determine taxes, do not consider the structural typology of the buildings. Therefore in the last years the determination of the structural typology of the new buildings of Barcelona is not clearly included in a data base of the government of Barcelona. Fortunately, the major part of the new buildings in Barcelona has been built with reinforced concrete. Therefore, it is possible to assume a small and negligible error that occurs, if it is considered that all the new buildings of Barcelona have been built with reinforced concrete (Lantada et al. 2009b).

**Table 13.15** Classification by structural typology of the dwelling buildings in the 10 districts of Barcelona

| | | *Typologies* | | | | | | | |
|---|---|---|---|---|---|---|---|---|---|
| *District* | *No. buildings.* | *M31* | *M32* | *M33* | *M34* | *RC32* | *S3* | *S5* | *W* |
| 1. Ciutat Vella | 5675 | 4069 | 112 | 690 | 151 | 459 | 101 | 47 | 46 |
| 2. Eixample | 8723 | 1624 | 57 | 3990 | 384 | 2309 | 182 | 155 | 22 |
| 3. Sants-Montjuïc | 7410 | 2288 | 44 | 1974 | 816 | 1874 | 166 | 243 | 5 |
| 4. Les Corts | 2587 | 428 | 29 | 539 | 352 | 1155 | 49 | 33 | 2 |
| 5. Sarrià-Sant Gervasi | 8152 | 1426 | 206 | 1923 | 1773 | 2539 | 124 | 140 | 21 |
| 6. Gràcia | 6976 | 2049 | 32 | 2023 | 1003 | 1635 | 80 | 113 | 41 |
| 7. Horta-Guinardó | 9762 | 1321 | 216 | 2324 | 3354 | 2289 | 48 | 195 | 15 |
| 8. Nou Barris | 6912 | 1025 | 75 | 1613 | 2169 | 1761 | 51 | 194 | 24 |
| 9. Sant Andreu | 7000 | 1728 | 157 | 981 | 2002 | 1890 | 101 | 133 | 8 |
| 10. Sant Martí | 6785 | 2240 | 12 | 1270 | 353 | 2443 | 201 | 234 | 32 |
| Total | 69,982 | 18,198 | 940 | 17,327 | 12,357 | 18,354 | 1103 | 1487 | 216 |

The data base of the Town Council of Barcelona include a code to identify each building according to a constructive typology. Therefore, it was necessary to define an equivalence between the typologies considered by the Town Council of Barcelona and the structural typologies that were proposed in the Risk-UE project (Lantada 2007). According to this equivalence, the major part of the buildings of Barcelona can be classified into some of the typologies of the Risk-UE project included in Table 13.7.

Table 13.15 shows the classification of dwelling buildings by district of Barcelona according to the structural typologies that were defined in the Risk-UE project (Milutinovic and Trendafiloski 2003). According to the data of this table from the total dwelling buildings of Barcelona 69.76% are masonry buildings, and 26.23% are reinforced concrete buildings. At the same time, it is posssible to observe that from the total dwelling buildings of the Ciutat Vella district (Old Town) 88.49% are masonry buildings, and only 8.09% are reinforced concrete buildings.

#### 13.4.4.2 Results of Seismic Vulnerability

We applied the software USERISK2015 to obtain the seismic vulnerability curves for each dwelling building of Barcelona. These individual curves are the vulnerability curves used to compute seismic risk. However, to determine an average of the seismic vulnerability of a group of buildings, it is possible to estimate an equivalent seismic vulnerability curve that represents the seismic vulnerability of a group of buildings. To determine this equivalent and simplified curve, it is possible to compute the geometric mean of the values of $\alpha$ and $\beta$ that define the curve of seismic

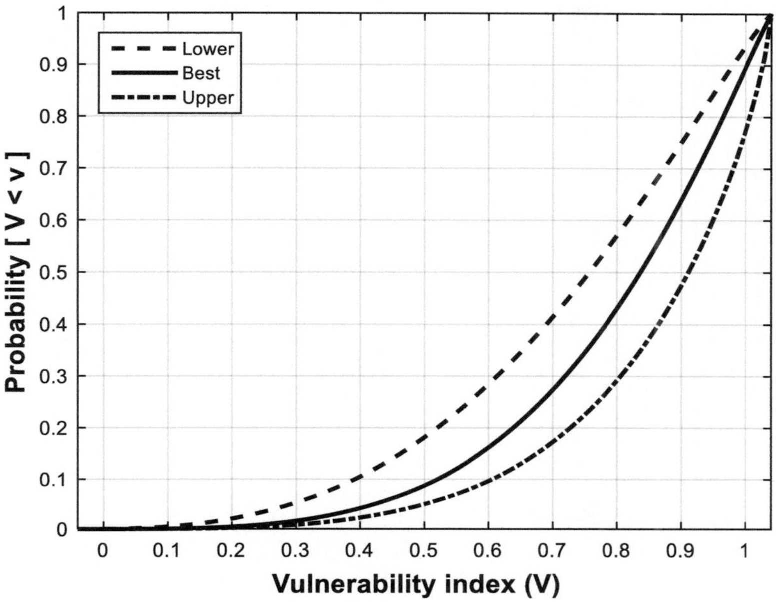


**Fig. 13.12** Seismic vulnerability curves of the dwelling buildings of Barcelona. Case 1 (Table 13.16)

**Table 13.16** Values that define the equivalent vulnerability curves of the 69,982 dwelling buildings of Barcelona (Case 1, where the regional modifiers of the seismic vulnerability are considered)

| Seismic vulnerability curve for a range of values of $V = (-0.04, 1.04)$ | $\alpha_m$ | $\beta_m$ | Mean | Standard deviation |
|---|---|---|---|---|
| Lower | 2.77 | 1.18 | 0.72 | 0.22 |
| Best | 3.73 | 1.10 | 0.79 | 0.19 |
| Upper | 3.75 | 0.76 | 0.86 | 0.17 |

vulnerability of each building of the group. For this purpose, the Eqs. 13.8 and 13.9 can be applied.

$$\alpha_{g-mean} = \sqrt[n]{\alpha_1 \cdot \alpha_2 \cdots \alpha_n} \tag{13.8}$$

$$\beta_{g-mean} = \sqrt[n]{\beta_1 \cdot \beta_2 \cdots \beta_n} \tag{13.9}$$

Figure 13.12 shows the equivalent curves that represent in a simplified way the seismic vulnerability for the 69,982 dwelling buildings of Barcelona for Case 1, where the regional modifiers of vulnerability of Eq. 1 are considered. Meanwhile, Fig. 13.13 also shows the equivalent curves but for Case 2, where the regional

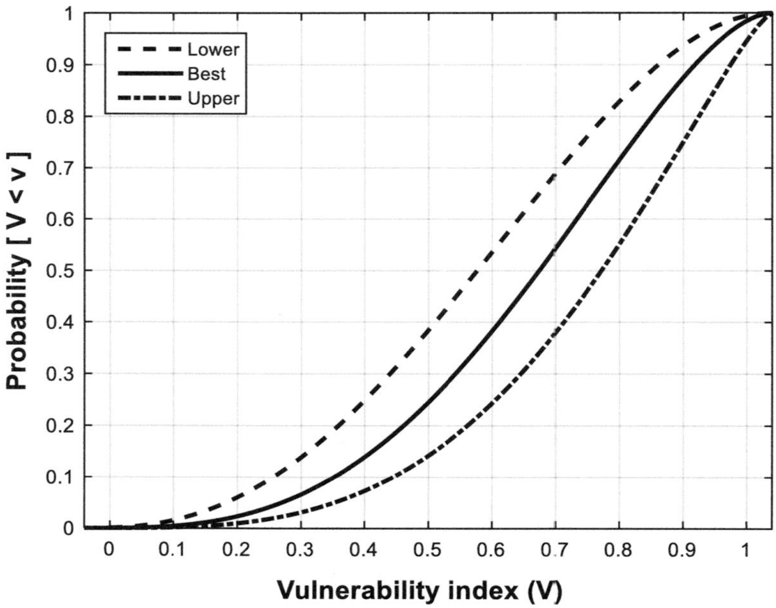


**Fig. 13.13** Seismic vulnerability curves of the dwelling buildings of Barcelona. Case 2 (Table 13.17)

**Table 13.17** Values that define the equivalent vulnerability curves of the 69,982 dwelling buildings of Barcelona (Case 2, where the regional modifiers of the seismic vulnerability are not considered)

| Seismic vulnerability curve for a range of values of $V = (-0.04, 1.04)$ | $\alpha_m$ | $\beta_m$ | Mean | Standard deviation |
|---|---|---|---|---|
| Lower | 2.65 | 2.07 | 0.57 | 0.22 |
| Best | 3.19 | 1.79 | 0.65 | 0.21 |
| Upper | 3.43 | 1.34 | 0.74 | 0.20 |

modifiers of Eq. 1 are not considered. On the other hand, Fig. 13.14 shows the equivalent vulnerability curves for each district of Barcelona, for Case 1.

According to the results of seismic vulnerability shown in Fig. 13.14 the district with the major seismic vulnerability of the city is Ciutat Vella, and the district with the lowest seismic vulnerability of the city is Nou Barris. The values that define each vulnerability curve of Fig. 13.14 are shown in Table 13.18.

The important differences between the seismic vulnerability curves of Barcelona of Fig. 13.12 and the seismic vulnerability curves of Fig. 13.13 are only due that in the first case the regional modifiers of the seismic vulnerability were considered (Case 1), and in the second case, the regional modifiers were not considered (Case 2). According to these results, it is possible to conclude that the

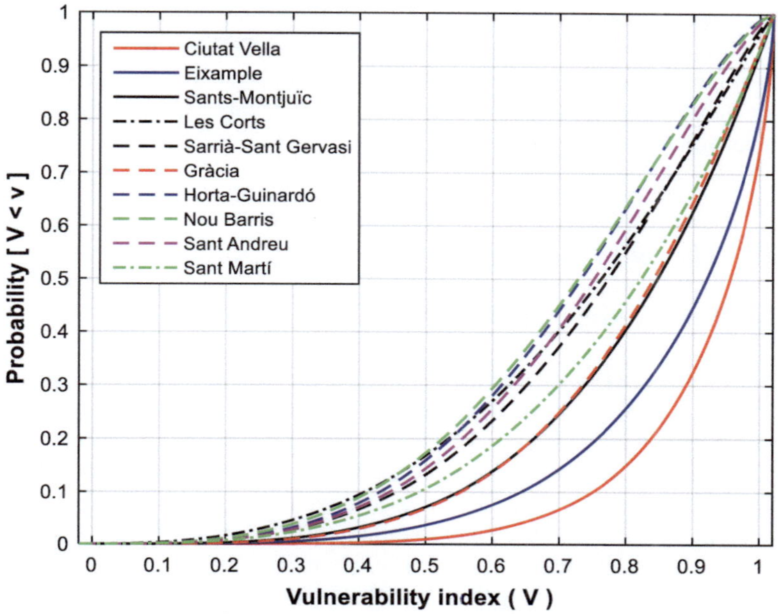


**Fig. 13.14** Representative best curves of the seismic vulnerability of the dwelling buildings of the districts of Barcelona, defined by the values of Table 13.18 (Case 1)

**Table 13.18** Parameters that define the *pdf* beta type that define the seismic vulnerability of the dwelling buildings of each district of Barcelona. Seismic vulnerability curves for a range of values of $V = (-0.04, 1.04)$ (Case 1, where the regional modifiers of the seismic vulnerability are considered)

| | *Lower seismic vulnerability* | | | | *Best seismic vulnerability* | | | | *Upper seismic vulnerability* | | | |
|---|---|---|---|---|---|---|---|---|---|---|---|---|
| | V = (−0.04,1.04) | | | | | | | | | | | |
| *District* | $\alpha_L$ | $\beta_L$ | $\overline{V_{I\_L}}$ | $\sigma_L$ | $\alpha_m$ | $\beta_m$ | $\overline{V_I}$ | $\sigma_{\overline{V}}$ | $\alpha_U$ | $\beta_U$ | $\overline{V_{I\_U}}$ | $\sigma_U$ |
| 1. Ciutat Vella | 3.85 | 0.63 | 0.89 | 0.19 | 5.62 | 0.64 | 0.93 | 0.12 | 5.65 | 0.49 | 0.95 | 0.11 |
| 2. Eixample | 2.69 | 0.69 | 0.82 | 0.16 | 3.92 | 0.67 | 0.88 | 0.16 | 4.1 | 0.53 | 0.92 | 0.14 |
| 3. Sants-Montjuïc | 2.7 | 1.04 | 0.74 | 0.21 | 3.84 | 1.01 | 0.82 | 0.18 | 3.83 | 0.68 | 0.88 | 0.16 |
| 4. Les Corts | 2.12 | 1.26 | 0.64 | 0.22 | 2.78 | 1.12 | 0.73 | 0.22 | 2.89 | 0.75 | 0.82 | 0.20 |
| 5. Sarrià-Sant Gervasi | 2.55 | 1.37 | 0.66 | 0.25 | 3.43 | 1.28 | 0.75 | 0.20 | 3.42 | 0.86 | 0.82 | 0.19 |
| 6. Gràcia | 2.91 | 1.13 | 0.74 | 0.23 | 4.07 | 1.09 | 0.81 | 0.18 | 3.99 | 0.72 | 0.87 | 0.16 |
| 7. Horta-Guinardó | 2.94 | 1.79 | 0.63 | 0.22 | 3.67 | 1.6 | 0.71 | 0.20 | 3.65 | 1.08 | 0.79 | 0.19 |
| 8. Nou Barris | 2.81 | 1.77 | 0.62 | 0.22 | 3.4 | 1.52 | 0.71 | 0.21 | 3.47 | 1.04 | 0.79 | 0.19 |
| 9. Sant Andreu | 2.79 | 1.6 | 0.65 | 0.22 | 3.61 | 1.46 | 0.73 | 0.20 | 3.61 | 0.98 | 0.81 | 0.19 |
| 10. Sant Martí | 2.3 | 1.04 | 0.70 | 0.22 | 3.19 | 0.98 | 0.79 | 0.20 | 3.15 | 0.63 | 0.86 | 0.18 |

influence of the regional modifiers in the seismic vulnerability of the buildings is very important for the buildings of Barcelona. At the same time, it is possible to highlight that the values of these modifiers are a proposal of experts, therefore they are values that can suffer important changes according to the criteria of the expert consulted.

## *13.4.5 The Seismic Risk of Barcelona*

### 13.4.5.1 Seismic Risk of the Dwelling Buildings of Barcelona

According to the RISKBUA-E methodology, the results of the PSHA and the results of the PSVA can be used to perform a PSRA. For this purpose, we applied USERISK2015. Figure 13.15 shows average seismic risk curves of the 69,982 dwelling buildings of Barcelona for Case 1. According to the results of seismic risk for Case 1 it is possible to affirm the following: 34.29% of the dwelling buildings of Barcelona have a probability equal or greater than 1% of reach or exceed the damage grade 4 in the next 50 years. In other words, 34.29% of the dwelling buildings of Barcelona have a probability equal or greater than 1% of suffer some kind of partial collapse of their structure in the next 50 years. But at the same time, no building in Barcelona has a probability equal or greater than 1% of experiencing the damage grade 5. However, if the seismic risk is computed for Case 2, then it is possible to obtain the results that are shown in Fig. 13.16. According to these results it is possible to affirm that the whole dwelling buildings of Barcelona have a probability lower than 1% of reach or exceed the damage grade 4 in the next 50 years.

When the seismic risk results are analysed in terms of the districts of the city, it is possible to observe that in the Eixample District for the Case 1 the seismic risk results indicates that 62.17% of the dwelling buildings of the Eixample District have a probability equal or greater than 1% of reaching or exceeding the damage grade 4 in the next 50 years. In other words, 62.17% of the dwelling buildings of the Eixample District have a probability equal or greater than 1% of experiencing some kind of partial collapse of their structures in the next 50 years (Fig. 13.17). But at the same time, no dwelling building of the Eixample District has a probability equal or greater than 1% of suffering the damage grade 5. It is important to remember that in the damage state 4, partial collapse of the building can occur. On the other hand, according to the seismic risk results for the Case 2, all the dwelling buildings of the Eixample District have a probability lower than 1% of reaching or exceeding the damage grade 4 in the next 50 years (Fig. 13.18).

In order to facilitate the comparison of the results of the present work with the results of previous works, we considered that the total economic value of about 70000 dwelling buildings of Barcelona that were analysed is of 31522.8 million of euros. This economic value was chosen for comparative purposes of a work of

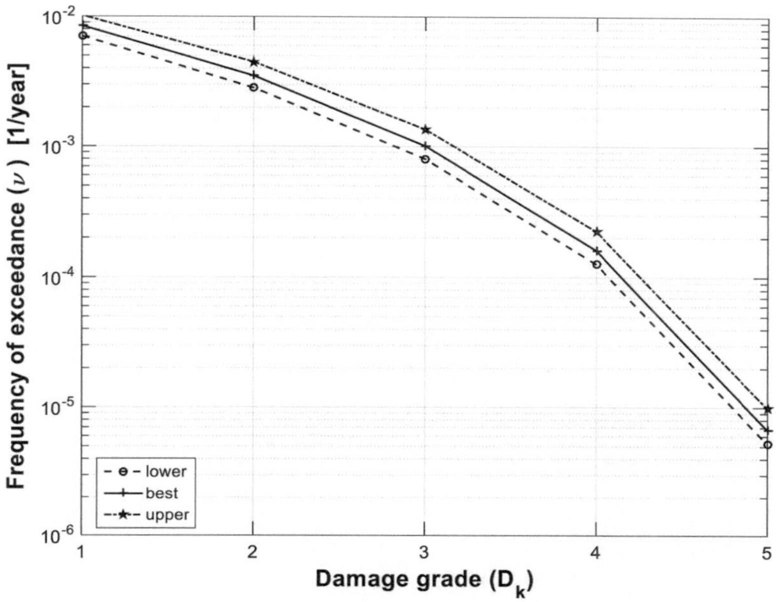


**Fig. 13.15** Seismic risk curves of the dwelling building of Barcelona computed with the seismic hazard of Barcelona truncated to 475 years (Case 1)

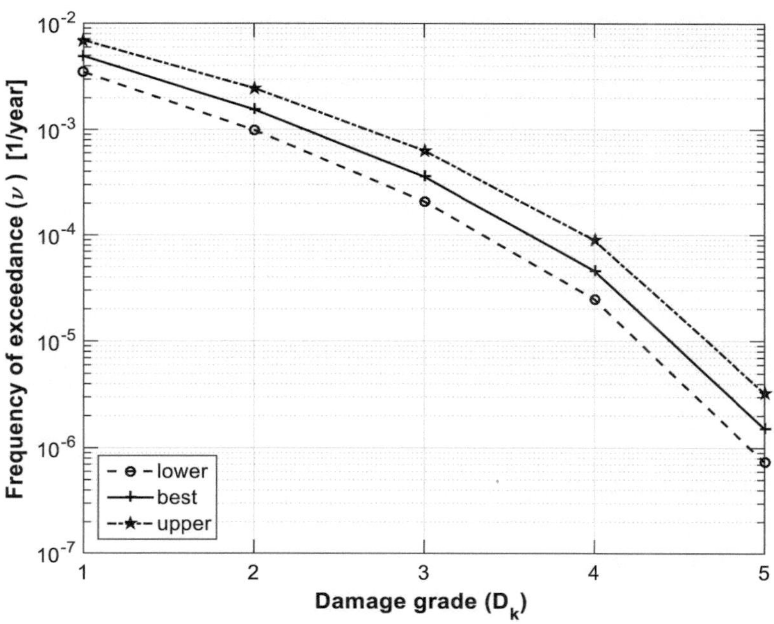


**Fig. 13.16** Seismic risk curves of the dwelling building of Barcelona computed with the seismic hazard of Barcelona truncated to 475 years (Case 2)

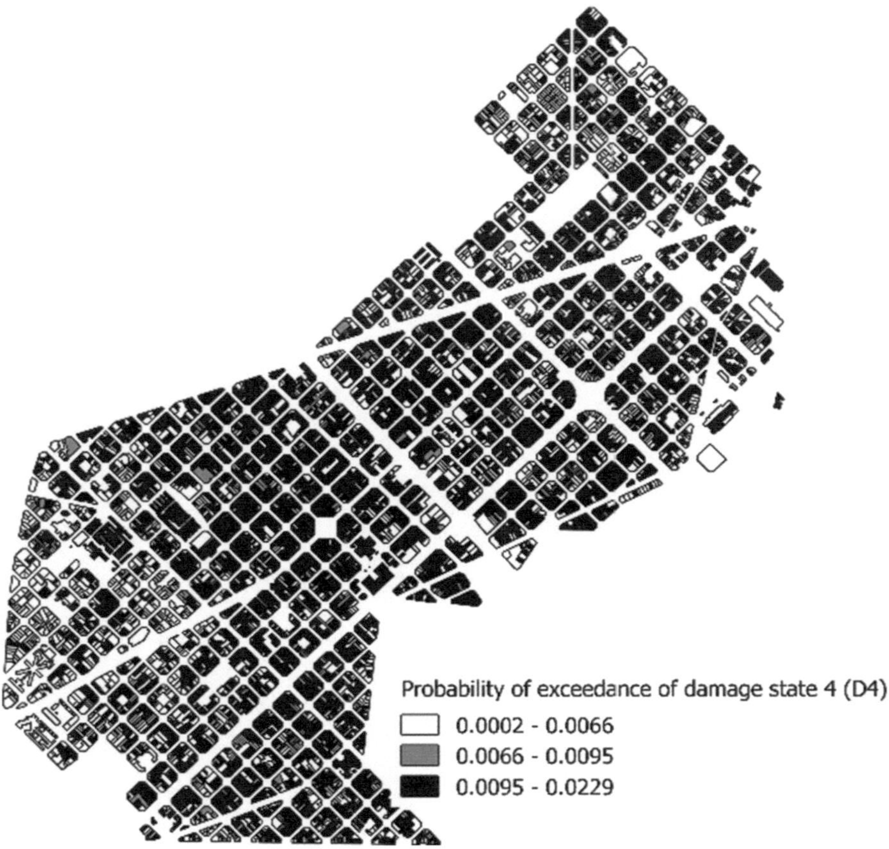


**Fig. 13.17** Map of the Eixample District of Barcelona that shows the probability that the damage state 4 occurs over the next 50 years in each building, considering a seismic hazard curve truncated to 475 years (Case 1)

Marulanda et al. (2013). At the same time we considered the damage factors that were proposed by Dolce et al. (2006). It is possible to observe in Fig.13.19 that the losses computed by Marulanda et al. (2013) are in good agreement with the losses computed in the present work for the Case 2.

## 13.5 Possible Uses of the Results of Seismic Risk of Barcelona

The results of seismic risk of Barcelona computed through the RISKBUA-E methodology can have different uses. Some of these possible uses are the following:

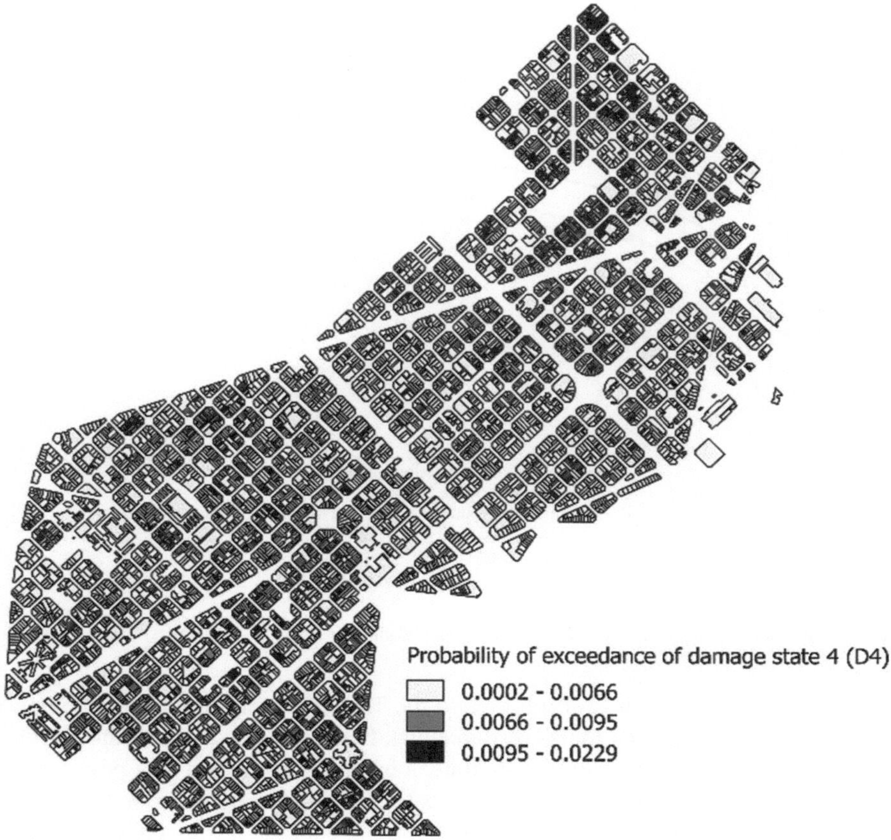


**Fig. 13.18** Map of the Eixample District of Barcelona that shows the probability that the damage state 4 occurs over the next 50 years in each building, considering a seismic hazard curve truncated to 475 years (Case 2)

- The type of maps of seismic risk computed in the present work can be used by the Civil Protection Department of the Town Council (*Ajuntament*) of Barcelona, to define strategies to improve the emergency attention to the population during possible scenarios of damage due to earthquakes;
- The knowledge of the seismic losses of the dwelling buildings of Barcelona is information that can be used by the Town Council of Barcelona as a reference to define if it is necessary to change the amount of money that must be assigned in the annual public budget to attend emergencies;
- The seismic risk results can be used by the Town Council of Barcelona to evaluate the convenience of promote as mandatory that all the buildings of the city must to have a damage insurance that includes damage due to earthquakes;
- The maps of seismic risk computed can be used by the Town Council of Barcelona to highlight the importance of defining a program, norm, or law that could

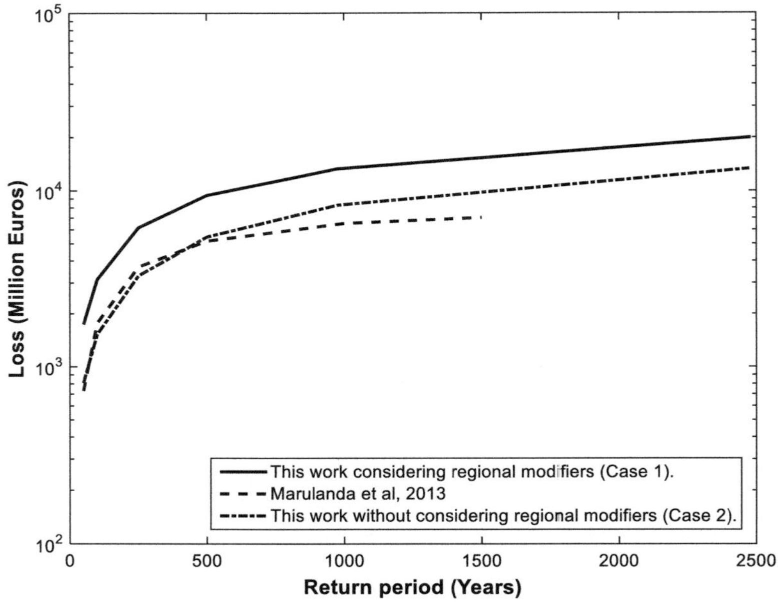


**Fig. 13.19** Seismic loss curves of buildings of Barcelona

be established as mandatory to give to the Town Council the main data about the structures of the buildings of Barcelona. Because nowadays, the data of the structures of the new buildings of Barcelona are not included in a public data base. Therefore, if the Town Council does not obtain the basic structural information of the new buildings of Barcelona, the data required to compute the seismic risk of buildings of Barcelona will be incomplete in future assessments. It is convenient to note that the structural data of the buildings are also essential data to assess other kind of risks as risk by hurricane, or risk by fire;

- The results of seismic risk can be used as reference to define a program to assess with more detail the buildings, which were identified in the present study as buildings with significant levels of both seismic vulnerability and seismic risk. This is especially relevant because of the age of numerous buildings of Barcelona. For instance, the average age of the dwelling buildings of the Eixample District is about 80 years. Therefore, the results of the present work can be also used as a reference to generate an integral program to define which buildings must to be evaluated with priority, in order to determine both their structural conditions and probable behavior due to different types of loads: live, dead and accidental (earthquake, wind, snow, tsunami, etc.). This is especially relevant because numerous buildings have exceed their original lifetime;
- The results of seismic risk of the present work can be used as reference to determine an appropriate criterion to define the level of seismic risk, which can be considered by the authorities of Barcelona and Catalonia as an acceptable level of seismic risk for the existing dwelling buildings of the city. As an example of

decision criterion it is possible to remember that the Standard ASCE/SEI 7–10 (ASCE 2010) states that the buildings are allowed to have a probability of collapse of 0.01 (1%) due to an earthquake that has a return period of 475 years.

## 13.6 Conclusions

### *13.6.1 Conclusions about the Seismic Risk of Barcelona*

In the last fifteen years the seismic risk of dwelling buildings of Barcelona has been assessed. For this reason, nowadays, there are important results about the levels of seismic risk of dwelling buildings of Barcelona. However, with the purpose of applying part of the recommendations of the Sendai Framework to increase the resilience of the city of Barcelona, it is convenient to execute actions, as the following:

- To publish the seismic risk results available for Barcelona in an official document or web site, where the information about the different hazards and risks that affect to the city can be observed;
- To include as a technical requirement the assessment of the seismic risk and others risks related to any building of Barcelona, which will be rented or sold. A summary of the results of the assessment will be published;
- To perform a project to divulgate the seismic risk results of the dwelling buildings of Barcelona, in order to contribute to increase the knowledge of the citizens about the seismic risk of the city. This is relevant due that recent studies confirm that it is necessary to increase the actions oriented to communicate the seismic risk that exists in the cities, especially in cities with a larger seismicity than Barcelona (Marincioni et al. 2012);
- To create a program to verify or improve the connections of the nonstructural elements of the buildings of Barcelona with their main structure. The execution of this program can contribute to reduce the risk of damage due to earthquakes, but also the damage due to excessive winds or other similar perturbations. For this purpose, it is necessary to recognize that in the recent earthquake of Lorca (Table 13.1), the few deaths related to this earthquake were mainly due to the collapse of nonstructural elements (Carreño et al. 2012).

In spite of the fact that the emphasis of this work is on the assessment of the seismic risk of buildings, some data, procedures, and results of the present work can be incorporated into a multi-hazard and multi-risk analysis. For instance, part of the seismic hazard results computed for Barcelona in the present work can be considered as a starting point to assess the hazard of fire that can be triggered due to earthquakes. Similarly, it is possible to assess the hazard for tsunami for Barcelona based on the data and results of the seismic hazard assessment. On the other hand, the seismic risk curves of economic losses computed in the present work can be aggregated to other similar curves due to different hazards that are not related to earthquakes, in order to express the global risk in a curve that incorporates risks due to different hazards. Finally, we underline that we are agree with the idea that states that both single hazard approach and multiple hazard approach are nowadays relevant tools to assess the levels of risk in the cities, in order to use the results that can be generated by both approaches to increase the resilience in the cities.

### *13.6.2 Conclusions About the RISKBUA-E Methodology*

The RISKBUA-E methodology is a valuable methodology to assess the seismic risk of buildings in urban zones. The results of seismic risk computed by this last methodology in the present work are in reasonable agreement with the seismic risk results, which were computed for Barcelona with the CAPRA methodology that has been widely validated. The RISKBUA-E methodology can be easily applied to assess the seismic risk of any other city. The seismic risk results obtained by this methodology have important applications and they can be communicated in versatile ways.

During the application of the RISKBUA-E methodology it is convenient to put especial attention to the regional vulnerability modifiers, because these modifiers could have a significant influence in both the seismic vulnerability results and the seismic risk results. Also it is important to highlight that the regional modifiers were originally proposed for the assessment of seismic scenarios according to the VIM method. Therefore, with the results obtained in the VIM_P method is reasonable to consider a revision of these regional modifiers. The RISKBUA-E methodology allows to obtain seismic risk results in a relative short time due to the fact that both codes CRISIS2015 and USERISK2015 are versatile and powerful tools.

Finally, it is convenient to allocate resources and efforts to define standard procedures to incorporate the RISKBUA-E methodology in multi-hazard and multi-risk approaches. These procedures must allow executing both single-hazard and multi-hazard assessments, and also to perform both single-risk and multi-risk assessments.

**Acknowledgements** Thanks to University of Veracruz, Barcelona Supercomputing Center, CONACYT and PRODEP. This research has been partially funded by the Ministry of Economy and Competitiveness (MINECO) of the Spanish Government and by the European Regional Development Fund (FEDER) of the European Union (UE) through projects referenced as: CGL2011-23621 and CGL2015-65913 -P (MINECO /FEDER, UE).

## References

Abramowitz, M., & Stegun, I. A. (1964). Handbook of mathematical functions: with formulas, graphs, and mathematical tables (Vol. 55). Courier Corporation.

Aguilar-Meléndez, A., Pujades, L., Barbat, A., & Ordaz, M. (2008). Probabilistic assessment of seismic risk in urban areas. In *Proceedings of the 14th world conference on earthquake engineering* (pp. 12–17).

Aguilar-Meléndez, A., Pujades, L., Barbat, A., & Lantada, N. (2010). *A probabilistic model for the seismic risk of buildings: application to assess the seismic risk of buildings in urban areas*. A: US National and Canadian conference on earthquake engineering. "9th US National and 10th Canadian conference on earthquake engineering". Toronto, pp. 1–10.

Aguilar-Meléndez, A. (2011). *Evaluación probabilista del riesgo sísmico de edificios en zonas urbanas* (297 pp). Tesis doctoral. Universidad Politécnica de Cataluña.

Aguilar-Meléndez, A., Pujades, L., Barbat, A., & Lantada, N. (2011). *USERISK2011. Software for computing seismic vulnerability and seismic risk of buildings in urban areas*.

Aguilar-Meléndez, A., Pujades, L. G., Barbat A. H., Ordaz, M. G., Lantada, N., & García-Elías, A. (2012). *Probabilistic assessment of the seismic risk of Barcelona*. 15 WCEE Lisboa 2012.

Aguilar-Meléndez, A., Pujades, L. G., De la Puente, J., Barbat, A. H., Lantada, N., & Campos-Rios, A. (2015a) *USERISK2015. Software for computing seismic vulnerability and seismic risk of buildings in urban areas*. Last accessed 11 Feb 2016. https://sites.google.com/site/userisk2015/

Aguilar-Meléndez, A., Pujades, L. G., Barbat, A., Ordaz, M. G., & Lantada, N. (2015b). *Estimación del Peligro Sísmico de Barcelona y su Aplicación en la Determinación del Riesgo Sísmico de la Ciudad. Capítulo 3*. Libro: Mora, I. Coordinador, Metodologías aplicadas a las Ciencias de la Tierra. ISBN: 978-607-9091-49-1.

Aguilar-Meléndez, A., Ordaz, M., De la Puente, J., González-Rocha, S. N., Rodríguez-Lozoya, H. E., Córdova-Ceballos, A., García-Elías, A., Calderón-Ramón, C., Escalante-Martínez, J. E., Laguna-Camacho, J. R., & Campos-Rios, A. (2017). Development and validation of software CRISIS to perform probabilistic seismic hazard assessment, with emphasis on the recent CRISIS2015. Computación y Sistemas (In press).

Ajuntament de Barcelona. (2016). *Touris statistics*. Barcelona: city and surroundings. Last accessed 2 July 2017 http://www.diba.cat/documents/74348/78145163/1+Estad%C3%ADstiques+de+turisme+2015.+Barcelona_ciutat+i+entorn.pdf/d913b713-ba01-4ac9-abad-047df5a85405

ASCE. (2010). ASCE Standard ASCE/SEI 7–10. In *Minimum design loads for buildings and other structures* (658 pp).

ATC-13 (1985). *Earthquake damage evaluation* (492 pp). Redwood City: Data for California; Applied Technology Council.

Barbat, A. H., Lagomarsino, S., & Pujades, L. G. (2006). Vulnerability assessment of dwelling buildings. In C. S. Oliveira, A. Roca, & X. Goula (Eds.), *Assessing and managing earth-quake risk* (pp. 115–134). Dordrecht: Springer.

Benedetti, D., Benzoni, G., & Parisi, M. A. (1988). Seismic vulnerability and risk evaluation for old urban nuclei. *Earthquake Engineering and Structural Dynamics, 16*(2), 183–201.

Cardona, O. D., Ordaz, M. G., Reinoso, E., Yamín, L. E., & Barbat, A. H., (2012) *CAPRA-Comprehensive approach to probabilistic risk assessment: International initiative for risk management effectiveness*. 15 world conference on earthquake engineering. Lisboa.

Carreño, M. L., Lantada, N., Irizarry, J., Valcarcel, J. A., Barbat, A. H., & Goula, X. (2012). Comportamiento Sísmico de los Edificios de Lorca/Seismic behavior of the buildings in Lorca. *Física de la Tierra, 24*, 289–314. http://revistas.ucm.es/index.php/FITE/article/view/40142.

Cid, J., Figueras, S., Fleta, J., Goula, X., Susagna, T., & Amieiro, C. (1999). *Zonación Sísmica de la Ciudad de Barcelona* (pp. 263–271). Murcia: Primer Congreso Nacional de Ingeniería Sísmica.

Cornell, A. (1968). Engineering seismic risk analysis. *Bulletin of the Seismological Society of America, 58*(5), 1583–1606.

Department d'Estadística. (2016). *Cifras oficiales de población. 1 enero 2015*. Last accessed 22 Dec 2016. http://www.bcn.cat/estadistica/castella/dades/tpob/ine/a2015/sexe/bcn.htm

Dolce, M., Kappos, A., Masi, A., Penelis, G., & Vona, M. (2006). Vulnerability assessment and earthquake damage scenarios of the building stock of Potenza (Southern Italy) using Italian and Greek methodologies. *Engineering Structures, 28*, 357–371.

Ellingwood, B. R. (2006). Mitigating risk from abnormal loads and progressive collapse. *Journal of Performance of Constructed Facilities, 20*(4), 315–323.

EM-DAT (2015). *Disaster list*. EM-DAT. The International Disaster Database. www.emdat.be. Université Catholique de Louvain, Brussels (Belgium). Last accessed 19 Jan 2016.

ERN-AL. (2010). *CAPRA, Comprehensive approach for probabilistic risk assessment.* http://www.ecapra.org/. Accessed 2 May 2012.

Esteva, L. (1970). *Regionalización símica de México para fines de ingeniería.* Mexico: Institute of Engineering Series-246, UNAM.

Faccioli, E. (2006). Seismic hazard assessment for derivation of earthquake scenarios in Risk-UE. *Bulletin of Earthquake Engineering, 4*, 341–364. https://doi.org/10.1007/s10518-006-9021-2.

Faccioli, E., Pessina, V., Pitilakis, K., & Ordaz, M. (2003). *WP2: Basis of a handbook of earthquake ground motions scenarios. An ad-vanced approach to earthquake risk scenarios with applica-tions to different European towns.* Contract: EVK4-CT-2000-00014, 93 pp.

FEMA. (2015a). *HAZUS® MH 2.1. Multi-hazard loss estimation methodology. Earthquake model. User manual.* Washington, DC, 718 pp. Last accessed 10 Mar 2016. http://www.fema.gov/media-library-data/20130726-1820-25045-1179/hzmhs2_1_eq_um.pdf

FEMA. (2015b). *HAZUS® MH 2.1. Multi-hazard Loss Estimation Methodology. Earthquake Model. Technical Manual.* Washington, DC, 718 pp. Last accessed 10 Mar 2016. http://www.fema.gov/media-library-data/20130726-1820-25045-6286/hzmh2_1_eq_tm.pdf

Gardoni, P., & LaFave, J. M. (2016). Multi-hazard Approaches to Civil Infrastructure Engineering: Mitigating Risks and Promoting Resilence. In *Multi-hazard approaches to civil infrastructure engineering* (pp. 3–12). Springer International Publishing.

Giovinazzi, S. (2005). *The vulnerability assessment and the damage scenario in seismic risk analysis.* Doctoral thesis, Technical University of Braunschweig, and University of Florence, 222 pp.

Goula, X., Susagna, T., Secanell, R., Fleta, J., & Roca, A. (1997). *Seismic hazard assessment for Catalonia (Spain)* (pp. 173–177). Barcelona: Proceedings Second Congress on Regional Geological Cartography and Information Systems.

Grünthal, G. (1998). *European Macroseismic Scale 1998* (Vol. 15, pp. 1–99). Luxemburg: Cahiers du Centre Europ'een de G'eodynamique et de S'eismologie.

Irizarry, J., Goula, X., & Susagna, T. (2003). *Evaluación de la peligrosidad sísmica de la ciudad de Barcelona en términos de aceleración espectral* (pp. 389–399). 2° Congreso Nacional de Ingeniería Sísmica, Málaga, España.

Irizarry, J.. (2004). *An advanced approach to seismic risk assessment. Application to the cultural heritage and the urban system of Barcelona.* Doctoral thesis, Universitat Politècnica de Catalunya, Barcelona, 406 pp.

Irizarry, J., Lantada, N., Pujades, L. G., Barbat, A. H., Goula, X., Susagna, T., & Roca, A. (2010). Ground-shaking scenarios and urban risk evaluation of Barcelona using the Risk-UE capacity spectrum based method. *Bulletin of Earthquake Engineering, 9*(2), 441–466.

ISDR-UN. (2005, March). Hyogo framework for action 2005–2015: building the resilience of nations and communities to disasters. In *Extract from the final report of the world conference on disaster reduction.* (A/CONF. 206/6) (Vol. 380).

Lantada N. (2007). *Evaluación del riesgo sísmico mediante métodos avanzados y técnicas GIS. Aplicación a la ciudad de Barcelona* (Vol. 1, 338 pp). Tesis doctoral, Universitat Politècnica de Catalunya, Barcelona.

Lantada, N., Pujades, L. G., & Barbat, A. H. (2009a). Vulnerability index and capacity spectrum based methods for urban seismic risk evaluation. A comparison. *Natural Hazards, 51*, 501–524.

Lantada N., Pujades L.G., & A. H. Barbat (2009b). *Escenarios de riesgo sísmico para la ciudad de Barcelona* (Vol. II. 76 pp). Informe técnico para el Servicio de Protección Civil. Ayuntamiento de Barcelona.

Lantada, N., Irizarry, J., Barbat, A. H., Goula, X., Roca, A., Susagna, T., & Pujades, L. G. (2010). Seismic hazard and risk scenarios for Barcelona, Spain, using the Risk-UE vulnerability index method. *Bulletin of Earthquake Engineering, 8*(2), 201–229.

Lizarralde, G., Johnson, C., & Davidson, C. (2009). *Rebuilding after disasters: From emergency to sustainability.* New York: Routledge.

López Casado, C., Molina, S., Delgado, J., & Peláez, J. A. (2000). Attenuation of intensity with Epicentral distance in the Iberian peninsula. *Bulletin of the Seismological Society of America, 90*, 34–47.

Marincioni, F., Appiotti, F., Ferretti, M., Antinori, C., Melonaro, P., Pusceddu, A., & Oreficini-Rosi, R. (2012). Perception and communication of seismic risk: The 6 April 2009 L'Aquila earthquake case study. *Earthquake Spectra, 28*(1), 159–183.

Marulanda, M. C., Carreño, M. L., Cardona, O. D., Ordaz, M. G., & Barbat, A. H. (2013). Probabilistic earthquake risk assessment using CAPRA: Application to the city of Barcelona, Spain. *Natural Hazards, 69*(1), 59–84.

McGuire, R. K. (2004). *Seismic hazard and risk analysis* (221 pp). Earthquake Engineering Research Institute. MNO-10.

McGuire, R. K. (2008). Probabilistic seismic hazard analysis: Early history. *Earthquake Engineering and Structural Dynamics, 37*, 329–338.

Milutinovic, Z. V., Trendafiloski, G. S. (2003). *WP4: Vulnerability of current buildings. RISK-UE. An advanced approach to earthquake risk scenarios with applications to different European towns* (109 pp). Contract: EVK4-CT-2000-00014.

Olivera, C., Redondo, E., Lambert, J., Riera, A., & Roca, A. (2006). *The earthquakes of the XIV and XV centuries in Catalonia (NE Spain). First European conference on earthquake engineering and seismology*. Ginebra, Suiza, 10 pp.

Ordaz, M., Martinelli, F., Aguilar, A., Arboleda, J., Meletti, C., & D'Amico, V. (2015). *CRISIS2015. Program for computing seismic hazard*. Last accessed 18 Feb 2016. https://sites.google.com/site/codecrisis2015/

Ordaz, M., Martinelli, F., D'Amico, V., & Meletti, C. (2013). CRISIS2008: A flexible tool to perform probabilistic seismic hazard assessment. *Seismological Research Letters, 84*(3), 495–504.

Rockefeller Foundation. (2017). *100 resilient cities*. Last accessed 2 Dec 2017. http://www.100resilientcities.org/cities#/-_/

Secanell, R., Goula, X., Susagna, T., Fleta, J., & Roca, A. (2004). Seismic hazard zonation of Catalonia, Spain, integrating random uncertainties. *Journal of Seismology, 8*, 25–40.

Susagna, M. T., & Goula, X. (1999). *Atles Sísmic de Catalunya. Volum 1: Catàleg de sismicitat* (p. 413). Barcelona: Institut Cartogràfic de Catalunya.

UNISDR. (2015). *Sendai framework for disaster risk reduction 2015–2030*. http://www.unisdr.org/files/43291_sendaiframeworkfordrren.pdf. Last accessed 3 Mar 2017.

United Nations. (2012, March). *How to make cities more resilient. A handbook for local government leaders*. Geneva. http://www.unisdr.org/files/26462_handbookfinalonlineversion.pdf. Last accessed 20 Jan 2016.

United Nations. (2013). *Plan of action on disaster risk reduction for resilience*. United Nations System. Chief Executives Board for Coordination. 14 pp. http://www.preventionweb.net/files/33703_actionplanweb14.06cs1.pdf. Last accessed 19 Jan 2016.

UN-Habitat. (2016). *City resilience profiling programme*. http://unhabitat.org/urban-initiatives/initiatives-programmes/city-resilience-profiling-programme/. Last accessed 19 Jan 2016.

USGS. (2016a). *2010 significant earthquake and news headlines archive*. http://earthquake.usgs.gov/earthquakes/eqinthenews/2010/

USGS. (2016b). *2011 significant earthquake and news headlines archive*. http://earthquake.usgs.gov/earthquakes/eqinthenews/2011/

Villani, M. (2010). *High resolution SHA in the vicinity of earthquake sources*. Ph. D. thesis, Instituto Universitario di Studi Superiori, Pavia, ROSE School.

Villani, M., Faccioli, E., & Ordaz, M. (2010). Verification of CRISIS2008 code.

Zschau, J. (2017). *MATRIX (New multi-hazard and multi-risk assessment methods for Europe) Report Summary*. Project ID: 265138. Funded under: FP7-Environment. Germany. http://cordis.europa.eu/result/rcn/156166_en.html. Last accessed 8 Mar 2017.

# Chapter 14
# Urban Resilience and Post-Disaster Reconstruction. Evidences from Mexico and France

**Aleyda Reséndiz-Vázquez**

**Abstract** This paper relates resilience to post-disaster reconstruction phase. The concept of resilience (in the urban-architectural field and in the field of disaster risk) is understood as the ability of a complex system to maintain or rapidly return to desired functions in the face of impacts of natural or human origin. The reconstruction process of a given territory is one of the post-disaster stages, in which the above societal conditions must be restored. The reconstruction is then a phase of the "disaster risk" cycle where exist the opportunity to decrease risk and therefore, to increase resilience. Resilience through reconstruction is recognizable in two cases: the later French reconstruction after World War II (governmental and centralized) and the Mexican reconstruction after the earthquakes of September 1985 (also vertical and centralized, while at the same time spontaneous and solidary).

**Keywords** Urban-architectural resilience · Post-disaster reconstruction · Industrialized construction · French reconstuction · Mexican reconstruction

## 14.1 Resilience through Reconstruction

France, after the Second World War, faced a crisis housing and public services: shelling, *baby-boom*, rural exodus and migration. When the war got to an end, the main difficulty was the necessity to build fast and at a low-cost. In this context, many building systems and organizational forms were invented, and some others modernized to thereby fulfill the reconstruction needs. The "solution" was the industrialization of the prefabrication (Reséndiz 2010). French reconstruction reveals an administrative and technical process where the government set the foundation, proposed technological systems and controlled the reconstruction policies. Nevertheless, years later, the social and economic debate in the late 1960's strongly questioned the architectural and urban production of the preceding three decades.

---

A. Reséndiz-Vázquez (✉)
Instituto Politécnico Nacional, Mexico City, Mexico

 
G. Brunetta et al. (eds.), *Urban Resilience for Risk and Adaptation Governance*, Resilient Cities, https://doi.org/10.1007/978-3-319-76944-8_14

In Mexico, in September 1985, two earthquakes strongly shook Mexico City. Regarding the participation of "civil society"(Mexican term for citizens), the reconstruction process led by the Federal Government, was a success (Monsiváis 2005). The 1985 earthquake has become a research accepted model for several disciplines - including urban-architecture – and highlights the importance of historical studies in investigations about risks. Mexican reconstruction, on one side, is characterized by the participation of civil society or citizens, and on the other, is a testimony of the influence of physical and socioeconomic vulnerability in risk management.

The concept of resilience, and especially that of urban resilience, in the current debate of the last decade, discusses and complements the concept of reconstruction. Meerow et al. (2015) define urban resilience from six conceptual contradictions (i.e., characterization of 'urban', notion of equilibrium, resilience as positive concept, pathway to resilience, understanding of adaptation, timescale of action). The textual definition is:

> Urban resilience refers to ability or an urban system-and all its constituent socio-ecological and socio-technical networks across temporal and spatial scales- to maintain or rapidly return to desired functions in the face of a disturbance, to adapt to change, and to quickly systems that limit current or future adaptive capacity" (Meerow et al. 2015, p. 39 y 45).

This definition recognizes urban systems as dynamic, multifactorial, adaptable and whose ability to absorb disturbances does not necessarily mean going back to the previous situation but concentrating on important functions. Change, to achieve resilience, is not static and contains varying degrees (persistence, transition and transformation) that depend on the conditions of time and space. Therefore, we agree with Meerow, Newell and Stults that the "resilience for whom, what, when, where and why needs to be carefully considered" (Meerow et al. 2015, p. 46). The mexican and french cases precisely illustrate the importance and divergence of the answer to these questions.

Both reconstruction models (the French and the Mexican ones) possess advantages and disadvantages; consequently, they are important to be considered in the development of a resilient reconstruction referential plan. Accordingly, the first tracks of this research are based on the question: What are the resilient elements of the French and the Mexican reconstruction processes that could be implemented in the event of massive and urgent constructions, particularly after a disaster?

### *14.1.1 Resilience and Reconstruction*

Reconstruction is one of the phases of a post-disaster recovery. The recovery process involves the "restoration" of spaces and activities of the affected territory. For Oliver-Smith (1994), "of all stages of a disaster, reconstruction is probably the longest, the most expensive and most complex" (p. 3). The stages of a disaster can be defined in relation to the time that each phase occupies in the post-disaster period; so, you can define an emergency phase (Asfi 2012; and Oliver-Smith 1994), followed by a recovery phase (USAID n.d.), which includes the rehabilitation and reconstruction (Oliver-Smith 1994; USAID, n.d). The recovery stage "is one in which the process of restoring normal conditions of life of a community affected by a disaster" (USAID, s. F: MR-9). The process of "back to normal" that "begins immediately after it has completed the emergency phase" (UNISDR 2009, p. 23) is divided, in turn, into two sub-stages: the rehabilitation and the reconstruction. These two sub-stages are defined by the U. S. Agency for International Development (USAID) as:

> [Recovery] covers two aspects, the first aimed at restoring in the short term and transiently essential basic services [rehabilitation] and the second moving towards a permanent solution and long-term, which seeks to restore normal conditions life of the affected community [reconstruction] (USAID n.d.: MR-9).

On this ground, the reconstruction phase should be regarded as the phase that allows a "return to the normal" situation, as expressed by the USAID, which may mean the return of the risk, or the reconstruction, considered as the post-disaster stage. In this phase, the principle of " a better re-building" should be applied (UNISDR 2009, p. 26); this is the process in which new risks should be mitigated. Furthermore, this one should be considered as an "opportunity development" stage with respect to the affected territories (Cuny 1983; Pantelic 1991; Oliver-Smith 1994; and Lavell 1999). The USAID defines the reconstruction phase as the "process of reparation of the medium and long term, physical, social and economic damages, to a higher level than the existing development before the event" (USAID, n.d.: MR-10). Hence, the reconstruction incorporates actions that reduce vulnerability, which are simultaneously linked to development efforts that seek to improve the life quality, jobs, resources, social and cultural values of the community (USAID n.d.; Pantelic 1991). The idea of this concept of reconstruction is to create new systems that incorporate prevention and risk mitigation.

The international organization *Architectures sans Frontières* (ASFI 2012) divides the post-disaster stage in the following sub-stages: early recovery, reconstruction and renewal stage, depending on the time to be considered for each one of these concepts. The idea of early recovery may be associated with Davis (1980), when referring to the "accelerated reconstruction", defined as building strategy that is carried out in a few days in Third World countries, situation that in other countries would take several months. Therefore, as Oliver-Smith (1994) defines, the notion of accelerated reconstruction can also be related to the stage of rehabilitation, yet emphasizing that each stage implies both time and the social conditions that were affected by a disaster.

The General Civil Protection Act states that "this process [the reconstruction] should look as much as possible to reduce the existing risks, ensuring the no generation of new risks and the improvement of the preexisting conditions" (2012, p. 5); nonetheless, it is an approach that - as the ones previously mentioned - no longer sees the disaster as a linear process in which there is "a before and an after" regarding a disastrous event.

The comprehensive risk management model exposed by USAID (n.d.: MR-6), place the reconstruction phase within an analysis of areas and components, where the "recovery area" with its "rehabilitation components and reconstruction", are located within the same surface as the risk analysis areas, risk reduction and management of adverse events, with their respective components ranging from the study of threats and vulnerabilities to the reconstruction. Thereby, this scheme associated an *a priori* knowledge of the risks (and not disasters) with reconstruction.

This leads to a scheme of analysis of the process of reconstruction within a system that not only observe the post-disaster, but also the previous stage; leads to consider the society's conditions before the disaster, as they may determine the magnitude of the disaster as well as the degree of change that it is possible to achieved by means of the reconstruction.

The post-disaster rebuilding process can be analyzed as a multi-factorial system which involves almost all the actors of society. Social, psychological, economic and physical factors (Cuny 1983, and Oliver-Smith 1994) are the ones that must be reestablished by "(…) a social process through a complex interaction between institutions, groups and individuals who have to do with the allocation and form of material and non-material resources towards the achievement of goals that are culturally derived for society" (Oliver-Smith 1994: 4).

Under this conceptual scheme, the history of post-disaster reconstruction can be a source of inspiration from various aspects. In this case, we have referred to some of its characteristics related to vulnerability and reconstruction public policies and both these factors directly related to social aspects.

### 14.1.2 Building Reconstruction

Understood as one of the stages of recovery, the reconstruction phase or the post-disaster reorganization phase is analyzed from: a) the physical parts involved in the process of urban and architectural reconstruction (i.e. materials, construction systems, hand construction or urban planning); b) Government participation in the reconstruction process; and c) the relationships between citizens and the governmental institutions engaged in urban management and housing production that originate from the disaster.

A first question on materials and construction systems is related to the possibility of reducing the risks in respect to the building components. For example, after the earthquake of Lisbon in 1755, the wooden frame was introduced in order to make structures more rigid; after the earthquake in Guatemala in 1976, many of the heavy materials were substituted with lighter ones (Pantelic 1991). Construction rules may also be included in reducing physical vulnerability, such as the existence of these norms and their relation to the risks, together with their correct application.

A second question looks at the origin of the materials and the construction systems; to whether they may or may not be unconnected to the affected society. There are many examples of the use of prefabrication; a system that is usually oblivious to the affected communities and is applied mainly as a solution to massive reconstructions. In the Turkey's 1970 earthquake, the authorities introduced a program of (permanent) prefabricated houses, which were built in 8 weeks; this is faster than the emergency constructions (Davis 1980: 106–107). The solutions provided by this prefabrication system may be applicable in seismic areas as shown by the prefabricated system "Camus", applied in 1966 in the reconstruction of Tashkent, capital of Ouzbékistan. Nevertheless, the system of "industrialization of prefabrication" supposes a technologically complex organization and management (Reséndiz 2002); and therefore, requires long periods of time to be implemented. Additionally, prefabrication may be associated with centralized decisions that are unrelated to the affected community. Also, as noted by Pantelic "One important measure of a reconstruction program's success is the extent to which a community can preserve the cultural identity and lifestyle" (1991: 346).

The idea of using prefabrication for reconstruction may oppose the evolutionary housing raised by Davis (1980); for him it is a myth that a temporary accommodation is needed before the definitive one, especially in the Third World where reconstruction usually begins immediately [after the disaster]" (p. 111). The notion of evolutionary housing proposes that the initial structure of the house is small and grows throughout time adapting to the needs of its inhabitants. This means that when facing any sort of structure, whether is an emergency shelter, a temporary or a long-term structure, this one must be very strong from the beginning because people will continue using the house in its original form and it will evolve with the addition of new rooms, thus becoming a long-term structure. This informs us that an emergency shelter or a temporary home should be designed considering the last phase of its evolutionary process (ibid: 106).

The notion of evolutionary housing should not be confused with temporary housing, with which several cases were rebuilt and, after several years, are still used. This is the case, for example, of the period of prefabricated emergency houses that followed the end of World War II in France or the Quonset huts in the reconstruction of Sicily in1968. Another question on physical reconstruction may refer to the source of labor. Pantelic (1991) argues that the locals can be involved in the reconstruction, as well as being trained to do so. In the reconstruction of Guatemala, after the earthquake of 1976, residents were trained in building techniques, learning to identify the safest places to construct their new homes (Cuny 1983, and Pantelic 1991). This example shows the importance of citizen participation in the reconstruction process, in which the technical and organizational experience of the locals becomes fundamental. It is possible that a controversy arises regarding the place in which the reconstruction must be carried out. This one must be in the same territory where the disaster took place -, in spite of the possibility of this one being a place exposed to a constant threat - or in a different place, likely to be safer. For some authors (Pantelic 1991, and Davis 1980) the complete relocation of settlements has been one of the long-standing operations. Nonetheless, as suggested by Davis (1980) this solution has not been successful in the long term neither from an economical nor from a social perspective. The classic example is Antigua in Guatemala, where the Spanish authorities at the time officially relocated the population in 1779, only for the inhabitants to return a few years later (Tobriner 1980).

Finally, one more question may refer to urban and architectural preservation during reconstruction. An example of a radical transformation of urban heritage was the case in Skopje, a medieval city before the earthquake in 1963 that was rebuilt as a linear city 24 kilometers long (Davis 1980).

### *14.1.3 Reconstruction by the Government and the Society*

Governmental participation in a reconstruction policy could be analyzed from at least three different scenarios:

One in which the strong centralization implies that the decision-making process takes place away from the social conditions, therefore away from the needs and the local culture (Oliver-Smith 1994). According to Davis (1980), it is a "myth" that governmental reconstruction programs are effective in solving housing problems. He says: "It is true, otherwise"; meaning that the response of the local population is an effective solution.

The second scenario is opposite to the previous one: corresponds to a decentralized process that because of its flexibility may produce a faster response than the centralized process, together with a greater awareness of the local culture and needs. Notwithstanding, as pointed by Oliver-Smith paraphrasing Alcira Kreimer (1994), decentralized processes "lack of capacity to offer comprehensive plans for the development of infrastructure and social facilities for communities" (1994:6). In this second scenario, the participation of local existing organizations and of those formed by the disaster itself is essential; these are the ones that Audefroy denominates as regulator agents that shall bring solutions, governmental and non-organizations, programs, actions and rules designed to protect the affected agents, especially the population (2003: 54).

The third scenario refers to a government control that encourages social participation and facilitates the reconstruction process. Wilches-Chaux, referring to his participation as an officer in the reconstruction of Popayan (Colombia) says: "At the time, our role was to support the people that with their own hands were going to rebuild their homes, enabling them to do it as good as possible" (2012:248).

## 14.2 The French Reconstruction

The years 1945–1975 correspond to the French Reconstruction after World War II. In this period, the government ensured the economic conditions for the implementation of a suitable reconstruction system, which comprised the industrialized construction.

From 1944 to 1954, the Construction and Architecture Plan included the studying of a "housing policy", which would be carried out by the State and implemented by the Ministry of Reconstruction and Urbanism (MRU, for its acronym in French).

Shortly after, the state is forced to define an economic and technological policy to solve the shortage of materials, machinery and skilled labor. In this manner, reconstruction became a testing ground, which will result in operations run by the state in collaboration with building professionals (architects, engineers and companies).

In 1944, the Ministry of Reconstruction and Urbanism (MRU, for its acronym in French), is created. This ministry was responsible for developing this centralized policy: "After 1945, the habitat and the city became one national project, such as the energy and transport supplies are" (Vayssière 1988). Since 1945, the Administration procedures were put into progress to solve technical problems of reconstruction such as materials, transport, energy and infrastructure. As suggested by Strobel, the modernization ideas will be "[...] retaken by the central State and its technicians in order to promote an economic policy (reorganize and concentrate the building industry), one technical policy (industrialization) and one social policy (HLM housing)"(1983: 6).

Under the tutelage of the State, the technical authorities launched different mechanisms of productivity to cope with reconstruction: construction aggregation procedures, experimental works and contests. Also, the MRU, ensured the conditions and guarantees of a long-term market, to encourage companies to invest. "In less than a year, the MRU will dictate all modes of control and markets granting, all directives and circulars, all recommendations [...]" (PCA: 10).

In 1947, the MRU launches "experimental works." The first experimental operation in Noisy-le-Sec, fifty houses, was useful in testing some industrialized construction procedures, standards and building materials. In 1950, 800 homes were built in Strasbourg.

Since 1952, the "industrial sector program" (Simon 1950) began its application. Since the mid-1950s, the continuity of construction, necessary for the investment in machinery and the incentive of the prefabrication developments, is sought by large construction operations, such as the *Grands ensembles* (Great Housing Series). In the mid-1950s, in the Courneuve (a French commune in the Paris region) was built a large set of 4000 homes.

In the mid-1980s, one of his "bars" - the "Debussy" - was dynamited (Vayssière 1988). These are the "towers and bars" - as named by Faure (1996) - that composed these large housing assemblies; they aroused debates of possible social, urban and architectural interventions today and even lead to consider its destruction, considering them as one of the "mistakes of history."

From 1967 to 1973 the amount of constructions decreased (Monnier 2000). Administration's interventions that favored innovation began in 1967, It is a period that brings into question the feasibility of mass construction in comparison with other principles that were developed in this decade: quality, comfort, flexibility, variability, new cities.... From these factors, become deduced a redefinition of all forms of architecture industrialization. In 1973, the “Guichard” circular officially prescribed the construction of the “Grands Ensembles”.

The history of the development of the industrialization of construction in France, is also the story of its failure. “Rarely a subject has been so unanimously discredited, or even worse, collectively rejected: firstly, by the media and after by the university that have strongly criticized the Great Housing from 1963–1964” (Vayssière 1988: 9).

## 14.3 The Mexican Reconstruction

The earthquake of September 19th, 1985 in Mexico City, is one of the events that has caused major destruction in the country, between 1970 and 1999 (Perló Cohen 1999). This earthquake and its aftershocks seriously damaged the central area of the city (80% of the Cuauhtémoc Borough), (Audefroy 2008; Salcido 2010) generating the collapse of about 30 thousand buildings; in addition, another 60 thousand constructions were partially damaged (Perló Cohen 1999).

Following this disaster, in Mexico various reconstruction programs (Delgadillo 2007; Mecatl et al. 1987; Pliego 1994) were established such as the Emerging Housing Phase I and Phase II, the Democratic Reconstruction of the Nonoalco Tlatelolco Unit, the Popular Housing Renovation (RHP) and the Program for Affected Workers from the Textile Sector (“Seamstresses”). These reconstruction programs launched by the central authorities are privileged sources: on one hand, of the reconstruction problems and on the other hand of the reconstruction potentiality.

Mexican reconstruction seems to correspond to a strange combination of a centralized process with citizenship participation. “The condition of the citizens and the presence of social organizations of different kinds victims were key elements of the public housing reconstruction policy” (Mecatl et al. 1987: 23).

The relationship of disaster risk with physical vulnerability became manifest in the earthquake of 1985. Unfortunately, many of the collapsed buildings unmasked the level of corruption and poor planning of our country. Also, the earthquake revealed the physical vulnerability of the constructions of Mexico City. The new building regulation, considered to improve the seismic resistance of physical infrastructure, was ready 5 weeks after the earthquake of 1985 (Pantelic 1991).

Besides, Mexico City’s reconstruction is an example of how the life conditions of the population can be improved as a result of a disaster. Before the earthquake, most homes had an average of 22.25 m2, often sharing bathrooms and kitchens. With the process of reconstruction, housing gained an average of 40 m2 with two bedrooms, kitchen, dining room and bathroom (Pantelic 1991); “The new, the different, is the fact that the central neighborhoods of the city, from the 1985 earthquake, began to be the protagonists of an improvement process of the quality of their popular housing” (Mecatl et al. 1987: 13).

Historical cultural context in which reconstruction was carried out in Mexico according to López and Veduzco, prior to the September 1985 earthquakes was expressed in "a degraded environment, a population that is largely impoverished and with high unemployment, a serious housing deficit, chaotic and saturated services and on the verge of collapse "(1986: 25). In addition, as a result of the migration to Mexico City, on the one hand the marginal colonies on the outskirts of the city expand, land invasions and clandestine subdivisions occur; on the other hand, the city center continues the physical and social degradation of the neighborhoods. The inhabitants most affected by the earthquake were in these old houses of the center of the city, rented to families of low income.

Before 1985 many homes had already been displaced from the center in order to save this area to commercial investments. The destruction of downtown buildings opened up a new opportunity for speculation.

The main players of the Reconstruction were the inhabitants of damaged and collapsed buildings but also the tenants victims of real estate speculation represented by increased rents and evictions. In this way, the popular mobilization that was born with the earthquakes involves not only the direct victims but also other sectors already behind and in need of housing. Faced with real estate speculation, the tenants were resisting in a scattered manner, what was generated in the reconstruction was the organization in defense of the popular housing that agreed the housing demands.

The processes of organization of the various sectors affected are developed in an environment of governmental incapacity to face the need for housing. The government imposes a separation between the victims thus officially recognized of the rest of the population. After the emergency phase, in October 1985, the neighborhood organizations set up the Single Coordinator of Victims (SCV). The SCV was formed from the movement of the victims, grouped very heterogeneous social groups but also collaborated in the segregation of the other sectors in struggle. Tlatelolco, a central housing complex, with the earthquakes of 1985, was one of the most affected areas; one of its buildings the "Nuevo Leon" collapsed. The survivors of this building led the mobilizations of victims throughout the city. Another leader and pioneer group of the organization was constituted by the tenants of the damaged neighborhoods in the Historical Center.

A first government response given to the victims of the neighborhoods was to offer housing on the outskirts of the city. In response, the tenants of the neighborhoods struggled to rebuild their homes in the same place they inhabited before the earthquake.

By governmental decision was ordained the expropriation of the properties where there were damaged neighborhoods. "The expropriations were concentrated in those areas where the neighborhood organizations were more belligerent" (López and Veduzco 1986:36). In order to undertake the reconstruction in the expropriated estates, the Popular Housing Renewal Program was created.

The collective reactions of the reconstruction exposed "resilient" characteristics that manifested themselves beyond the moments of emergency: solidarity, neighborhood coexistence, popular inventiveness, rapidity, popular defense of neighborhoods, collective memory. Characteristics with which damaged and lagged demanded their rights before owners of owners and public officials.

Besides the victims, who led the movements for reconstruction, other governmental and non-governmental organizations participated in the process. Almost immediately after the earthquakes, the government, through the Secretary for Urban Development and Ecology (SUDE), constituted the National Reconstruction Commission (NRC) (which later became known as the Social Aid Committee). The NRC was in charge of overseeing some reconstruction programs and the management of economic resources.

The government's response to the organized victims was first the expropriation and later the agreement of Democratic Concertación for the Reconstruction. The agreement was signed by the federal government, foundations civil associations, universities and organizations of victims. The Convention becomes an emblem of the movement as it is the first time that the government recognizes organized citizens as interlocutors (Ramírez Cuevas 2005). Reconstruction programs were funded mainly through World Bank loans and NGO funds. The World Bank collaborated in the reconstruction of hospitals, schools and housing. According to Kreimer and Echeverria (n.d.), the World Bank pushed the policy of on-site reconstruction instead of a relocation. On-site reconstruction depended on the expropriation of the land which took place in October 1985. The expropriation order stated that the construction of new housing would be for people who already lived on expropriated lots.

The organization of the programs, such as the RHP, gave great importance to the supervision and control of construction by creating several modules that managed the construction activities. At the same time, organized social and technical groups were in charge of reviewing the design and construction process of the new houses.

The organization of the actors in the reconstruction process depends on the program in which the various sectors were organized. The Emergency Housing Program Phase I, which was the first to be implemented, was commanded by SUDE and the Federal District Department (FDD). This program was for the victims that had the accreditation of the property or were owners. In this case, the existing housing institutions were entrusted with the management of resources for reconstruction.

The second program implemented was the low-income (popular) Residence Renovation Program (RRP). Through this program was responded to the demands of the tenants of the damaged neighborhoods expropriating buildings. In this program, the actors who participated in the design and planning of the new housing were the government, universities, technical support groups, groups of private and religious volunteers. The actors who participated in the approval of the plans and in the construction of the works were the committees formed in the various programs, the low-income Housing Fund (FONAPHO), the office of the FDD and the National Bank of Works and Public Services (BANOBRAS). Civil associations already existing before the earthquake, such as the Housing, Environment and Social Development Institute (HESDI), the Operational Center for Housing and Population (OCHP), the House and the City and the Institute of Communication and Popular Education (ICPE) participated.

Housing reconstruction programs are an example of analysis where the participation of the affected population is unable to be relocated "The status of citizen and the presence of social organizations of victims of different types were key elements of the reconstruction policy of popular housing" (Mecatl et al. 1987: 23). Contrary to new forms of urbanization, the Mexican reconstruction was a *in situ* reconstruction. Mexico City's historic center economic activity includes formal and informal trade; for this reason, this *in situ* reconstruction allowed the local population to maintain their economy.

The urban-architectural reconstruction entails as well the loss, recovery, and rehabilitation of heritage buildings. Among the rehabilitation actions, the reconstruction of several of these buildings is found. Nevertheless, not all buildings were restored; there are still buildings with significant structural damages that are inhabited. Such is the case of a four level multi-family building, located on the Republic of Venezuela street in Mexico City downtown, Cuauhtémoc City District (Garcia-Lopez 2013). In this building, as in many others, despite the risk, their inhabitants recognize their heritage:

> [...] It is a privilege to have in this place a home, because it is a historic area, not anyone could stand firm in such a place [...] is a great advantage for me! it's something important, having a little house, a home in this place (Interview with Ms. Lucila Garcia Nicolas, resident) (García López 2013).

## 14.4 Conclusions: Ideas for a Reconstruction

The concepts of resilience and reconstruction were fundamental to analyze reconstruction processes in Mexico and France. The reconstruction of buildings carried out by the state and by the society in two different historical and cultural contexts showed how the factors of urban resilience are dynamic and depend on the questions of who, for whom, what, when, where and why.

Thus the first purpose was to refer to the concepts of reconstruction and resilience with special emphasis on the partition of the state and society in the process of reconstruction. Second, from the conceptual scheme derived from these concepts, the cases of Mexican reconstruction following the earthquakes of 1985 and French after the Second World War were presented.

In the later French reconstruction after the World War II a state centralized phenomenon is appreciated, where the State industrializes the prefabrication to rebuild the country. In the case of the reconstruction after the 1985's earthquake in Mexico City, a similar phenomenon is to be found, vertical and centralized at the same time, in which an spontaneous and solidary citizen participation existed.

The review of the French and Mexican experiences of post-disaster reconstruction suggests the elaboration of an urban-architectural reconstruction plan under a formal and hierarchical structure (as in the French case) that includes social participation systems and strategies. (as in the Mexican case).

The review of the French and Mexican reconstruction experiences allows us to distinguish two reconstruction stages. The first suggests a reconstruction based on a centralized government that provides massive and urgent housing production. The second presents the need to create "emerging" reconstruction models inspired by the ideal of fairer cities; other forms of production are possible and their inspiration may be the citizen participation that the 1985's earthquake showed.

A first reconstruction model suggests state control, under the control of a few real state monopolies, as the case of the French Reconstruction. As we have noted, this model has proved to be a failure. A second reconstruction model suggests the analysis of social participation, including informal and self-construction. In the informal construction, Mexican culture, we can find a number of possibilities to face the global problem of lack of habitable spaces. What is remarkable about the post-earthquake 1985 reconstruction is the alliance emerges between governmental authorities, citizens and the private sector.

The objective of a historical analysis of different reconstructions processes is to generate hypotheses that allow us to "invent" new structures, based on new modes of organization and management.

It is possible, for example, to develop a reconstruction urban-architectural plan, under the tutelage of governmental institutions (such as the French case), using a formal and perfectly hierarchical structure, and at the same time using systems and strategies that include social participation (like the Mexican case).

It would be possible to insert in a governmental reconstruction program, strategies, organizations and performances invented by the inhabitants: flexible and less expensive.

Models in which the residents' autonomy is restored and citizens participate in the elaboration of their homes' projects, making them suitable to their needs. It is about learning to negotiate with governmental agencies, to preserve and apply the "know-how" (*savoir faire*) of the local construction techniques; and to collaborate with the invention and innovation of technologies that enable massive, urgent and sustainable reconstructions.

Based on the question posed initially: What are the resilient elements of French and Mexican reconstruction that could be implemented in the event of massive and urgent constructions, particularly after a disaster? We could point out the conclusions obtained through the resilience factors detected in the cases studied: (i) urban planning of reconstruction is important in the same site, (ii) building materials and constructions systems should be socially and institutionally accepted, (iii) the act of the state together with the society as main actors in the planning, design, organization and control of the reconstruction process.

## References

Architectes Sans Frontieres Internacional (ASFI). (2012). *Challenging practice (essential for the social production of habitat)*. In http://asfint.or/Challenging-Pactice.html [November 2013].

Audefroy, J. (2003). La problemática de los desastres en el hábitat urbano en América Latina in *Boletín INVI*, vol. 18. Santiago de Chile, p. 52–71.

Audefroy, J. (2008). *Riesgos y vulnerabilidad en la ZMCM. Construcción de modelos geoespaciales*. México, D.F.: Coalición Internacional para el Hábitat.

Cuny, F. (1983). *Disasters and Development*. New York: Oxford University Press.

Davis, I. (1980). *Arquitectura de emergencia*. Barcelona: Gustavo Gili (Tecnología y Arquitectura).

Delgadillo, V. (2007). De habitantes suicidas y simulacros. In *Ciudades* (Vol. 74, pp. 40–46). México, D.F., April–June: RNIU.

Faure, A. (1996). *Entre les tours et les barres (Restructurer les espaces publics des grands ensembles)*. Lyon: Centre d'études sur les réseaux, les transports, l'urbanisme et les constructions publiques, CERTU.

Gustavo, G. -L. (2013). *La historia de las reconstrucciones en México: el caso de los edificios dañados por sismos en la ciudad de México*. Proyecto de investigación para obtener el título de ingeniero arquitecto. Instituto Politécnico Nacional, Escuela Superior de Ingeniería y Arquitectura, unidad Tecamachalco. Directora Reséndiz, A.

Kreimer, A., & Echeverria, E. (n.d.). Case Study: Housing Reconstruction in Mexico City, in *Disaster Risk Management*, The World Bank in http://web.worldbank.org/WBSITE/EXTERNAL/TOPICS/EXTARBANDEVELOPMENT [April, 2013].

Lavell, A. (1999). The impact of disaster on development gains: clarity or controversy. Génova, July. In http://www.desenredando.org/public/articulos/1999/iddg/IDDG1999_mar-1-2002.pdf [November 2012].

López, A., & Veduzco, C. (1986). Vivienda Popular y Reconstrucción. In *Cuadernos políticos* (Vol. 45, pp. 25–37). México D.F., January-Mars.

Mecatl, J. L., Michel, M., & Ziccardi, A. (1987). *Casa a los damnificados: dos años de política habitacional en la reconstrucción de la ciudad de México*. México: UNAM/ Instituto de Investigaciones Sociales.

Meerow, S., Joshua, N. P., & Stults, M. (2015). "Defining urban resilience: A review" 147 (2016) p. 38–49. In www.elsevier.com/locate/landurbanplan [March 2017].

Monnier, G. (2000). *L'architecture moderne en France (Tome 3: de la croissance à la compétition1967–1999)*. Paris: Picard.

Monsiváis, C. (2005). *No sin nosotros: los días del terremoto, 1985–2005*. México: Ediciones Era.

Oliver-Smith, A. (1994). "Reconstrucción después del desastre: Una visión general de secuelas y problemas" in Lavell, A. (compilador), *Al norte del rio grande*. La Red. In <http://www.desenredando.org/public/libros/1994/anrg/anrg_cap02-RDDD_oct-8-2002.pdf> [March 2012].

Pantelic, J. (1991). "The link between reconstruction and development" in *Land use policy*, October.

Perló Cohen, M. (1999). Desastres en las grandes ciudades. México: Dirección General de Protección Civil/Red Mexicana de Estudios Interdisciplinarios para la Protección Civil de Desastres. In http://www.cridlac.org/digitalizacion/pdf/spa/doc12803/doc12803.htm [November 2012].

Plan Construction et Architecture (PCA), Institut Français d'Architecture Fresnais, Jacques. (coord.), Vayssière, Bruno. (Editor). (1995). *Une politique de logement Ministère de la Reconstruction et de l'Urbanisme* (pp. 1944–1954). Paris: PCA/IFA.

Pliego, C. F. (1994). *Hacia una sociología de los desastres urbanos: un balance del proceso de reconstrucción en la ciudad de México*. México: UNAM/ Instituto de Investigaciones Sociales.

Ramírez Cuevas, J. (11 de septiembre de 2005). Repercusiones sociales y políticas del temblor de 1985 -Cuando los ciudadanos tomaron la ciudad en sus manos. La Jornada. http://www.jornada.unam.mx/2005/09/11/masjesus.html. Consultado el 14 de junio de 2016

Reséndiz, A. (2002). Los sistemas constructivos mixtos (Nuevo reto para el control de obra). Tesis para obtener el grado de maestra en ciencias con especialidad en arquitectura, México, Escuela Superior de Ingeniería y Arquitectura, Sección de Estudios de Posgrado e Investigación, unidad Tecamachalco, Instituto Politécnico Nacional. Dir. F. Luna.

Reséndiz, A. (2010). La industrialización de la construcción. El caso de la prefabricación en la construcción escolar en Francia (1951–1973). Tesis de doctorado, Centro de Historia de Técnicas (de la Arquitectura) y del Medio Ambiente, Conservatorio Nacional de Artes y Oficios de Paris. Dir. A. Guillerme.

Reséndiz, A. (2014). Las reconstrucciones: francesa post-guerra y mexicana post-sismo in Audefroy, Joel (coordinador), *Arquitecturas en riesgo*. México: Ediciones Navarra.

Salcido, I. (2010). El terremoto de 1985. In *25 años en nuestra memoria*. México: Martín Adame Editor.

Simon, E.H.L. (1950). La préfabrication dans la construction (emploi actuel et possibilités d'avenir) in Le Moniteur des Travaux publics et du Batiment, no. Hors série, March, p. 50.

Strobel, Pierre (1983). Les politiques d' industrialisation de la construction en France depuis la Libération en France in Centre Georges Pompiduou, Architecture et industrie (Passé et avenir d'un mariage de raison). Paris: Centre de Création Industrielle/Centre GeorgesPompidou.

Tobriner, S. (1980). Les politiques d'industrialisation de la construction en France depuis la Libération en France in Centre Georges Pompidou: Architecture et Industrie (Passé et avenir d'un mariage de raison). Paris: Centre de Création Industrielle/Centre Georges Pompidou.

U.S. Agency for International Development (USAID). (n.d.). *Evaluación de los daños y análisis de necesidades nivel toma de decisiones (EDAN-TD)* in http://transition.usaid.gov/our_work/humanitarian_assistance/disaster_assistance/publications/prep_mit/RDAP_training/files/DANADM/MR_EDANTD_2007.pdf [February 2013].

UNISDR. (2009). Estrategia Internacional para la Reducción de Desastres de las Naciones Unidas. *Terminología sobre reducción del riesgo de desastres*. Suiza: ONU.

Vayssière, B. (1988). *Reconstruction - déconstruction: le hard French ou l'architecture française des Trente Glorieuses*. Paris: Picard.

Wilches-Chaux, G. (2012). La gestión del riesgo: del deber de la esperanza a la obligación del milagro. En Briones, F. (coordinador), *Perspectivas de investigación y acción frente al cambio climático en Latinoamérica*, número especial de *Desastres y Sociedad* en el marco del XX aniversario de la Red. Venezuela: La Red/CIGIR.

# Chapter 15
# Operational Mitigation – Transient Housing: The Case of Jammu, India

**Faiz Ahmed Chundeli and Lakshmi Visakha**

**Abstract** Continual rains that led to intense flooding of the Jhelum River in Jammu during September 2014, affected a population of 1, 75,000 people over 19 districts. Urban poor and middle class families faced severe crisis in accessing basic amenities like communication, road access, hygiene and drinking water facilities. With the rising water level and lack of road transport, people were stranded on the rooftops. The main challenge of the project was the rehabilitation of the people considering lack of transportation facilities and resources needed for construction. The post disaster management project was an operational mitigation developed to address crucial aspects of urban life such as to provide permanent housing, restoring livelihoods, education and skill building to 'build back better', create better medical facilities and improve the scenario to resist the disasters in the future. In this case it is observed that the community participation and usage of available resources helped building temporary shelters in the aftermath of the disaster as part of rehabilitation measures taken. While the transient housing addressed the immediate needs of the victims, the need for permanent housing stands as the main concern.

**Keywords** Resilient building   Floods · Vernacular architecture · Transient housing · Community participation

F. A. Chundeli (✉) · L. Visakha
School of Planning and Architecture, Vijayawada, India



G. Brunetta et al. (eds.), *Urban Resilience for Risk and Adaptation Governance*, Resilient Cities, https://doi.org/10.1007/978-3-319-76944-8_15

| **Image or graphic** | <br>*Source*: *Transcending the Tragedy Together(2015)*[(1)] |
|---|---|
| **Location** | Region: Asia,India, Jammu and Kashmir |
| **Duration** | 2 months |
| **Time Frame** | [Start date: 09/2014] [End date: 11/2014] |
| **Consortium** | Seeds' India |
| **Cost** | Estimated Cost of 1 shelter - 2091.083 USD |
| **Resources** | Human resource, Local knowledge, Sustainable Technology |
| **People affected** | 1000 families |
| **Donors** | TATA Relief Committee, Trafigura, UNICEF and United Way of India |
| **Urban Sectors** | Water, Waste, Building, Energy |
| **Resilience dimension** | Disaster resistant housing and Post disaster Management |
| **Strengths** | Socially Driven Mitigation |
| | Environmentally appropriate materials and design |
| | Safe and comfortable housing |
| **Weaknesses** | Durability |
| | Vulnerability of the Transient structures in the event of sudden impact. |
| | Lack of a permanent solution to the floods of Jammu. |
| **Impact** | Disaster Management |
| | Operational Mitigation |
| | Post Disaster Transient Housing |
| **Stakeholders List** | District Government/Jammu & Kashmir Police |
| | Local Village Administration |
| | House Owners |
| | Local trained masons and carpenters |
| | United Way India and NGO Partner. |

<table>
<tr><td>Challenge</td><td>Continual rains that led to intense flooding of the Jhelum River in Jammu during September 2014, affected a population of 1, 75,000 people over 19 districts. Affecting the rural areas to a greater dimension, the flood left 145 villages submerged and 2500 villages affected. Urban poor and middle class families also faced crisis in accessing the basic amenities like communication, road access, hygiene and drinking water facilities. With the rising water level and lack of road transport, people were stranded on the roof tops. The main challenge was the rehabilitation of the people considering lack of transportation facilities and resources needed for construction.</td></tr>
<tr><td>Project description</td><td>The post disaster project addressed crucial aspects such as to provide permanent housing, restoring livelihoods, education and skill building to ‘build back better’, create better medical facilities and improve the scenario to resist the next disaster with good force are taken into consideration. Satisfactory outcomes were a result of coordinated community participation and trust between the local people and the NGO. Adequately proportioned houses based on the humanitarian shelter standards for family of 7 members resulted in a 24′ x 12′ shelter plan in a 30′ x 30′ site area. The A shaped roof structure with 9 triangular frames at equal intervals of 3′ is sloped at an angle of 45°was designed to handle the extreme snow load along with 0.4 mm thick corrugated galvanized iron (CGI) roofing sheet, and a 2′6″ plinth in stone masonry as shown in the figure. For increased stability, the A frame is further fixed into the ground with the foundation of timber members in PVC pipes that are packed with cement slurry.</td></tr>
<tr><td></td><td>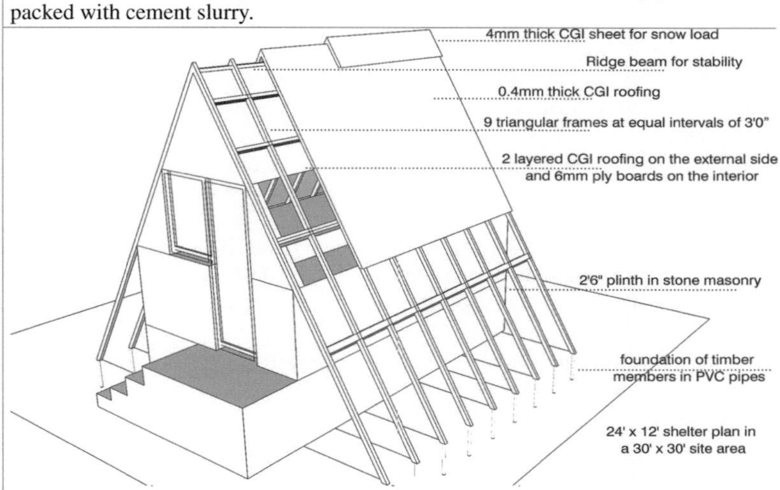
</td></tr>
<tr><td></td><td>Stability of the roof is rendered by the ridge beam that runs across the frame and the timber posts at each corner and the center. Also, seismic safety is provided by using cross- bracings in galvanized iron (GI) wire. Thermal Comfort during the extreme conditions in winter is provided by the two layered CGI roofing on the external side and 6mm ply boards on the interior. The air - gap between the CGI sheet and the ply board was proposed to be sandwiched using Egg crates/Paper/Cardboard for insulation. Sanitary needs were satisfied by soak-pit based 4′ x 4′ toilet for each house. Reduced material procurement was achieved by the use of salvaged materials in the brick wall over the stone plinth, and the door/ window shutters which contained wood from damaged houses.</td></tr>
</table>

| | |
|---|---|
| **Urban resilience elements** | In the case of Jammu, the disaster management in the aftermath of the floods in September 2014 was an "*Operational Mitigation*[2]" pertaining to post disaster management and recovery. It was observed that the community participation and usage of available resources helped building temporary shelters in the aftermath of the disaster as part of rehabilitation measures taken. |
| **List of innovation elements** | Salvaged and low cost materials used for construction to overcome the hassles of transportation and material procurement |
| | Thermal comfort assessed and insulation for about sub-zero temperatures was provided using low-cost materials, the structural design was done considering seismic safety and the plinth heights were considered as per snow level. |
| | Easily Replicable and locally adapted design. |
| | People centric process promoting Community Natural Resource Management on a larger scale. |
| **Lessons learned** | The initiatives taken during the mitigation have opened the doors to rethink on the idea of sustainability in the following lines |
| | Strengthening of design process through public involvement, thus making it more people centric. |
| | Low cost construction technology by re- use of salvaged materials in order to promote environmentally sound design and reduce time required for construction and transportation needs. |
| | The need for environmentally sound design in order to overcome the effects of the disaster in the future. |
| **Up scaling and replication** | While the transient housing addressed the immediate needs of the victims, the need for permanent housing stands as the main concern. |
| | In the aftermath of the disaster however, the replication of the design proved scalable and replicable. However, it is observed that the house cannot be considered incremental as it addresses only temporary needs of people. |
| **Web page** | *Resilient Cities - ICLEI: Urban resilience planning. Resilient-cities.iclei.org*. Retrieved 23 August 2016, from http://resilient-cities.iclei.org/resilient-cities-hub-site/resilience-resource-point/resilience-library/urban-resilience-planning |
| | *Home \| ICLEI Global. Iclei.org*. Retrieved 24 August 2016, from http://www.iclei.org/ |
| | *United Way India. Unitedwayindia.org*. Retrieved 24 August 2016, from http://www.unitedwayindia.org/ |
| **Reference** | SEEDS India. (2015). *Transcending the tragedy together*. New Delhi: SEEDS. Retrieved from http://www.seedsindia.org/final-photo-documentation-J&K-Flood-response2014-15-documentation.pdf |
| | Sphere India. (2014). *Joint Rapid Needs Assessment Report: Jammu and Kashmir Floods 2014* (pp. 1–37). Sphere India. Retrieved from http://reliefweb.int/report/india/humanitarian-appeal-kashmir-flood-response |
| | United Way India. (2016). *Assessment of Relief interventions and Plans for Jammu and Kashmir*. United Way India Response Team J&K. Retrieved from http://www.unitedwayindia.org/.../UWI%20Team-Report%20on%20visit%20to%20JK-Nov |
| | United Way India. (2014). *Project for Reconstruction of Shelters in Affected Communities of Jammu and Kashmir PHASE-I: First Plan for Utilization of Funds Received under Hum Hain Ummeed E Kashmir By United Way of India*. United Way India Response Team- J&K. Retrieved from http://www.zeetv.com/humhain/HHUEK-%20Shelter%20Reconstruction-%201st%20Utilizat |

# Chapter 16
# Planning Recovery and Reconstruction After the 2010 Maule Earthquake and Tsunami in Chile

**Stephen Platt**

**Abstract** This chapter analyses the urban planning process and the disaster recovery strategies adopted, both at the national and local level, after the Maule Chilean earthquake of 27 February 2010. In particular it focuses on how well Chile balanced the need for speed with building back better and how effective was the transition from temporary relief to long-term resilience.

In part, the analysis is based on a field trip conducted by the author eighteen months after the disaster (Platt S, Reconstruction in Chile post 2010 earthquake. CAR, Cambridge (2012a)). The author visited three cities: Concepci6n, Viña del Mar and Valparaiso, that were subject to earthquake damage and three coastal settlements, Tumbes, Dichato and Tubul in the Region Bio-Bio, that were hit by the subsequent tsunami. He interviewed senior people in the Ministry of Transport and the Ministry of Housing and Urbanism responsible for coordinating national reconstruction, people in Bio-Bio responsible for reconstruction in that region and residents in the coastal settlements.

This study of disaster recovery in Chile is one of a series of 10 case studies of places that have suffered major earthquake related disasters in the recent past. The performance and effectiveness of the strategies and solutions in Chile are compared with those in these other places (Platt S, So E, Speed or deliberation - a comparison of post disaster recovery in Japan, Turkey and Chile. Disasters (Online; forthcoming in print): https://doi.org/10.1111/disa.12219 (2016); Platt S, Factors affecting the speed and quality of post disaster recovery and resilience. In: Olafsson S, Rupakhety R (eds) Recent developments in earthquake engineering and structural dynamics: in memory of Prof. Ragnar Sigjörnsson. Spinger. (forthcoming) (2017)).

Most significantly Chile did better, in terms of the speed and quality of recovery, than any of the other countries. The chapter explores the reasons underlying this relative success and offers important lessons for planning recovery after all types of major natural disaster.

**Keywords** Disaster recovery · Reconstruction · Earthquake · Tsunami · Chile

S. Platt (✉)
Cambridge Architectural Research Ltd, Cambridge, UK
e-mail: steve.platt@carltd.com

 
G. Brunetta et al. (eds.), *Urban Resilience for Risk and Adaptation Governance*, Resilient Cities, https://doi.org/10.1007/978-3-319-76944-8_16

## 16.1 Introduction

An important aspect of resilience in a city or city region is the ability to recover quickly after a major disaster. Recovery means people who were displaced getting back in permanent homes, businesses getting back to full capacity, schools hospitals and infrastructure repaired or rebuilt, roads cleared and life in general getting back to normal (Merriam-Webster; Quarantelli 1999; Bruneau et al. 2003). But recovery is not only about speed. As well as the timely restoration of normal life, post-disaster planning also involves enhanced protection against the impacts of future hazards (Alexander 2013).

This chapter analyses the urban planning process, both at the national and local level, after the Maule Chilean earthquake of 27 February 2010. The analysis is based on a field trip conducted by the author eighteen months after the disaster (Platt Platt 2012a). The author visited three cities that were subject to earthquake damage: Concepción, Viña del Mar and Valparaiso; and three coastal settlements, Tumbes, Dichato and Tubul in the Region Bio-Bio, that were hit by the subsequent tsunami. He interviewed senior people in the Ministry of Transport and the Ministry of Housing and Urbanism responsible for coordinating national reconstruction, people in Bio-Bio responsible for reconstruction in that region and residents in the coastal settlements.

The case study was part of a larger 10 year project investigating recovery after major earthquake-related disasters in 10 countries: Chile, China, Iran, Italy, Japan, New Zealand, Pakistan, Thailand, Turkey, and the United States (Table 16.1). The aim of the research was to identify commonalities and to define the critical factors in increasing resilience and improving the speed and quality of recovery.

In the main the affected areas were visited, typically 12–18 months after the incident, and the principal actors were interviewed, including: politicians and local governors; architects, engineers, planners, and scientists; personnel in local government offices, central government ministries, special disaster authorities, and civil defence and emergency response entities; business people; members of chambers of commerce, voluntary organisations, and local NGOs; and community leaders and

**Table 16.1** Data from 10 major earthquakes

| Country | Name | Year | Mw | Displaced | Deaths | Loss US$ bn |
|---|---|---|---|---|---|---|
| USA | Northridge | 1994 | 6.7 | 125,000 | 61 | 44 |
| Iran | Bam | 2003 | 6.6 | 75,000 | 26,271 | 1.5 |
| Thailand | Indian Ocean | 2004 | 9.2 | 1,690,000 | 276,025 | 14 |
| Pakistan | Kashmir | 2005 | 7.6 | 3,500,000 | 100,000 | 2.3 |
| China | Wenchuan | 2008 | 8.0 | 1,940,000 | 90,000 | 75 |
| Italy | L'Aquila | 2009 | 5.8 | 67,000 | 309 | 16 |
| Chile | Maule | 2010 | 8.8 | 800,000 | 550 | 30 |
| New Zealand | Christchurch | 2011 | 6.3 | 25,000 | 185 | 16 |
| Japan | Tohoku | 2011 | 9.0 | 130,927 | 18,499 | 235 |
| Turkey | Van | 2011 | 7.1 | 50,000 | 604 | 1 |

residents. It is important to note that interviews elicit both facts and subjective opinion and that these individual accounts, like all interpretations of reality, may be partial and biased. This issue was addressed by careful choice of informant, by interviewing at least 30 people in each place, by asking probing questions, and by verification through observation, crosschecking between informants, and conducting a comprehensive review of published sources.

## 16.2 Impact of the Maule Earthquake

The Maule Chile earthquake of 8.8Mw occurred at dawn (3:34 am) on the 27 February 2010 200 miles southwest of Santiago (USGS 2011a). The earthquake is believed to be the fifth most powerful since seismic measurements began. It triggered a tsunami whose waves hit the Chilean coast and caused severe damage to many coastal settlements (Lubkowski et al. 2010). The earthquake affected more than 2 million people, but despite the magnitude of the event, there were few major structural collapses and a relatively low death toll of 500–600 (USGS 2011b). Most buildings performed as they had been designed, to protect life, but eco-nomic losses were huge, totalling approximately USD 30 billion or 17 per cent of gross domestic product (GDP) (Franco and Siembieda, 2010). Despite the magnitude of the earthquake there were few major structural collapses and a relatively low death toll: 525, with another 25 missing. The standard of construction in Chile is generally good (Booth and Taylor 1988). Fortunately, the city avoided the tsunami that followed the earthquake. Nevertheless, 260,000 homes, 4000 schools and about 70% of hospitals in the quake-affected area were severely damaged, as well as roads, highways, churches, airports, stadiums, courthouses, electricity and telecommunications networks and military installations.

One-fifth of the population in Maule was made homeless by the earthquake. Displaced persons were given two options: (i) to stay on their property; or (ii) to move to an *Aldea*, or temporary housing village. In both instances, displaced persons were housed in *Media Aguas*, or temporary timber housing. Grants for tools and supplies were distributed to those working on repairing their homes (Hinrichs et al. 2011). Temporary housing was constructed from local timber. These homes were one-room, 18 square metre timber chalets, smaller than either the Japanese or Turkish equivalents, and the camps comprised 250–500 units plus sanitary blocks. Some families incorporated the shelter into their rebuilt homes.

### *16.2.1 Impact in Concepción and Other Cities*

Concepción is the capital of the Biobío Region or Region VIII. With 889,725 inhabitants (2002 census) Gran Concepción (which includes Talcahuano, San Pedro de la Paz, Hualpén, Chiguayante, Penco, Tomé, Lota, Coronel, Hualqui and Concepción) is the second-largest conurbation in the country. The city itself has a population of 212,003.

Previous earthquakes and tsunamis razed the town in 1570, 1657, 1687, 1730, 1751, 1822, 1835, 1868, 1877, 1906, 1918, 1922, 1937 and 1960. During the1570 earthquake of magnitude 8.3 a delay between the earthquake and the tsunami allowed the population to escape to higher ground and no lives were lost, but every house was destroyed. The 1751 Concepción earthquake was one of the strongest and most destructive recorded quakes in Chilean history. It struck the Central Valley, destroying the cities of Concepción, Chillán, Cauquenes, Curicó and Talca. During the earthquake and the subsequent tsunami, all of the buildings in the city of Concepción were destroyed. The records indicate that the earthquake was so intense that "the residents could not remain standing." Fourteen years after the quake the city was relocated to the Valle de la Mocha alongside the Bio-Bio River where Concepción presently lies.

In September 2011 there were clear signs of recovery. Shops and restaurants were open in the city centre, the port of Talcahuano was in operation and weekend trippers were eating in local restaurants in Tumbes, a local fishing village of that had been inundated. Of the three bridges across the River Bio-Bio, only the railway bridge survived unscathed. One of the two road bridges collapsed and had been replaced with a temporary metal bridge, nominated the Mecano. The other bridge survived and was in use but the carriageway was distorted.

In the commercial centre there was severe earthquake damage to multi-storey buildings. The most notorious collapse, the fifteen-story newly completed residential building, the Alto Río, fell horizontally, trapping many of the residents. This site had just been cleared. The eighteen-storey Torre O'Higgins building pancaked at the 10th floor and demolition work was still blocking the main avenue of O'Higgins. Two other tall buildings, the Plaza del Rio and the Centro Mayor had significant damage. The possible causes of these failures have been analysed elsewhere (Lagos and Kupfer 2012) (Fig. 16.1).

Several factors contributed to the low casualty rate and to rapid recovery, especially the robust building code and its comprehensive enforcement. In particular, Chile has a law that holds building owners accountable for losses in a new building for 10 years. There were few fires after the earthquake, because the electricity grid was shut down immediately. In many areas, the emergency response was very effective and there was close local coordination between emergency managers and fire and police personnel, without a need for communication with the capital. Finally, the high level of awareness meant that most people evacuated in good time and were able to manage living in temporary accommodation.

### *16.2.2 Impact on Coastal Settlements*

Talcahuano is a principal port. The earthquake and tsunami caused extensive damage, especially in the town centre, the naval base and in the barrio of El Arenal and the small fishing ports of El Morro and Tumbes. The tsunami flooded the Canal Ifarle and inundated Villamar, Las Salinas, San Marcos and Vegas Perales

**Fig. 16.1** Torre O'Higgins, Concepción, still blocking main avenue 18 months after the earthquake (Photo by Author)

(Robertson, Roberson 2010). The damage to road, rail and port infrastructure caused severe access problems in the days following the disaster. New social housing was being constructed in the barrios on hills above Talcahuano and informal development of timber shacks on the more difficult steep slopes (Muñoz 2010).

Tumbes extends along the seafront and up a wooded valley perpendicular to the coast. On the 8 March, just over a week after the tsunami, 2010 Chilean TV reported that Tumbes was left completely destroyed, that the residents had found shelter in tents in the hills in fear of another wave and that the future of the settlements was uncertain. From the video shot at the time it certainly looked like that. But eighteen months later it was clear that more than half the homes in Tumbes had survived the wave, the economy was recovering and the school site had been cleared and a new school built.

Dichato is a seaside town of 4046 inhabitants 37 km north of Concepción. It sits at the head of an enclosed bay of calm water suitable for water sports and popular with tourists in summer. The horseshoe bay and the flat estuary make the site vulnerable to tsunami. In total 80% of built structures including 1343 homes were destroyed by a water column of 6.5 m that reached an acceleration of 3.5 meters per second. The tsunami came at dawn and, warned by the earthquake, people had time to get to high ground.

**Fig. 16.2** Dichato after the wave retreated 27/3/2010

Much of the damage was caused by ships and solid debris being thrown against buildings. This was compounded by poor quality construction and inadequate structures. The destruction extended upstream of the estuary, affecting various residential sectors, the primary school, police headquarters, fire station that were wholly or partially destroyed. Three bridges were destroyed, seriously affecting connectivity and accessibility (Mackintosh 2010). Residents were temporarily relocated in four emergency villages, the largest being "El Molino", with a total of 450 shacks. This location, on high ground above any future tsunami risk represents an option for a future urbanization (Fig. 16.2).

I spent one day with Ivan Cartes, the Dean of the Faculty of Architecture at Bio Bio University and the person in charge of coastal settlemnent planning. We visited people in the Municipal planning authority in Tomes, walking around Dichato and talking to business people and restaurant owners. It was a beautiful sunny day and although it was still early spring one could imagine that this was a very popular resort. There were still many signs of the devastation. Most of the homes within 500 metres of the shoreline had been carried away and all that remained were the sanitary blocks poking out of the rubble and weeds like rotten teeth. Most poignantly, each family had painted their name, address and mobile number in large block letters on the concrete stump (Fig. 16.3).

Tubul is a small fishing village at the estuary of the River Tubul, 70 km south of Concepción, of 2031 inhabitants at the time of the last census in 2001 (see below). 246 buildings were destroyed, principally by the tsunami. The settlement was

**Fig. 16.3** Bathroom block, all that's left of the Mendez family home (Photo Author)

founded about 30 years ago by an invasion of poor fishermen and their families. The site is a triangle of flat wetland of about 1 $km^2$ bounded by the estuary to the south, the sea to the east and a line of hills to the northwest. It was these hills that people climbed in the moonlight to escape the tsunami. The two sides of the L-shaped settlement, by the river and along the sea front, form two distinct communities. San Jose, along the river, housed slightly more families than Nueva Esperanza on the sea frontage. San Jose was flooded to a depth of 1–2 metres but over half the homes in Nueva Esperanza were destroyed. There was a particular problem accessing Tubul because the road bridge from Arauco and the north collapsed when all eight steel girder superstructures became unseated. The main road south was closed, which, together with a landslide meant that relief operations and food supplies had to divert to a dirt road through the coastal mountains.

## 16.3 Post Disaster Planning Process in Chile

In Chile the central government recognised it would be unable to reconstruct everything or even control the process, and national coordination was limited to defining the scale of the problem and allocating resources. Planning and implementation was delegated to the regional government and specialist teams of experts.

Claudia González Muzzio (2011 described the urban planning framework in Chile. She focused on how risks and hazards are (or not) taken into account and analysed the shortfalls of the regulations and zonation in the light of the difficulties faced by authorities and practitioners during the reconstruction after the Chaitén Volcano of May 2008 and the Maule Earthquake 2010. The Chilean Urban Planning Framework comprises the General Law of Urbanisation and Construction (LGUC) and its Ordinance (OGUC). This law consists of four hierarchical planning instruments, or types of plan.

- Regional Urban Development Plan
- Intercommunal Plan
- Communal plan
- Plans for specific sectors of cities

These plans are implemented through the application of 14 norms, or regulations, managed through the building control process and the granting of building permits, that cover land use, plot size, plot occupation and building height. The planning framework also seeks to define high risk or 'hazardous' areas where construction is restricted, but not prohibited. However, the concept of vulnerability to hazards is not well articulated by existing planning law, which also does not cover pre-disaster planning for earthquakes or tsunamis. This means that plans already in force fail to adequately address risk from natural hazards such as earthquakes, tsunami and volcanic eruption.

### *16.3.1 National Plan for Reconstruction*

The most significant initial measure the Government took was to establish the main principles of recovery and to allocate the funds for reconstruction. Pablo Allard was appointed National Coordinator of Urban Reconstruction at the Ministry of Housing and Urban Development. The slogan of the Plan de Reconstrucción (MINVU 2010) was: "Our challenge: turn a catastrophe into an opportunity. .. for better cities, better life" (MINVU 2010). The plan underlined the principle that the state was unable to reconstruct everything or even control the process of recovery centrally from Santiago. It established that it was the responsibility of each region, town council, and community to develop its own plans, with the support of the state. It also highlighted that although the authorities in each locality face huge problems that they will want to address immediately, it is important that they have a long-term strategic vision and that they proceed methodically in dealing with important issues concerning cultural identity, the environment, and citizen engagement if opportunities to build back better are to be realised. The local authorities took a central role in setting budgets and in coordinating the actions of the government ministries, although the exchequer had the final say in approving plans and budget assignments.

The plan was to build 70,000 dwellings and the goal was to have everyone back in permanent housing by the end of 2012. Allard is quoted as saying: "This is one of the most complex reconstruction efforts anyone has ever undertaken. . .

Reconstruction means not only rebuilding what was there before, but rebuilding it better, much better. .. We can't just build short-term ghettoes. We have to rebuild these cities to mitigate the [future] risk and exposure to earthquake and tsunami hazards" (Padgett 2011). In fact the majority of people (>90%) were back in their homes 2 years after the disaster and most aspects of the economy and society were back to normal within 5 years (Platt 2017).

Comerio (2013) analysed housing recovery after the Maule earthquake of 2010. The disaster damaged or destroyed some 370,000 housing units (approximately 10 per cent of the housing in six regions). Yet, within six months, the Ministry of Housing and Urban Development had published a plan to repair or rebuild, with governmental assistance, 220,000 low and middle-income housing units within four years. In October 2012, at the midpoint of the programme, 54 per cent were complete and occupied and a further 30 per cent were under construction.

### *16.3.2 Coastal Settlement Plan*

The responsibility for master planning was passed from the National Government in Santiago to local authorities. Almost immediately after the earthquake a group of 10 architects/planners was seconded to the Regional Government from the two main universities in Concepción, the Universidad Bio-Bio and Universidad Católica de Chile, and from the Ministerio de Vivienda y Urbanismo, MINVU to assist the local authorities. Of the 36 coastal settlements in the Region, 18 were severely damaged and the team decided to concentrate their efforts on these. Interestingly all these settlements were exposed to tsunami waves from the north. They divided into three teams (Cartes 2011).

The first thing the planners did was to try to gather the data they needed for planning and to assess the scale of the damage. The teams worked closely both with the planners in the various municipal authorities, with academics and with local residents and business people in the affected settlements to develop master plans for each settlement. This process was called the Plan de Reconstrucción del Borde Costero – PRBC18 (2011). The plan had the following overall objective: to plan the reconstruction of coastal settlements affected by the earthquake and tsunami ensuring a high quality restoration that is inclusive and integrated. It was based on four general principles:

**Safety** planning for mitigating the level of vulnerability of inhabitants living on the coastal margin and improving the resilience of settlements to cope with a similar event.

**Sustainability** limit occupancy of urban sensitive areas to allow natural systems to recover; implement building systems that provide greater energy efficiency and transport systems that deliver more sustainable urban mobility.

**Quality of Life** stakeholders in each locality identified the factors affecting quality of life and these have been translated into indicators that are reported in each Master Plan.

**Fig. 16.4** Planner briefing to a residents association Tubul, Chile (Author)

**Platform for the Future** each plan should provide conditions that encourage opportunities for growth and economic development and a platform for developing new production and service activities, according to potential of each locality.

Project design and planning comprised defining potential project interventions, costing these and securing a budget using the damage assessments collected by each municipality. This was followed by extensive engagement with residents and local businesses through focus group meetings and face-to-face engagement (González Muzzio 2012). Maps, images and sketches were used to communicate and share ideas. This process of consultation, of seeking opinion, establishing priorities and keeping people informed about progress was extremely impressive and effective (Fig. 16.4).

The recovery manuals that the reconstruction team consulted (FEMA 2008a, b, c; NTMP 2010) stressed that, after the immediate relief effort, it was important to reinstate livelihoods as quickly as possible, both to take people's minds off the tragedy and to make them independent of state support. Consequently, the main focus of the master planning teams was to get business and commerce up and running. In addition, the team designed and built temporary restaurants set back from the beach and designated land for a park and the planting of trees.

The main components of the master plans in Dichato, Tubul and other coastal settlements was to move buildings back from the beach, plant trees to break the

**Fig. 16.5** Master plan for Tubul, Chile (Plan Maestro Tubul, 2010)
KEY: *1* health centre; *2* school; *3* police; *4* fire brigade; *5* wetland park; *6* housing (266); *7* main street; *8* waterfront; *9* info centre; *10* park; *11* cycleway; *12* port infrastructure; *13* new road; *14;* evacuation assembly

force of the waves and to canalise rivers (Fig. 16.5). The sea defences proposed involve a much lower sea wall than those in Japan, a setback of 50–80 metres for buildings, and a 20–30 metre band of trees. In Tubul, the coastal strip where people had been living was unsafe and the plan moved the majority of homes to higher ground at the other end of the village, which necessitated the purchase of land from the *latifundista* landowner. The government expropriated existing homes and paid people the commercial value for the house and improvements, independent of the subsidy for the new housing. The new homes are 114 square metres, slightly smaller than the existing houses. Homes built in the old centre of the village were tsunami-resistant. Residents were closely involved in planning the move and in decisions about the design of the home. Interestingly, it was the Great East Japan Earthquake of 2011 that helped to convince the undecided in Chile about moving.

## 16.4 Issues with the Planning System

### *16.4.1 Coordination*

A major issue facing the planning team was coordinating the efforts of the many agencies involved in reconstruction. It was an immensely complicated process involving weekly meetings, emails and site visits but was eventually successful in coordinating decision-making. The master plan involved three distinct types of work. The first involved defining a set of interventions, costing them and securing a budget. The second involved extensive and repeated engagement with residents and local businesses and the third was to coordinate the national, regional and local public institutions – the ministries, municipalities and local associations to deliver the project. Initially, there had been resistance from some sectors of the community and from parts of government, but finally by dint of talking they had achieved consensus.

Masters plans were not part of Chilean urban planning legislation. This meant that they were indicative rather than normative. To action them Regulation Plans had to be modified in line with the strategy in the master plan. Although the Housing Ministry was in charge of this process, the 'Contraloría' or exchequer has the final say in approving plans and this created problems and delay. There was, however, no political support to change the normal urban planning law to allow a faster rebuild.

### *16.4.2 Rebuild in Same Location or Move?*

The key question the country needed to address in planning for future tsunami was whether to rebuild in the same area, taking as many precautions to limit risk as possible, or whether to relocate development to areas of less or no risk. Chile adopted the first strategy. In April 2011 The Government declared its intent to incorporate tsunami risk mitigation into Chilean urban planning law, but there was little planning experience of how to consider this kind of risk and determine its impact on land use standards and building codes (Bustos Erwenne 2011). Current urban planning law did not define minimum building height and materiality and it was impossible to insist that only high concrete buildings were constructed in tsunami run-up areas. Many critical facilities were located in vulnerable areas and it was difficult to move these facilities to safe areas.

However, there is no technical certainty that the new tsunami housing and proposed mitigation works will withstand large magnitude tsunami and the barriers. Chile needed to rebuild homes faster than the planning process could keep up with. The fear factor was not enough to make people move and people started to rebuild their homes in the same disaster prone areas.

### 16.4.3 *Difficulty Obtaining Information*

There were major problems in obtaining the data necessary for planning at a national level. The planning process involved using a Geographic Information System (GIS) to integrate land cadastral information with 2002 Census data, various ministerial and local authority databases and satellite imagery, as well as managing the data. The relevant ministries: Housing, Public Works, National Assets etc. worked with different data and graphics systems and there was no time to coordinate information. National legislation would have been necessary to create a single information system used by all the relevant authorities and to define a clear procedure for sharing data. The ministries, as well as the regional and local authorities also lacked trained GIS personnel to manage the data.

The Chilean team working on coastal settlement reconstruction tried and failed to build this database in time to be useful to the master planning process. On the one hand master planners would like detailed information at an individual plot scale, but they can, and have to, make design decisions based on much sketchier aggregate information and local knowledge. In practice most countries facing the impacts of a major disaster will lack the time and resources to build a systematic database. The key is to define the minimum needs and to be prepared in advance of an event.

## 16.5 Information for Planning

The immediate problem the planners faced was a lack of cartography. Obtaining maps of the settlements prior to the disaster was fundamental to the planning process. The planners even lacked maps showing access and the location of buildings, although there was some aerial imagery. They also faced problems getting reliable statistics and cadastral information about damaged properties. The coastal planners described how they immediately started to build a GIS database but were overwhelmed by the size of the problem and had to plan without this information. Crowd-sourcing data was discussed and the possibility of training large groups of university students to help with this task, but the team was skeptical about whether there were resources available to coordinate this type of data collection.

There was no centralized national system of impact assessment using geospatial technology. Local damage was assessed for each Municipality by young architect volunteers overseen by the coastal planning team in the month after the earthquake. The results were then aggregated by the Housing Ministry to get a national figure. The results of this impact assessment were used in the first stage of the planning process to quantify the amount of damage, to determine the financial support needed for reconstruction and how long the work was likely to take.

Aerial imagery (privately acquired) was used to identify different land uses as residential, education, health, etc., and also to identify areas damaged or not damaged by the tsunami. Data from the Census 2002 was used to estimate the number of households in each location. Topographic maps, containing building footprints and land uses at 1:5000 scale were digitized and added to the GIS. Urban Regulation Plans (2000–2005) were also available for some of the bigger localities. Finally risk mapping was done by Bio Bio University alongside the Master Planning process. Maps of flood, tsunami, liquefaction and landslide risk were produced by the National Service of Geology and Mining (SERNAGEOMIN) and Bio Bio University. Combining the risk and topographic maps allowed planners to identify features and facilities at risk and to find suitable relocation points for them. This analysis also defined the locations of coastal defenses, mitigation forest and tsunami-resistant buildings.

Considerable thought was given in to developing quality of life indicators organized into: urban character, environment, mobility, public spaces, equipment, basic services and employment, that would show the level development of a locality and the welfare of its inhabitants. … First and foremost they are defined by the priorities of the community but they also include universal indicators that establish parameters for comparison and analysis. Defining indicators can be a tool that represents a picture of the current state of the urban situation. With it, you can observe, measure, compare and track the time, how growth affects quality of life. As a result it should be possible to focus public policy and investment decisions about urban interventions in terms of improved continuous habitat.

One of the insights from this study in Chile is that two types of information are needed after a major disaster. Immediately after the event there is a need for information about damage to buildings, roads and bridges and about the level of relief and shelter required. This information is needed to make decisions about rebuilding or relocating infrastructure and buildings and about providing support and compensation to families and businesses. Almost simultaneously teams of people begin planning the process of recovery at an urban scale and they need a different type of aggregate information about all aspects of the places affected.

In reality these are two types of need rather than two distinct types of information, since, theoretically, the information could be derived from the same data. For practical reasons of timing, they may have to be distinct. They key issue is how quickly can information of either type be provided. Information, particularly in the form of maps, aerial imagery and photographs both immediately before and after the event is a first requirement. Ideally there would also be base line socio-economic information about the community and cadastral information. In practice, in many countries at risk from natural disaster, this base-line data is not available.

## 16.6 Discussion

*Speed and Deliberation*

As in all recovery planning after major natural disasters there was a need to balance speed and deliberation (Olshansky 2006; Platt and So 2016). In Chile there was intense pressure from residents to rebuild homes, to restore facilities, and to get the economy moving but there was also a desire on the part of the authorities to develop new urban plans that would improve these communities and make them safer.

Chile achieved a better balance between speed and deliberation than any of the other 9 countries. The central government recognised that it would be unable to reconstruct everything or even control the process. National coordination was therefore limited to defining the scale of the problem and allocating resources. Responsibility for planning and implementation was allocated to the regional government and specialist teams of experts.

*Land Use Change and Safety*

There are three land use planning choices: (i) rebuild in the original place; (ii) move to a safer adjacent neighbourhood; or (iii) relocate to a new place. There is also the issue of whether to move whole communities collectively or whether individuals should make their own arrangements (Iuchi 2015). This decision depends largely on the degree of damage, the willingness of the inhabitants to move, the difficulty of mitigating future risk, and the economic implications of the move (Ye 1996). In Japan, for example, the policy was to define a hazard zone and to move people up and away. In Turkey, between one-third and one-half of displaced persons were rehoused in new apartment blocks (Platt and Durmaz 2016). The rest of the population plus new migrants to the city were housed on redeveloped demolition sites or in repaired houses. In Chile, commercial and industrial facilities were moved back from the coast to behind a new promenade and a treed margin that will protect against storm surge and break the force of a tsunami. Housing in the hazard zone will be tsunami-proof.

*Economic Recovery*

Economic recovery is quite likely the most serious issue facing most communities in a post-disaster period, and almost certainly the central issue confronting national authorities in every major disaster. Bolton (1996) points out that a major disaster spawns an urgency to decide many things at once. In Chile the main focus of both the National Plan and the local planners was to reinstate the local economy as swiftly as possible. As well as restoring infrastructure quickly, considerable efforts where made to house businesses in temporary accommodation and to provide subsidies for rebuilding permanent accommodation and to keep people in business. Within 6 weeks, nearly 60% of public infrastructure damaged by the quake had been repaired, and 11,000 emergency housing units had been built.

*Planning System*

A large number of authorities and organisations are involved in different aspects of recovery. In many countries it is unclear which agency, department, or organisation is responsible for planning post-disaster recovery. There may also be

a) range of pre-existing and special earthquake-related plans. There is a pressing need after major disasters to coordinate decision-making, land availability, the reconstruction programme, and service provision, but often there is tension between local, regional, and national authorities.

The architects heading the three groups visited their areas regularly, at least once a fortnight, meeting residents and business people and just walking the streets. The author accompanied two of them on various trips to the affected area and attended resident association meetings and ad hoc meetings with restaurant owners and shopkeepers, witnessing the quality of this consultation in action.

*Insurance and Government Subsidy*
Insurance can be a powerful ex-ante strategy in an earthquake risk mitigation framework. Its objective is to provide monetary compensation for damaged assets or lost income, but also it can help to achieve other important goals for society, such as the establishment of safer building practices, the dissemination of risk information, and the promotion of financial responsibility (Franco 2014).

In New Zealand the level of household cover was almost too high in that it seemed to have an almost debilitating effect on people and their psychological capacity to recover (Platt 2012b). In contrast in Pakistan insurance was almost non-existent. In Chile, insurers paid out approximately $8billion in losses, 95% of which were reinsured. Claims were settled relatively quickly with a low level of litigation.

The government assessed damages and losses at $30 billion, and estimated total public spending for reconstruction would come to $12 billion. The government funded this spending through moderate tax increases, reserves, budget reallocations, the sale of assets and, most significantly, through concessionary schemes with the private sector, such as new highways linking new peripheral shopping malls and housing developments with city centres. Although premiums jumped significantly after the earthquake, insurance and rapid government funding were key factors in speedy recovery (Useem et al. 2015).

*Political Issues*
Davis (2006) points out that all aspects of disaster management, including longer-term recovery, occur within political contexts, that disasters place immense demands on government officials, and that the public, especially affected persons, has high expectations of leaders and public officials. However, the brevity or superficiality of media coverage tends to underplay the challenges faced by governments. In Chile, all those involved in reconstruction had a personal commitment to involve local people in decision-making and to keep them informed about progress.

## 16.7 Conclusions and Main Lessons

What really distinguished recovery in Chile was the balance between the speed of recovery and the quality of reconstruction. In Chile cities and settlements returned to ‘normality’ within about five years of the disaster. Normality is defined as access being fully restored; power full restored; temporary housing completely cleared; 90% of displaced rehoused; 90% of children back in school and 90% of people back to work (Platt 2017). But as well as being quick, Chile also managed to 'build back better' and achieved 84% perfect performance in terms of indicators measuring an improvement in safety, amenity ecology housing and local economy (Table 16.2).

Both urban planning in the major cities and master planning of the 18 coastal settlements was completed by December 2010, only 10 months after the earthquake, reconstruction was well advanced after two years and things were 'back to normal' within 5 years.

The key recovery measures which increased safety and improved amenity were re-establishing economic activity, moving key public services and some housing to safer locations, building tsunami resistant housing in risk prone areas and mitigating the effect of future tsunamis through retaining walls and forest planting. Overall this represents a significant improvement in the quality and resistance of these settlements. There are however, many other settlements along the Chilean coast that were not badly affected by this event but remain vulnerable to future events. The main reasons for this success and the lessons for other places may be summarised as follows.

*Authority, Leadership*
There was clear demarcation of responsibility between the central government that provided the oversight, passed the necessary legislation, set the planning parameters and allocated the resources, and the regional government that devised and implemented the plans.

*Governance*
One of the really distinctive aspects of the recovery process in Chile was the quality of participation and consultation. Local communities were kept informed and involved in decision-making through choosing options and setting priorities, through residents associations, meetings with business people and casual encounters in the street throughout the planning and implementation process.

*Decision-Making and Window of Opportunity*
The regional government lost no time in seconding enthusiastic and highly talented architects and planners to the planning teams. These people realised that they had a 'window of opportunity' of about 18 months to devise the plans and get them started. They made full use of this limited time by working efficiently and hard.

*Urban Planning*
Master planning of the urban environment had the aim not of of restoring homes and businesses it also aimed to improve safety, amenity and the quality of life. Chilean planners, contractors and building owners also achieved a sensible balance of repair and rebuild.

*Finance and Resourcing*
About 20$bn was allocated for recovery (12$bn from central government and 8$bn from the insurance industry). Bureaucratic hurdles were minimised and the financing of recovery and reconstruction was speedy and effective.

**Table 16.2** Speed and quality of recovery

| Name | Speed (Years to return to normality) | Quality (Percentage of perfect performance) |
|---|---|---|
| Northridge, USA | 3 | 64% |
| Bam, Iran | 25 | 36% |
| Indian Ocean | 5 | 52% |
| Kashmir, Pakistan | 15 | 32% |
| Wenchuan, China | 5 | 52% |
| L'Aquila, Italy | 20 | 40% |
| Maule, Chile | 5 | 84% |
| Christchurch, NZ | 10 | 68% |
| Tohoku, Japan | 12 | 60% |
| Van, Turkey | 5 | 56% |

*Science and Information*
This is the only area in which the planners felt Chile could have done better in terms of having the information systems, in terms of cadastral data and up-to-date maps, in place before the disaster.

Recovery in Chile after the Maule earthquake can be judged as exemplary when compared to the other 9 events in the study. Specifically Chile did better in terms of the quality of the community consultation, the desire to rebuild as quickly as possible from the bottom up and to build back better. What distinguished recovery in Chile, for me personally, was the community consultation and the desire from the bottom up to rebuild as quickly as possible but also to build back better (Platt and So 2016).

## References

Alexander, D. (2013). *Planning for post-disaster reconstruction.* http://www.grif.umontreal.ca/pages/papers2004/Paper%20-%20Alexander%20D.pdf. Last accessed on 26 Oct 2016.

Bolton, P. A. (1996). The integration of housing recovery into reconstruction planning. In F. Y. Cheng & Y. Y. Wang (Eds.), *Post-earthquake rehabilitation and reconstruction* (pp. 163–178). Oxford: Pergamon Press.

Booth, E., & Taylor, C. A. (1988). *The Chilean earthquake of 3 March 1985*. A Field report by EEFIT Ove Arup and Bristol University June 1988.

Bruneau, M., Chang, S., Eguchi, R., Lee, G., O'Rourke, T., Reinhorn, A., Shinozuka, M., Tierney, K., Wallace, W., & von Winterfelt, D. (2003). A framework to quantitatively assess and enhance the seismic resilience of communities. *Earthquake Spectra, 19*(4), 733–752.

Bustos Erwenne, C. (2011). *Methodology for urban planning of coastal settlements with rsik of tsunami based in modeling of future scenes*. PhD Thesis outline. cberwenne@gmail.com

Cartes, I. (2011). The reconstruction process of coastal cities in the north of the Biobio region: Sustainability, participation and resilience. In *Department of planning and urban design*. Chile: Universidad del Bío-Bío. Unpublished paper.

Comerio, M. C. (2013). *Housing recovery in Chile: A qualitative mid-program review*. PEER Report 2013/01. Pacific Earthquake Engineering Research Center, University of California, Berkeley, CA.

Davis, I. (2006). *Learning from disaster recovery: Guidance for decision makers*. Kobe: International Recovery Platform.

FEMA. (2008a). *Earthquake publications for community planners and public policy makers*. www.fema.gov/library/viewRecord.do?id=3553. Accessed 15 Dec 2011.

FEMA. (2008b). *Tsunami*. http://www.fema.gov/hazard/tsunami/index.shtm. Accessed.

FEMA. (2008c). *Recovering from Disaster.* www.fema.gov/rebuild/recover

Franco, G. (2014). Earthquake mitigation strategies through insurance. In M. Beer et al. (Eds.), *Encyclopedia of earthquake engineering*. ISBN: 978–3–642-36197-5 (online).

Franco, G., & William Siembieda, W. (2010). Chile's 2010 M8.8 earthquake and Tsunami: Initial observations on resilience. *Journal of Disaster Research, 5*(5), 577–590.

González Muzzio, C. (2010). *Exploring community resilience – The social-urban aftermath of the Biobío earthquake*. MSc Thesis in Environment, Science and Society University College London. ccgonzalezm@gmail.com

González Muzzio, C. (2011). *Planning back better – Urgent and important*. PowerPoint Presentation University College London. ccgonzalezm@gmail.com

González Muzzio, C. (2012). *Planning back better, urgent and important*. Salzburg Congress on Urban Planning and Development 45th Annual SCUPAD CongressMay 9th-12th, 2012 Salzburg, Austria. www.scupad.org/web/sites/.../2012_Congressppt_Muzzio_Chile.pdf

Hinrichs, R., Jones, L., Stanley, E. M., & Kleiner, M. (2011). *Report on the 2010 Chilean Earthquake and Tsunami Response*. U.S. Geological Survey Open-File Report 2011–1053, version 1.1. http://pubs.usgs. gov/of/2011/1053/. Last accessed on 26 Oct 2016.

Iuchi, K. (2015). Planning resettlement after disasters. *Journal of the American Planning Association, 80*(4), 413–425.

Lagos, R., & Kupfer, M. (2012) Lessons Learned from the 2011 Great East Japan earthquake, March 1–4, 2012, Tokyo, Japan. *Proceedings of the International Symposium on Engineering.*

Lubkowski, Z. et al. (2010). *The MW8.8 Maule Chile earthquake of 27th February 2010*. A preliminary field report by EEFIT 27th February 2010.

Mackintosh, C. (2010, April 30). Letters from Chile: Visiting Dichato – The town that was. *Permaculture Review.* http://permaculture.org.au/2010/04/30/letters-from-chile-visiting-dichato-the-town-that-was/

MINVU. (2010). Plan de Reconstrucción "Chile Unido Reconstruye Mejor" Ministerio de Vivienda y Urbanismo Agosto 2010 © Ministerio de Vivienda y Urbanismo, Gobierno de Chile 3era Edición Octubre 2010.

MINVU. (2011). *Ministry of housing and urban development reconstruction plan*. First English Edition May 2011 ISDN: 978–956–7674-53-4.

Muñoz, Y. (2010). *Información para la gestión de ayuda humanitaria Comuna de Tacahuano Cruz Roja Chilena Julio 2010*.

NTMP. (2010). *National Tsunami Hazard Mitigation Program 2009–2013 Strategic Plan June 21, 2010*. www.nthmp.tsunami.gov/publications.html. Accessed 3 Nov 2011.

Olshansky, R. B. (2006). Planning after hurricane Katrina. *Journal of the American Planning Association, 72*(2), 147–153.

Padgett, T. (2011, March 4). '*Rebuilding Chile: Harder than rescuing miners?*'. http://world.time.com/2011/03/04/rebuilding-chile-harder-than-rescuing-miners/. Last accessed on 27 October 2016.

Plan Maestro Tubul. (2010). *Plan de Reconstrucción del Borde Costerro – PRBC18*.

Platt, S. (2012a). *Reconstruction in Chile post 2010 earthquake*. Cambridge: CAR.

Platt, S. (2012b). *Reconstruction in New Zealand post 2010–11 Christchurch earthquakes*. Cambridge: CAR.

Platt, S. (2017). Factors affecting the speed and quality of post disaster recovery and resilience. In S. Olafsson & R. Rupakhety (Eds.), *Recent developments in earthquake engineering and structural dynamics: In memory of Prof. Ragnar Sigjörnsson*. Spinger. (forthcoming).

Platt, S., & Durmaz, B. (2016). Disaster management decision making in Turkey: Case studies of Van and İzmir. *IJDRR International Journal of Disaster Risk Reduction, 17*, 220–237.

Platt, S., & So, E. (2016) *Speed or deliberation - a comparison of post disaster recovery in Japan, Turkey and Chile*. Disasters (Online; forthcoming in print): https://doi.org/10.1111/disa.12219.

Quarantelli, E. (1999). *The disaster recovery process - what we know and do not know from research*. Preliminary Paper #286. University of Delaware Disaster Research Center.

Roberson, I. (2010). *Tsunami Effects of the February 27, 2010 Chile Earthquake Univ. of Hawaii, March 20, 2010 EERI Survey Debrief*.

Useem, M., Kunreuther, H., & Michel-Kerjan, E. (2015). *Leadership dispatches: Chile's extraordinary comeback from disaster*. Stanford: Stanford University Press.

USGS. (2011a). *Sobreviviendo a un tsunami: lecciones de Chile*, Hawai y Japón Compiled by Brian F. Atwater, Marco Cisternas, Joanne Bourgeois, Walter C. Dudley, James W. Hendley y Peter H. Stauffer USGS Circular 1218.

USGS. (2011b). *Report on the 2010 Chilean earthquake and tsunami response. In cooperation with American Red Cross Multi-Disciplinary Team*. U.S. Geological Survey Open-File Report 2011–1053, v. 1.1, 68 p. Available at http://pubs.usgs.gov/of/2011/1053/

Ye, Y. (1996). Decision-making for recovery and reconstruction following a strong earthquake. In F. Y. Cheng & Y. Y. Wang (Eds.), *Post-earthquake rehabilitation and reconstruction* (pp. 56–68). Oxford: Pergamon Press. 01/02/2016 17:16.

MIX
Papier aus verantwortungsvollen Quellen
Paper from responsible sources
FSC® C105338

If you have any concerns about our products,
you can contact us on
**ProductSafety@springernature.com**

In case Publisher is established outside the EU,
the EU authorized representative is:
**Springer Nature Customer Service Center GmbH**
**Europaplatz 3, 69115 Heidelberg, Germany**

Printed by Libri Plureos GmbH
in Hamburg, Germany